Women Working
in the Environment

Women Working in the Environment

Edited by
Carolyn E. Sachs
Department of Agricultural Economics and Rural Sociology
The Pennsylvania State University

USA	Publishing Office:	Taylor & Francis 1101 Vermont Avenue, NW, Suite 200 Washington, DC 20005-3521 Tel: (202) 289-2174 Fax: (202) 289-3665
	Distribution Center:	Taylor & Francis 1900 Frost Road, Suite 101 Bristol, PA 19007-1598 Tel: (215) 785-5800 Fax: (215) 785-5515
UK		Taylor & Francis Ltd. 1 Gunpowder Square London EC4A 3DE Tel: 0171 583 0490 Fax: 0171 583 0581

WOMEN WORKING IN THE ENVIRONMENT

Copyright © 1997 Taylor & Francis. All rights reserved. Printed in the United States of America. Except as permitted under the United States Copyright Act of 1976, no part of this publication may be reproduced or distributed in any form or by any means, or stored in a database or retrieval system, without prior written permission of the publisher.

1 2 3 4 5 6 7 8 9 0   E B E B   9 8 7

This book was set in Times Roman. The editors were Heather Worley and Christine Winter. Cover design by Michelle Fleitz. Cover photographs courtesy of Carolyn Sachs.

A CIP catalog record for this book is available from the British Library.

∞ The paper in this publication meets the requirements of the ANSI Standard Z39.48-1984 (Permanence of Paper)

Library of Congress Cataloging-in-Publication Data

Women working in the environment/edited by Carolyn E. Sachs.
    p. cm.
    Includes bibliographical references.

    1. Women in development—Environmental aspects. 2. Ecofeminism.
3. Sexual division of labor. I. Sachs, Carolyn E., 1950-
HQ1240.W666   1997
305.42—dc21   97-21654
                                                                                                          CIP

ISBN 1-56032-629-8 (paper)

Contents

Contributors		xi
Chapter 1	**Introduction: Connecting Women and the Environment** *Carolyn E. Sachs*	1
	Ecofeminism	1
	Women, Environment, and Development	3
	Postmodern Shift	4
	Five Aspects of Gender Relationships	6
	References	10

Part One
Gender Divisions of Labor in Agriculture,
Mining, and Fishing Communities

Chapter 2	**Women's Forest Work in Laos** *Carol Ireson*	15
	Sources of Women's Influence	15
	Background and Setting	17
	Research Design and Procedures	18
	Results	22
	Notes	26
	References	27
	Epilogue	29
Chapter 3	**The Underground Proving Ground:** **Women and Men in an Appalachian Coal Mine** *Suzanne E. Tallichet*	31
	Introduction	31
	The Labor Process of Underground Coal Mining	32
	Women's Entry into the Male Culture of Mining	34
	Theoretical Framework	35
	Methodology	37
	From Red Hat to Miner: Women's Adaptations to Mining	38
	Conclusion	46
	Notes	47
	References	47

Chapter 4	**Gender, Culture, and the Sea: Contemporary Theoretical Approaches**	**49**
	Dona Lee Davis and Jane Nadel-Klein	
	Introduction	49
	Gender and Anthropology	50
	Women on Land and Sea	52
	Gender, Power, Production, and the State	53
	Critiquing Gender	56
	Conclusion	59
	Notes	59
	References	59
	Epilogue	62
Chapter 5	**Women's Status in Peasant-Level Fishing**	**65**
	James L. Norr and Kathleen F. Norr	
	Introduction	65
	The Community	66
	Women's Economic Activities	67
	Family Life and Daily Activities	69
	Leisure and Friends	70
	Marriage and Kinship Networks	70
	Women's Place: Behavioral Standards in the Fishing Village	71
	Comparative Analysis	72
	Discussion	76
	Acknowledgments	77
	Notes	77
	References	78
	Epilogue	80

Part Two
Property Rights: Access to Land and Water

Chapter 6	**Women and Irrigation in Highland Peru**	**85**
	Barbara Deutsch Lynch	
	Introduction	85
	Women's Work and Structural Subordination	87
	Women and Irrigation	87
	Class, Integration, and Participation in Irrigated Tasks	92
	Women and the Bureaucratic Transition in Highland Irrigation	93
	Conclusions and Implications	97

Contents   vii

	Notes	98
	References	99
	Epilogue	101

Chapter 7 **Rural Women and Irrigation: Patriarchy, Class, and the Modernizing State in South India** 103
Priti Ramamurthy

Theoretical Considerations	103
The Research Context	105
Implications for Women of Irrigation as a State Intervention	106
Transformations in the Sexual Division of Labor, Workloads, and Control of the Labor Process	108
Conclusions and Implications for Development Policy	115
Acknowledgments	117
Notes	117
References	118
Epilogue	121

Chapter 8 **Women as Rice Sharecroppers in Madagascar** 127
Lucy Jarosz

Sharecropping and Resource Access	127
Rice Sharecropping in Alaotra, Madagascar	128
Women Who Sharecrop	130
Gender Differences	133
Conclusions	135
Notes	136
References	136
Epilogue	138

Chapter 9 **Subsistence and the Single Woman Among the Amuesha of the Upper Amazon, Peru** 139
Jan Salick

Introduction	139
Background	140
Methods	140
Results	142
Discussion	149
Notes	151
References	152

Part Three
Women's Knowledge, Work, and Strategies for Sustainability

Chapter 10 **Women and Livestock, Fodder, and Uncultivated Land in Pakistan** 157
Carol Carpenter

Introduction	157
Women and Livestock Husbandry	158
Women and Fodder	160
Women and Uncultivated Land	161
Interdependence of Livestock Husbandry and Agriculture	162
Women and Livestock in Rainfed and Irrigated Farming Areas	164
Conclusions and Recommendations	167
Acknowledgments	169
Notes	169
References	170
Epilogue	172

Chapter 11 **Gender, Seeds, and Biodiversity** 177
Carolyn E. Sachs, Kishor Gajurel, and Mariela Bianco

Biodiversity and Crop Diversity	178
Seed Saving and Gender	179
New Crop Varieties	180
New Crops	181
Seed Saving: Gene Banks or Farmers' Fields	182
Seed Saving in the United States	183
Seed Saving in Peru	187
Conclusion	189
Notes	190
References	190

Chapter 12 **Women and Agroforestry: Four Myths and Three Case Studies** 193
Louise Fortmann and Dianne Rocheleau

Introduction	193
Myths and Realities	193
The Case Studies	196
Lessons for the Future	205
Summary	206
References	207
Epilogue	210

Part Four
The Gendering of Environmental and Social Movements

Chapter 13   **Men, Women, and the Environment: An Examination of the Gender Gap in Environmental Concern and Activism**   215
Paul Mohai

Introduction	215
Gender Differences in Environmental Concern	216
Gender Differences in Environmental Activism	218
Data and Method	220
Results	223
Discussion and Conclusions	230
References	232
Epilogue	234

Chapter 14   **Making a Big Stink: Women's Work, Women's Relationships, and Toxic Waste Activism**   241
Phil Brown and Faith I. T. Ferguson

Introduction	241
Experience, Knowledge, and Gender in Women's Toxic Waste Activism	244
Can Social Movement Theories Explain Women's Toxic Waste Activism?	257
Conclusion	261
References	262

Part Five
Policy Alternatives

Chapter 15   **Women and International Forestry Development**   267
Augusta Molnar

Introduction	267
Ongoing Development Work on Women and Forestry	268
How Women Participate in Forestry	268
Forestry Projects and Women's Participation	270
Restricted Access to Productive Resources	270
Restrictions on Decision Making Outside the Household	271
Restrictions on Women's Participation in the Labor Market	272
Lack of Skills and Incentives for Foresters and Project Staff to Involve Women	273
Gaps in Knowledge about Women and Forestry	274
Conclusions	274
References	275

Chapter 16	**Women and Community Forestry in Nepal: Expectations and Realities** *Irene Tinker*	277
	Two Problematic Assumptions	278
	Community Forestry in Nepal	280
	Summation	288
	Notes	288
	References	289
Chapter 17	**Integrating Gender Diverse and Interdisciplinary Professionals into Traditional U.S. Department of Agriculture–Forest Service Culture** *James J. Kennedy*	293
	Introduction	293
	Two Samples of Female and Male USFS Natural Resource Professionals	294
	Becoming Committed to a Natural Resource Profession and to the USFS	295
	Satisfaction with USFS Job and Career	297
	Career Counseling and Mentoring	299
	Conclusions	301
	Acknowledgment	303
	References	303
	Epilogue	305

Index   309

Contributors

Mariela Bianco is a Ph.D. student in rural sociology at Pennsylvania State University and is affiliated with Universidad de la Republica, Uruguay.

Phil Brown is a professor of sociology at Brown University. He earned his Ph.D. in sociology from Brandeis University.

Carol Carpenter teaches cultural anthropology and gender at Hawaii Pacific University. Dr. Carpenter received her Ph.D. in anthropology from Cornell University.

Dona Lee Davis is a professor of anthropology at the University of South Dakota and a visiting professor of anthropology at the University of Tromso, Norway. She earned her Ph.D. in anthropology from the University of North Carolina, Chapel Hill.

Faith I. T. Ferguson is completing her Ph.D. at Brandeis University in the Sociology Department. Her current work is on family arrangements among single mothers by choice. Her research interests include family, gender, work and policy, and research methods, particularly qualitative methods.

Louise Fortmann is a professor of natural resource sociology in the Department of Environmental Science and Management at the University of California, Berkeley. She earned her Ph.D. in development sociology from Cornell University.

Kishor Gajurel is a lecturer at the Institute of Agriculture and Animal Science at Tribuhuvan University, Nepal. He is currently a doctoral student in rural sociology and demography at Pennsylvania State University.

Carol Ireson is a professor of sociology at Willamette University in Salem, Oregon. She earned her Ph.D. in sociology from Cornell University.

Lucy Jarosz is an associate professor of geography at the University of Washington.

James J. Kennedy, a professor at Utah State University, teaches on and conducts research in the evolution of public land values and institutions. He received his Ph.D. in natural resource economics from Virginia Polytechnic University.

Barbara Deutsch Lynch is a visiting associate professor in the Department of City and Regional Planning and director of the Program in International Studies in Planning at

Cornell University. She received her Ph.D. in development sociology from Cornell University.

Paul Mohai is Associate Professor and Chair of the Resource Policy & Behavior Concentration in the School of Natural Resources & Environment, University of Michigan, Ann Arbor. Dr. Mohai received his Ph.D. in 1983 from Pennsylvania State University.

Augusta Molnar is an anthropologist who works for the World Bank. She is currently a project officer for NRM projects in Honduras and Mexico. Dr. Molnar received her Ph.D. in cultural anthropology from the University of Wisconsin.

Jane Nadel-Klein is an associate professor of anthropology at Trinity College. She received her Ph.D. in cultural anthropology from the City University of New York.

James L. Norr is Associate Head of the Department of Sociology at the University of Illinois, Chicago. He received his Ph.D. in sociology from the University of Michigan.

Kathleen F. Norr is a medical sociologist at the College of Nursing at the University of Illinois, Chicago.

Priti Ramamurthy is an assistant professor in the Department of Women's Studies at the University of Washington, Seattle. She received her Ph.D. in social science from the Maxwell School at Syracuse University.

Dianne Rocheleau is currently with The Rockefeller Foundation and International Council for Research in Agroforestry in Nairobi, Kenya.

Jan Salick is an associate professor of tropical ecology and ethnobotany in the Department of Plant Biology at Ohio University. She earned her Ph.D. in ecology from Cornell University.

Suzanne E. Tallichet is an assistant professor of sociology at Morehead State University. She received her doctorate in rural sociology from Pennsylvania State University.

Irene Tinker is a professor in the Department of City & Regional Planning and Department of Women's Studies at the University of California at Berkeley.

Chapter 1

Introduction:
Connecting Women and the Environment

CAROLYN E. SACHS

Department of Agricultural Economics and Rural Sociology
Pennsylvania State University
University Park, PA 16802-5600
USA

Increasing global and local concern with environmental degradation has focused attention on people's relationships with the environment. Ozone depletion, deforestation, decline in water availability and quality, land degradation, and pollution are among the environmental problems that affect and involve many people on the planet. Whereas many of the initial efforts to resolve environmental problems focused on technical and biological solutions, scholars and activists are increasingly looking to social causes and solutions for environmental problems. Feminist scholars have focused on whether women and men have different relationships with the environment and are concerned about the implications of these differences. Understanding the gendered nature of human relationships with the environment seems particularly critical for resolving environmental problems.

Theoretical discussions on women and the environment revolve around three major questions: (1) What are women's relationships with nature? (2) What are the connections between the domination of women and the domination of nature? and (3) What role do women play in solving ecological problems? Ecofeminists initially raised many of these questions in the late 1970s and early 1980s and set the stage for the ensuing debates concerning women and the environment. Versions of these questions on the relationship between women and the environment have been addressed by several major threads in feminist theory, which will be discussed in this introduction: ecofeminism, women and development, and postmodern feminism. Then, this chapter introduces a framework for understanding women's connections to the environment. This framework guides the organization of the book through considering gender divisions of labor, access and control over resources, knowledge and strategies for survival, participation in social movements, and policy concerns.

Ecofeminism

Ecofeminists and their critics are grappling with questions regarding women's relationships with nature: How do these differ from men's relationships to nature? To what extent are women more likely to be concerned and capable than men of solving environmental problems? and How can we end the domination of women and the environment? Feminist attempts to understand women's connection to nature begin with critiques of Western science and philosophy's assumptions concerning women's relationships with nature. Western philosophy associates women with nature, based on dualistic epistemologies that juxtapose culture/nature, male/female, reason/emotion, and mind/body. Women are associated with nature, emotion, and body whereas men are associated with culture, reason, and mind. Symbolic and ideological constructions of women as "closer to nature" than men set the stage for the domination of women and nature. Within this framework, men are positioned outside and above nature and women (Keller, 1985). The early promoters of science, led by the writings of Francis Bacon, used these symbolic and metaphoric constructions of women and nature as legitimation for men's control of nature and women's bodies. As Griffin (1989) points out, women become symbols of nature and are thereby transformed into objects of degradation. Ecofeminists challenge these hierarchical dualisms of nature/culture, male/female, and emotion/reason. Rather than rejecting women's association with nature, ecofeminists suggest this connection can be used as a vantage point for transforming the nature/culture distinction, revising the conceptualization of nature, and resolving environmental problems (King, 1989).

Ecofeminism refers to a plurality of positions relating to the connections between women and nature. Two of the major tenets of ecofeminism hold that the domination of women and the domination of nature are intimately connected and that women are particularly suited to lead ecological movements to save the planet. Merchant (1992) distinguishes between liberal, cultural, social, and socialist ecofeminism. Each of these versions of ecofeminism are concerned with improving the relationship between humans and nature, but their approaches and strategies for change differ. Liberal ecofeminists attempt to work within existing structures of government by changing laws and regulations related to women and the environment and providing equity for women in the workplace. Cultural ecofeminists critique patriarchy and emphasize the symbolic and biological connections between women and nature. From their perspective, women's bodies are closer to nature than men's bodies due to menstruation, childbirth, and pregnancy. According to cultural ecofeminists, these biological processes are the source of women's power and ecological activism. Social and socialist ecofeminists analyze the ways in which both patriarchy and capitalism contribute to men's domination of women and nature. Both of these perspectives explore and analyze social justice issues. Central to much of ecofeminism is the belief that women share an environmental "ethic of care" based on their biology, labor, or social position.

Ecofeminism has been subject to often well-deserved critiques from feminists and others for upholding an essentialist claim that women's nature is to nurture. Major criticism has been directed to ecofeminist assumptions that, due to their biology, women are closer to nature than men and are therefore privileged to think ecologically and positioned to be better caretakers of the earth. For many decades, feminists have battled "biology is destiny" arguments. The assumption that women are biologically inferior to men has often been used as a justification for the subordination of women. Ecofeminist attempts to change this relationship by emphasizing positive aspects of women's biological connection to nature seem particularly dangerous to other feminists.

In light of these essentialist critiques, many ecofeminists have rethought and revised their earlier positions. Biehl (1991), who was once an advocate of social ecofeminism, rejects ecofeminism altogether and bitterly accuses other ecofeminists of carrying the women–nature metaphor too far by rejecting rationalism and worshipping goddesses. Merchant (1996) suggests that the cultural baggage attached to gendering nature is quite problematic for social movements in the West and calls for a "partnership ethic" between humans and nature that does not endow women with a special knowledge of nature or ability to care for nature. In contrast to other ecofeminists, Shiva (1994) does not back down to the essentialist charge. Rather, she suggests that the charge itself is based on the critics' inability to transcend reductionist, dualistic epistemologies. Shiva replies that critics who suggest that she is essentializing women see difference as so "essential" that solidarity and commonality between women or between women and men seem impossible. Shiva insists that women acting together in ecological movements is not essentializing; there is not a divide between the environment and our bodies. Shiva's work also led the way in considering Third World women's connections to the environment.

Women, Environment, and Development

This approach to women and the environment emerges from scholars and practitioners working in the field of women and development. Beginning with the path-breaking work of Boserup (1970), many scholars working on women and development issues have critiqued the impact of Western development on Third World women's lives. Western development efforts have increased women's workload, decreased women's access to resources, and contributed to the feminization of poverty. As a response to problematic development projects, current efforts by women and development activists focus on creating alternative grassroots strategies and/or transforming government policies to empower women (Leonard, 1989, 1995; Moser, 1993). With the recent international attention on global environmental problems, feminists quickly recognized Western development practices as simultaneously detrimental to women and the environment. As van den Hombergh (1993) explains, "Environmental degradation on the one hand, and the feminization of poverty on the other are caused by or reinforced by male-biased development, based on a model of exploitation of resources mainly for the prosperity of Northern countries and Southern elites" (p. 20).

Shiva's (1989) work, *Staying Alive: Women, Ecology and Development*, forcefully connected the impact of development on women and the environment. Focusing on Indian women, she argues that ecological destruction and the marginalization of women have been the results of Western science and Western economic development paradigms. Rather than focus on women as passive victims of development, Shiva emphasizes that women's struggles for survival lead the way in illustrating ways to resist ecological destruction. She argues that Indian women have been at the forefront of struggles to preserve land, water, and forests (Shiva, 1993). "Because of their location on the fringes, and their role in producing sustenance, women from Third World societies are often able to offer ecological insights that are deeper and richer than the technocratic recipes of international experts or the responses of men in their own societies" (Shiva, 1994, p. 1). Third World women's "deeper" insights come from their participation in cultures that value the maintenance of life; in addition, the gender division of labor in these countries has forced women to provide subsistence for their families, while men seek profit-earning strategies. Thus, women's involvement in the environmental movement starts with the daily work and activities in their lives and the threats to the health of their families.

Also using Indian women's lives as the starting point for understanding the connection between women and the environment, Agarwal (1992) insists that men's and women's connections with the environment require a critique grounded in the concrete realities of their daily lives. However, from her perspective, ecofeminism fails to move beyond the symbolic and ideological associations of women and nature and falls into the trap of essentializing women. According to Agarwal, by remaining at the symbolic level, ecofeminists fail to account for differences between women by class, race, ethnicity, and national identity. Men's and women's relations with the environment must be understood in connection with the material reality of the division of labor, property, and power. Thus, Agarwal proposes a theoretical position she refers to as "feminist environmentalism" as an alternative to ecofeminism. Feminist environmentalism challenges notions about gender and the division of labor and resources between men and women, while also changing and transforming the appropriation of natural resources by the privileged. Agarwal's insistence on concretizing and materially locating women's relationships with the environment fuels arguments for closer studies of gendered relationships with the environment.

Recently, several scholars have critiqued the "women, environment, and development" perspective for narrowly focusing on women (Bradiotti et al., 1994; Leach, Joekes, and Green, 1995). Instead, they emphasize gender relations with the environment. From their perspective, grouping women together as a unitary category misrepresents women, makes men invisible, and clouds our understanding of human relationships with the environment (Leach, Joekes, and Green, 1995). These authors also critique dominant approaches that define women and nature as universally connected. Their critiques of ecofeminism and "women, environment, and development" approaches as universalizing and their subsequent calls for local and specific studies coincide with the move to postmodern approaches.

Postmodern Shift

Postmodernism represents another major theoretical shift that provides insights into understanding women's relationship with the environment. Postmodern theorists offer several useful insights for understanding women's relation with the environment, including the rejection of universalist claims, the focus on identity and difference, and the emphasis on local and subjugated knowledges.

The rejection of the search for universal truth and the profound questioning of the superiority of Western science and modernization lie at the heart of postmodernism. This rejection of universalism and critique of metanarratives open spaces for fresh understandings of human relationships with the environment. The postmodern focus on language and discourse provides tools for deconstructing processes, narratives, and concepts such as "women," "nature," "environment," and "development." Looking at these concepts through a lens that requires examination of how these terms are used to enhance the power of the privileged, one is able to critically evaluate assumptions concerning the connections between women, environment, and development. For example, Escobar (1995), Sachs (1993), and others have shown how "development" has become a powerful semantic term at the center of modern thought and behavior (Esteva, 1993). Development has served as the major guiding principle for the "mixture of generosity, bribery and oppression which has characterized the policies toward the South" (Sachs, 1993, p. 1). The concept of development simultaneously constructs "first world" countries at the height of an evolutionary scale of societies while portraying Third World countries as mired in poverty, hunger, overpopulation, environmental degradation, oppression of women, and traditional ways. This universal narrative of development is so pervasive that alternative scenarios for the Third World are difficult to envision.

The postmodern critique of modernist, universal metanarratives suggests a shift toward difference and identity politics. Within feminism, the shift from a unitary conception of women to an awareness of differences between women has given voice to African American women, Latinas, lesbians, and Third World women. Many of these women critiqued Western feminists for using the category "women" to represent all women, when in fact the "women" usually represented were white, upper middle class, heterosexual, and Northern. In understanding women's connections with the environment, this emphasis on difference establishes a theoretical framework in which women's relationships with the environment will differ by their class, racial, ethnic, sexual, and national identities. Rather than searching for the essential connections between "women" and "nature," a more useful approach is to turn toward the concrete situations of women.

Said's (1978) influential work on orientalism shows how colonial and postcolonial discourses have defined Third World people as the "other" in a move to perpetuate colonialist privilege while emphasizing the exotic and negative qualities of the "other." As early as deBeauvoir, feminists have battled male attempts to define women as "other." However, postmodern feminists such as Mohanty (1991) argue that feminists themselves have constructed Third World women as the "other." Mohanty indicts Western feminists for portraying Third World women as homogeneously poor, illiterate, overburdened, and victimized in contrast to the "liberated" Western woman. In fact, much of the discourse on women and the environment in the Third World has emphasized women as either victims of environmental degradation or uneducated exploiters of natural resources. In this way, Western women and development planners position themselves as experts to help solve the problems of Third World women and blindly proceed as if Third World women and men lack skills and strategies for resolving their problems.

By contrast, postmodernism's emphasis on local knowledge rather than universal truths provides new approaches to the study of women and development (Parpart, 1993). Development planners, feminists, and others should focus on local, concrete experiences of women and on their interpretations of their lives, needs, and goals (Parpart, 1993). This call for a more localized understanding of women's lives is particularly useful for exploring women's connections to the environment.

Haraway's (1991) work on "knowledge claims" proves particularly insightful for understanding women's connections to the environment. Haraway sees all knowledge claims as situated, embodied, and partial. She criticizes claims to knowledge that are universal and nonlocatable. Similarly to Agarwal, Shiva, and Parpart, Haraway's insistence on the situatedness of knowledge pushes us to look at women's daily activities and knowledge about the environment in specific places. Rather than claim that women are closer to nature or have a particular connection to the environment, researchers must look at the daily work, activities, and knowledge of women in locatable contexts such as forests, fields, or urban environments. Women's knowledge and understandings about the environment derive from their concrete, situated experiences. Although Haraway sees women's knowledge or any subjugated knowledge as different from and challenging to dominant perspectives, she cautions against the dangers of partial, subjugated knowledge. She emphasizes that "seeing from below" is difficult and may not result in women challenging dominant perspectives. In the case of women and the environment, Haraway might advise exercising caution in suggesting that women will necessarily attempt to protect the environment and/or challenge patriarchal authority and practices.

One of the most obvious and frequent feminist critiques of postmodernism holds that endless deconstruction results in the depoliticizing of women's issues. One must be careful, these women argue, not to devolve into an inability to write, act, or organize as women because of a focus on difference and locality (Zita, 1989). Focusing on the local, concrete experiences of women's lives in agricultural production, in forests, or in other activities need not preclude a more structural analysis that examines gendered divisions of labor and access to property and resources.

This book provides a framework for understanding women's connections to the environment. Each author(s) provides locally based, specific examples of men's and women's experiences and relations with nature and the environment. Rather than focusing on abstract connections between women and the environment, these studies examine women's relationships with the sea, the land, forests, and animals in particular places and localities. Using examples from Africa, Asia, Latin America, and North America, the contributions offer concrete pictures of difference as well as arguments about the importance of gender in defining human relationships with the environment.

Five Aspects of Gender Relationships

Based on theoretical insights from ecofeminism, women and development, and postmodernism, and the convincing empirical work of numerous scholars, I suggest and organize the book around five aspects of gender relationships that undergird theoretical and empirical understanding of women's relationships with the environment: (1) gender divisions of labor, (2) access and control over property, (3) knowledge and strategies for survival, (4) environmental movements, and (5) policies. Examining women's relationship with the environment using these five dimensions provides concrete, material examples of how women work with, control, know, and affect the environment and natural resources.

Gender Divisions of Labor in Agriculture, Mining, Forestry, and Fishing Communities

This section of the book focuses on gender divisions of labor in communities that are dependent on natural resources. In many of these communities, men control and dominate occupations directly involved with the use of land, plants, animals, forests, and the sea. Men's work as loggers, fishermen, farmers, and miners often eclipses women's contributions and work in these communities. While acknowledging that gender divisions of labor clearly exist, these authors provide concrete examples of women's work in fishing, forestry, and mining. The chapters in this section document the daily work lives and difficulties of women's employment in forests, mines, and fishing communities. Ireson's study (Chapter 2) of Laotian women emphasizes women's contributions to both household subsistence and family income. She depicts women's forest-gathering work as essential to the largely rice-based agricultural households. Her study documents the ways in which women's work differs depending on the quality of the forest. Women with access to old-growth forests make more visits to the forests and tend to have less of a commercial orientation to the forest than women who have access only to new-growth forests. Whether they are involved in subsistence and/or market activities, women's forest work contributes to their informal influence in their villages.

Whereas the majority of women in Laotian villages work in forests, few women in the United States work in mines. Tallichet's study (Chapter 3) of female miners in Appalachia reveals the extraordinary obstacles women face when they enter the mines. Using miners' own words, she vividly portrays the gendered nature of work in the mines as well as the problems female miners encounter in their workplaces, households, and communities. Tallichet's study examines their individual problems as they challenge the assumption that "only men are miners."

Two chapters on fishing (Chapters 4 and 5) address the problematic assumption that only men fish; each chapter attempts to unravel the cultural silences and popular myths about fishing and gender. Findings from international studies of fishing communities suggest that men typically go out to sea, whereas women's fishing-related activities vary immensely. In many regions, women market fish after the catch and their incomes contribute substantially to supporting their families. Norr and Norr's study (Chapter 5) of a fishing village in Tamilnadu, India, found that women did not fish, but met their husbands' or other male relatives' boats every day upon their return from sea and assumed responsibility for the marketing of the fish. Women in this village were not unique in their responsibility for marketing fish: fishermen's wives in villages in Thailand, Malaysia, and Taiwan also earned income selling fish. The authors also explore how women's status in fishing communities compares with the status of women in nearby agricultural communities. They found that women from the Indian fishing community had more independence and were less subservient than women in nearby agricultural communities. In an attempt to understand this finding, they analyzed studies from other countries and concluded that women in fishing communities have more independence and power than women in agricultural communities in India, Japan, and Taiwan, but not in Brazil, the Caribbean, or Malaysia. Where fishermen are lower in status than agriculturalists, their wives tend to have more independence than women married to farmers. Davis and Nadel-Klein (Chapter 4) critique Western gender models that view fishing as a male activity and suggest more complicated strategies for looking at gender and fishing. In criticizing maritime studies that associate women with land and men with the sea, Davis and Nadel-Klein suggest asking the following questions: How is fishing part of the symbolism of gender? How is this social construction of an activity connected to peoples' understanding of who can work? And where? And when? Their theoretical insights provide new strategies for studying gender in other natural resource–dependent communities.

These chapters describe the highly gendered nature of fishing, mining, and forestry work in many regions of the world. In examining and critiquing the association of fishing, mining, and logging with men, these writers emphasize the importance of women's work in communities that are heavily dependent on work with natural resources. These scholars attempt to understand how gender divisions of labor define women's status and power in their communities and households. Women's forest gathering in Laos, going to sea and selling fish in many fishing communities, and working in mines in Appalachia tend to improve their status and power in their households and communities. However, status and power derives not only from the work women do, but also from gender differences in control and access to resources.

Property Rights: Access to Land and Water

In this section, the writers address gender differences in access and control over land and water. Much of the scholarship on gender and natural resources has focused on the division of labor, but few scholars have examined women's access and control over resources. In agricultural systems, access to property, land, and water rights critically shapes people's relations with natural resources. Both state policies and local practices define women's and men's differential access to land and water. State-sponsored irrigation schemes notoriously overlook women's concerns, often increase women's workloads, frequently redistribute resources to men, and marginalize women's involvement in agricultural decision making. Lynch and Ramamurthy's chapters (Chapters 6 and 7) reveal the impact of state-sponsored irrigation systems on rural women in highland Peru and South India, respectively. Lynch documents women's key, but often invisible, role in irrigation work. She also delineates the consequences associated with women's exclusion from participation in irrigation management decisions. Compared to two decades ago, women in many parts of the Andes exercise more control over water in the fields, largely as a result of the declining ability of agricultural production to meet household subsistence needs and the out-migration of men. However, women play only token roles in the increasingly bureaucratized politics of irrigation. Ramamurthy's detailed study analyzes how a government-sponsored irrigation system in South India increased poor women's workloads and redistributed benefits to men. Both studies explain why irrigation policies remain largely uncontested by women and suggest how women's organizations can politicize, challenge, and shift irrigation policies.

In many regions of the world women have limited access to land, and, often, both state and local policies exclude women from access to land ownership. Jarosz's case study (Chapter 8) of a rice-growing region in Madagascar explores how gender and class shape access to land and resources. She provides a fascinating typology to show differences in men's and women's reasons for and benefits from sharecropping. For example, rich men use sharecropping to accumulate more land, in contrast to poorer women who sharecrop to gain access to male labor. Consequently, the young, poor, landless, and female-headed households suffer the greatest disadvantage from sharecropping.

Female-headed households in many parts of the world frequently lack access to property and land. Recognizing the problems experienced by female-headed households, Salick (Chapter 9) uses an ethnobotanical approach to study the particular subsistence strategies of single Amuesha women in Peru. Single women's fields are fewer, smaller, and more intensively cropped than other Amuesha fields. Compared to other households, these women rely more extensively on home gardens for their food supply because they have difficulty clearing fields without the assistance of men. These chapters on Madagascar and Peru reveal gendered patterns of access to land and water for agriculture and disclose the ways in which women suffer the consequences of uneven access and control over land and water rights. As state policies emphasize privatization and reduce the amount of common land, and governments tighten their control of natural resources, poor women lose access to natural resources and are further marginalized.

Women's Knowledge and Strategies for Sustainability

Gender divisions in labor and access to resources results in men and women harboring distinct knowledge about the environment. Women's knowledge about animals, plants, and land often goes unnoticed by agricultural development agencies and natural resource management personnel. However, evidence from many locations suggests that women's knowledge and strategies for survival may point to new directions for achieving environmentally sustainable agricultural and natural resources. Carpenter (Chapter 10) emphasizes women's centrality in maintaining balances among livestock, cropping systems, and fodder collection in Pakistan. Policies that attempt to bring marginal land into agricultural production overlook women's reliance on uncultivated land for fodder production; such policies eventually increase women's labor, decrease supplies of fodder, fuelwood, and manure, and endanger the sustainability of household agricultural systems. My chapter on seed saving and biodiversity (Chapter 11) documents women's central role in resisting the trend to plant fewer varieties of a given plant in many regions of the world. I discuss how women continue to improve crops and maintain biodiversity within species. Agricultural scientists rarely acknowledge women's information about plants and seeds in their attempts to develop new crops and preserve genetic diversity of crops. Fortmann and Rocheleau's chapter (Chapter 12) on women and agroforestry highlights the importance of women's work in and knowledge of agriculture and forestry systems. Using three case studies from the Dominican Republic, India, and Kenya, they detail the ways in which women's and men's knowledge, priorities, and access to resources differ. Written in 1985 to influence natural resource policies, this classic work on women and forestry provides insightful analyses of the myths of women's roles and targets issues that remain equally relevant today. Women's exclusion from access to resources pushes them to develop particular subsistence and income-earning strategies for their households and communities. In some cases, women's efforts to protect the environment result in organized movements.

The Gendering of Environmental and Social Movements

Women in many parts of the world are involved in grassroots-level environmental activism. Women's leadership in the toxic waste movement in the United States, the Chipko movement in India, and the greenbelt movement in Kenya exemplify grassroots-level organization by women. Although many theorists and research studies show women as more concerned about environmental problems than men, other studies refute the idea of gender differences in environmental activism. Mohai's study (Chapter 13) explores differences in women's and men's involvement in environmental issues in the United States. Based on a survey of 7,010 people, he found that women indicated somewhat greater concern than men for environmental issues, but, contrary to expectations, women had lower rates of environmental activism than men. He discusses possible explanations for these inconsistent gender differences in environmental concern and activism; moreover, he urges further research to investigate specific environmental issues that may be important to women, such as local health and safety concerns. Along similar lines, Brown and Ferguson's chapter (Chapter 14) on the toxic waste movement describes the central role played by women. Their insightful study of activists in Massachusetts reveals the ways in which predominantly working-class women with little previous political activist experience successfully mobilized to investigate the connection between toxic waste and health in their community. They conclude that the toxic waste movement is altering the environmentalist movement through focusing on local issues, emphasizing human well-being, introducing race and class issues, involving minorities and working-class people, and placing gender in a central role.

Policy Alternatives

National and international policies relating to natural resources have historically overlooked the ways in which gender shapes natural resource use. Perhaps this is changing. In the 1990s, international agencies, governments, and nongovernmental organizations are recognizing the importance of considering gender issues in natural resource policies. These chapters in the final section illustrate the ways in which policy issues relate to gender, forestry, and natural resources. Molnar (Chapter 15) draws on her experience in natural resource management and women and development programs in Asia and from evaluations of natural resource management projects in Asia, Africa, and Latin America. Tracing the increasing number of forestry programs targeting women, she highlights progress in documenting women's importance in forestry and in contributing to sustainably managed forests. Molnar provides numerous examples from around the world to explicate the problems faced by such projects. These include women's restricted access to resources, restricted roles in public decision making, and restricted access to labor opportunities, as well as project personnel's lack of skills in incorporating women. Rather than ending on a lamenting note, she provides examples and strategies for overcoming each constraint.

Tinker (Chapter 16) also highlights the constraints encountered in efforts to involve women in community forestry in Nepal. She argues that problematic assumptions that subsistence farmers are the major culprits in forest degradation and that women are closer to nature than men obscure the major family and class power relations that fundamentally contribute to deforestation. She documents constraints encountered in numerous programs and projects funded by international development agencies, the Nepalese government, and nongovernmental organizations. She warns that unrealistic expectations placed on women to solve deforestation may undermine Nepalese women's few gains. Expectations for women is also a central theme in Kennedy's study (Chapter 17). His study focuses on wildlife biologists entering the U.S. Forest Service as a result of federal mandates to hire both women and wildlife/fisheries biologists. The new women and men hired into the Forest Service as a result of these federal mandates encountered problems. The Forest Service was an organization dominated by traditional foresters who resented these new people who were hired to change their organization. Kennedy reports that compared to male wildlife/fisheries biologists, female professionals were more likely to enter the profession because of their concern for the environment. Female professionals also were more likely to acknowledge the importance of a mentor for their career and report greater satisfaction with their jobs. He concludes that federally mandated programs designed to integrate women and wildlife/fisheries biologists into the Forest Service placed many of the new, young professionals in difficult circumstances; they were ill prepared to change the dominant culture of the organization.

References

Agarwal, B. 1992. The gender and environment debate: Lessons from India. *Feminist Studies*, 18(1):119–158.
Biehl, J. 1991. *Rethinking Ecofeminist Politics*. Boston: South End Press.
Boserup, E. 1970. *Women's Role in Economic Development*. New York: St. Martin's Press.
Bradiotti, R., E. Charkiewicz, S. Hausler, and S. Wieringa. 1994. *Women, the Environment and Sustainable Development: Towards a Theoretical Synthesis*. London: Zed Press.
Escobar, A. 1995. *Encountering Development: The Making and Unmaking of the Third World*. Princeton, NJ: Princeton University Press.
Esteva, G. 1993. Development. In *The Development Dictionary*, ed. W. Sachs, pp. 6–25. London: Zed Press.
Griffin, S. 1989. Split culture. In *Healing the Wounds: The Promise of Ecofeminism*, pp. 7–19. Philadelphia, PA: New Society Publishers.
Haraway, D. J. 1991. *Simians, Cyborgs, and Women: The Reinvention of Nature*. New York: Routledge.
Keller, E. F. 1985. *Reflections on Gender and Science*. New Haven: Yale University Press.
King, Y. 1989. The ecology of feminism and the feminism of ecology. In *Healing the Wounds: The Promise of Ecofeminism*, eds. J. Plant, pp. 18–28. Philadelphia, PA: New Society Publishers.
Leach, M., S. Joekes, and C. Green. 1995. Gender relations and environmental change. *International Development Society Bulletin*, 26(1):1–8.

Leonard, A. 1989. *Seeds: Supporting Women's Work in the Third World*. New York: Feminist Press.

Leonard, A. 1995. *Seeds 2: Supporting Women's Work Around the World*. New York: Feminist Press.

Merchant, C. 1992. *Radical Ecology: The Search for a Livable World*. New York: Routledge.

Merchant, C. 1996. *Earthcare: Women and the Environment*. New York: Routledge.

Mohanty, C. 1991. Under Western eyes: Feminist scholarship and colonial discourses. In *Third World Women and the Politics of Feminism*, eds. C. Mohanty, A. Russo, and L. Torres, pp. 51–80. Bloomington: Indiana University Press.

Moser, C. 1993. *Gender Planning and Development: Theory, Practice, and Training*. New York: Routledge.

Parpart, J. 1993. Who is the "other"?: A postmodern feminist critique of women and development theory and practice. *Development and Change*, 24(3):439–464.

Sachs, W. 1993. *The Development Dictionary*. London: Zed Press.

Said, E. 1978. *Orientalism*. New York: Pantheon Books.

Shiva, V. 1989. *Staying Alive: Women, Ecology and Development*. London: Zed Press.

Shiva, V. 1993. Colonialism and the evolution of masculinist forestry. In *The Racial Economy of Science*, ed. S. Harding, pp. 303–314. Bloomington: Indiana University Press.

Shiva, V. 1994. *Closer to Home: Women Reconnect Ecology, Health and Development Worldwide*. Philadelphia, PA: New Society Publishers.

van den Hombergh, H. 1993. *Gender, Environment and Development: A Guide to the Literature*. Utrecht, the Netherlands: International Books.

Zita, J. 1989. The feminist question of the science question in feminism. *Hypatia*, 3(1):158–169.

PART ONE

GENDER DIVISIONS OF LABOR IN AGRICULTURE, MINING, AND FISHING COMMUNITIES

Photograph courtesy of Jean-Phillipe.

Chapter 2

Women's Forest Work in Laos

CAROL IRESON

Department of Sociology
Willamette University
Salem, OR 97301
USA

Abstract *Forest work is a significant part of the contribution of Lao rural women to the household economy. Women's forest work was studied by interviewing 120 rural women farmer/gatherers in eight villages in one province in central Laos. Women with access to old-growth forest as well as second-growth areas use forest products mainly for subsistence purposes, whereas women with access only to second-growth areas are more commercially oriented and are more likely to sell what they gather. Women's forest work in all cases contributes to the household economy and becomes even more important during poor crop years. It is suggested that women's forest activities, along with women's other work activities, foster their informal influence in household and village.*

Keywords: women, forestry, gathering, minor forest products, women's work, household economy, Laos, village.

The wealth of the lowland Lao is in their rice fields and animals, whereas their welfare system is in the forest. Lao women contribute substantially to both family wealth and welfare. In fact, Lao women's economic activities provide a basis for the informal influence women often enjoy in their families and villages. Village women regularly gather from the forest, providing supplementary food, medicine, and material for household items. In years of drought, flood, or disease, when rice yields are low or the rice crop is destroyed, forest food becomes even more essential for household survival. Most village women are foresters, but the forest work women do and the ways in which they utilize forest resources depend on the characteristics of the forest available to them and the kinds of economic resources available to their households. This study portrays the variations in the forest work of lowland Lao village women in a province of central Laos.

Sources of Women's Influence

Women's economic activities are often the basis for their relative social power. Peggy Sanday identified four factors that indicate the extent of female economic and political influence in a society: control over allocation of goods, demand for female-produced goods, female political participation, and the existence of female solidarity groups that

---

The larger study of which this is a part was sponsored by the National Institute for Social Sciences of the Lao People's Democratic Republic, approved by the Forestry Department of the Ministry of Agriculture and Forestry of the Lao People's Democratic Republic (PDR), and funded by the Swedish International Development Authority.

are politically and economically oriented (Sanday, 1974). In a later work, Sanday located women's power in a symbolic context and noted that the likelihood and meaning of female control over goods or participation in group decision making vary by the symbolic meaning of gender roles (Sanday, 1981). Leacock and Sacks both asserted that women's status is a function of women's control over economic resources and conditions (Leacock, 1978; Sacks, 1974, 1979). As Marxists, they focused on economic conditions and configurations rather than on symbolic arenas as basic to female power and influence. A number of other authors also have illustrated, with specific examples, this interaction between a woman's work and her status.[1] Women are often responsible for family welfare. It appears that women, perhaps more commonly than men, expend their efforts and resources on family nutrition, support for their children's schooling, and other unmet household needs (Babb, 1986; Charlton, 1984; Mueller, 1977; Sen and Grown, 1987).

Lao women, like women in a number of societies, are important economic actors. Most rural Lao women are active farmers and gatherers. The male household head, commonly a young or middle-aged man, organizes family paddy labor. Men sow the family rice seedbed and prepare paddy fields for planting, and the women and girls transplant seedlings. In the swidden or shifting cultivation system, the male head often selects the swidden field; both men and women clear, burn, and plant the field; and family females tend the growing field, weeding throughout the growing season and planting and harvesting subsidiary crops. Everyone, though, harvests and threshes the rice. Women and girls are mainly responsible for gardens, fruit orchards, and raising small animals. Men, on the other hand, build and repair houses and family implements and raise cattle and buffalo. Nearly all family members collect and utilize forest resources. Lao women and children regularly gather forest foods and other forest products, whereas males usually only enter the forest to hunt, gather, or cut timber for construction. Both sexes fish and gather water animals. In addition, some women produce goods, such as woven blankets and rice whiskey, to sell as well as marketing goods produced by others. Women usually market family produce, except large amounts of surplus rice or large animals. The division of labor by sex is somewhat flexible, though. For example, men sometimes transplant seedlings or tend gardens, and women occasionally plow or repair houses. Insurance for cases of illness or crop shortfall are the family animals, forest products for family use and for sale, and other income-earning activities, often pursued by family women with marketing experience.

Just as women's economic activities sometimes provide women with a resource base, traditional kinship and household organization sometimes provide a structural basis for female influence. Sacks noted that in societies emphasizing a woman's membership in the corporate kin group as a sister, rather than as an outside wife, women are more apt to have access to economic resources and social influence. On the other hand, women who are isolated from the support networks of their own natal families are more likely to be dominated by the men of their household.[2]

Women's informal influence in household and community is commonly observed among lowland Lao and Thai and is based on the structural significance of women's positions as inheriting daughters (Potter, 1977). Residence preference after marriage tends to be matrilocal, at least for a few years, with some likelihood that the youngest daughter and her husband will remain with her parents to care for them in their old age. No formal female solidarity groups have developed as a result of this pattern. However, daughters sometimes mediate communication between their husbands and their fathers, and female relatives sometimes work together. Lao women, especially those in middle age, have central economic roles as noted above. They traditionally hold the family purse strings, and although they are not considered to be equal to their men, they share financial decision making with their husbands (Ayabe, 1959; Condominas, 1970; Keyes, 1975; Potter, 1977).

Nearly all economic activity in Lao villages is carried out by individuals or households, which on the average have about seven members (National Committee of Plan, Lao PDR, 1985). In spite of Lao women's economic resources and significance in family structure, very few officials and no important religious leaders are women. Although women have some informal power, their lives are not easy. Besides working hard in fields and forests, the average Lao woman bears seven or eight children, losing two or three of them in childhood. She is either pregnant or nursing most of her adult life, suffers from chronic nutritional anemia, and has a life expectancy at birth of about 50 years (Stuart-Fox, 1986, p. 149; UNICEF, 1987, p. 17).

Rural women are often subsistence foresters (Fortmann, 1986; Fortmann and Richeleau, 1985; Hoskins, 1980). Poor farmers, particularly, count on forest products to enrich their hearths, their health, and their household income, especially during bad crop years. In one area of northeast Brazil, for example, forest products extraction is roughly equivalent to wage labor and to agriculture in its contribution to family income and is most important for poor households and for women (Hecht, Anderson, and May, 1988). In fact, many farming families in developing countries continue to be dependent for part of their livelihood on the forests near their farming plots (see also Barker, Herdt, and Rose, 1985; Belsky and Siebert, 1983; Kunstadter, Chapman, and Sabhasri, 1978; and Liemar and Price, 1987).

Although male dominance is evident in Lao political, religious, and social life, Lao women's economic activities and position in the family appear to enable them to enjoy some measure of informal influence in household and village. A village woman's forest work, which may serve both family welfare and family income functions, is one component of her power base. This work appears to vary, depending on the characteristics of the available forest and market. By studying the content and variations in this work in a specific time, place, and ethnic group, I hope to contribute to the understanding of linkages between inequality and the "particulars of women's lives, activities and goals" (Rosaldo, 1980, p. 417). Furthermore, the acceleration of forestry and other economic development efforts in Laos is now occurring with little information about village forestry and women's economic roles. This study was designed to better inform these efforts.

I begin by briefly presenting background information about Laos and describing the province studied. I then discuss the research design and procedures, including background information about the specific villages studied. Finally, I present and discuss study results.

Background and Setting

Economically, Laos remains a nearly undeveloped country. Encircled by Burma, Thailand, China, Vietnam, and Cambodia, it has been significantly affected by global politics and the world economy. Laos was a neglected part of French Indochina and subsequently experienced two wars of national liberation in the three decades after 1945, culminating in a relatively calm transition to communist control in late 1975.[3] The cooperative reorganization of the mainly subsistence village economy and the development of a state-run marketing network foundered in many parts of the country after 1979. Cooperativization removed incentives to produce goods for sale and made rice production problematic even in years with good weather. Moreover, it compelled Lao villagers to rely more on other traditional livelihood strategies—subsistence gardening, animal raising, crafts, and forest extraction.

Forest products, especially tropical hardwoods, have been an important source of income for the Lao nation and its people. These products account for about one-quarter of foreign exchange for the current government, which plans further development of commercial forestry. Laws reserving forests and limiting village logging are beginning to be enforced, and standing timber concessions have been allocated to various ministries, military units, and provincial governments in lieu of cash budgets.

The forest has, for generations, been the source of food, shelter, and safe haven for needy Lao. With Laos's historically low population density (currently 17 people per sq km) and low rate of forest resource exploitation, the mixed rural economy of paddy, swidden, gardening, forest gathering and hunting, and animal raising has been an adaptive one. Most of the chroniclers of rural life in Laos mention the subsistence forest activities of Lao villages as a minor source of income (Kaufman, 1956; Wulff, 1972), a dietary supplement or food source during lean years (Halpern, 1958; Orr, 1966), or as a regular part of the household economy (Branfman, 1978; Condominas, 1970; Maynard and Kraiboon, 1969). Little detail is available, however, on women's forest activities.

In the late 1980s, the Lao version of glasnost opened the country to foreign business, trade, and tourism, while facilitating the reestablishment of private trade networks and deemphasizing cooperatives. This had striking effects on the economies of a few large towns, but, by the time of this study in late 1988 and early 1989, these economic changes had not yet penetrated the countryside.

Lao gathering forests today are informally held as village commons and are protected by government policy against the clearing of large trees for swidden cultivation. Large trees, however, are not currently protected against commercial logging. Furthermore, the pace of present logging activities in Laos is likely to increase with recent government policies of economic openness, a strong world market in tropical hardwoods, and the rebuilding of several main roads.

At the village level, the effects of economic opening may be to reestablish opportunities for the sale of agricultural, craft, and forest products, while accelerating timber extraction from forests near some villages. But subsistence activities, including subsistence forestry, will continue to provide the safety net in years of poor crops and in politically difficult times.

Bolikhamsai Province, the site of the study, was newly established as a province in 1986. It is located in the narrow central part of the Lao People's Democratic Republic (PDR) just east of Vientiane Province, and it extends from the Mekong River border with Thailand on the south and west to Vietnam on the east and northeast. Land along the Mekong River and its tributaries tends to be flat or rolling, with the countryside becoming mountainous farther from these rivers. Some mountains, even along the rivers, rise steeply from the flat land below. All villages studied were located at lower elevations on flat or rolling land. The province has various forest types and three active logging companies.

Research Design and Procedures

To discover how women's forest work varies by household economic patterns and type of forest available, the author and a research team interviewed 120 women farmer/

gatherers from eight villages of two types in Bolikhamsai Province: four villages with access to old-growth forest and four villages with access only to second-growth forest and overgrown swidden fallows. Observations, forest visits, meetings, and conversations provided additional information. The research team, working with relevant provincial and local officials, followed a standard procedure in selecting village respondents and gathering information in each village. Data thus gathered were coded or summarized and analyzed using appropriate statistical techniques. This study was part of a larger study that also included respondents living in logging company towns.

Village Selection and Description

Village Selection. The larger study was designed to identify differences in living conditions and forest use between women of company towns and villages, and between women living in villages affected by logging company activities and women living in villages not affected by company activities. Representative villages were selected in consultation with provincial, district, and logging company officials, after villages that were located near logging company concession areas, seedling plantations, or company towns were identified.

Two incidents occurring early in the fieldwork period somewhat constrained village selection. One of the villages originally selected near a company town had to be abandoned when a band of antigovernment guerrilla soldiers surrounded a logging camp in the nearby forest and demanded rice in exchange for the hostages they had taken. While we were conducting interviews in another study village, two alleged drug merchants were murdered a few kilometers away. As a result of both incidents, we were forced to exclude villages from one entire district.

Village Descriptions. Lao distinguish between two main types of forest: old-growth forest with large trees and wild animals (*pa dong*) and second-growth forest or swidden fallows with only smaller trees and bushes and few wild animals (*pa lao*). Different terms are used to designate grassland with scattered trees and rainy-season swampland. Women in four of the study villages have access to forests containing large trees, whereas women in the remaining four villages have access only to second-growth forests and overgrown fallow fields. The *pa dong* formerly used by women in two of these latter villages had recently been logged or was being logged during the interview period.

Half of the villages studied are exclusively lowland Lao in ethnic origin. In the other half of the villages, midland Lao are a minority of the population, with most villagers being lowland Lao.

Survey villages range in size from 35 to 160 households. The newest village was first settled only 6 years ago, and the oldest was founded almost 150 years ago. Some villages were established by groups migrating from overpopulated or less agriculturally promising areas to their current location to clear land for new rice paddies; others relocated in response to the Indochina wars. All villages have schools, although only half of these village schools offer the full five years of elementary school. Five of the villages claim at least one nurse, though only two have functioning village clinics.

Village Differences in Living and Economic Conditions. Villagers with access to old-growth forest have slightly better living conditions than do other villagers, as demonstrated on several health- and education-related measures (Table 1). Villagers with access to old-growth forest are more likely to report that they gave birth most recently

Table 1

Living Conditions: *t*-Tests of Differences Between Villagers With and Without Access to Old-Growth Forest

Variable Name	Mean With	Mean Without	*t*-Test	Significance
Child survival (%)	0.65	0.73	—	n.s.
Boil water (scale 0–3)	2.31	2.56	−1.91	n.s.
Birth conditions (scale 0–3)	2.72	2.30	2.25	*
Education level (years)	2.57	1.68	2.91	**
Literacy (scale 0–3)	2.06	1.37	2.58	*
Diet quality (scale 1–3)	2.59	2.29	2.01	*
Children in school	0.91	0.93	—	n.s.

n.s. = not significant.
*$p < .05$.
**$p < .01$.

with the assistance of an experienced attendant or in a hospital and that they ate an animal protein food in at least one meal the previous day. Their educational level is barely higher than women without such access, though still quite low at an average of 2.5 years. Child survival rates, boiling of drinking water, and proportion of 7- to 15-year-old children attending school are similar for both groups.

The better living conditions of households with access to old-growth forest are probably due to settlement patterns over the last century. Three of the villages with access to old-growth forest are old, established villages with paddy as well as swidden fields. These three villages were founded well before the main road was constructed and as a result are not located near this road. Three of the villages with access only to second-growth areas were established in the last 30 years by villagers fleeing from war or forcibly removed from war zones. The villagers resettled along the new road to facilitate government control and delivery of rice rations during the war years. This same road provided easy forest access to Thai logging companies allowed entry by the Royalist government before 1975. By the time the more recent villages were established, the best potential paddy land had already been claimed and developed by earlier settlers.

Although women with access to old-growth forest have slightly better living conditions, the economic levels of the two groups appear similar. Income sources and agricultural production patterns of the two groups, though, are somewhat different. These differences are partly related to available forest type and are discussed below.

Operationalization of Variables

The research variables of interest fall into three general categories: descriptive information about households and their economies, description of the process and products of forest gathering, and differences in household economies and forest gathering by the type of forest available to villagers.

Household, household economy, and forest gathering variables were operationalized by one or several questions in the structured interview schedule administered to the women respondents. When possible, supporting data from other sources were used to verify interview statements. The main instrument, the structured interview schedule, was constructed, pretested, revised, and implemented completely in Lao.[4] Supplementary data were gathered during semistructured interviews with provincial, district, and village officials. In addition, members of the research team accompanied women skilled in forest-related activities into their gathering forest to observe the process of gathering and to see the products gathered. Team members also noted the types of forest available to the women of each village. Other data were gathered in the village from observations and informal conversations.[5]

Sampled Women and Their Households

One hundred twenty women aged 20–55 years from eight villages in Bolikhamsai Province were interviewed. Data from these interviews constitute the heart of this study. A sample of predetermined size based on village population was drawn by lottery from a list of all women aged 20–55 in each selected village. This list was compiled with the help of members of the village administrative committee and the village women's union. The drawing was held at the end of an exploratory meeting with village women. Women were eager to participate, so in two villages research team members felt the need to interview additional women who were disappointed that their names had not been drawn. These extra cases were not included in the data analysis.

The women sampled were interviewed in their homes. Interviews lasted 40 to 60 minutes. During the interview, the interviewer evaluated the respondent's housing conditions as well as conducting the interview. Most villagers received us warmly and responded well to interviewer questions. Some villages organized traditional welcoming ceremonies in our honor.

The average sampled household is like that described in national census statistics and ethnographic studies reported above. It usually contains a nuclear family of six to seven members but may include grandparents or other relatives, more commonly those of the wife. Only 10% of the sample households are headed by females. The typical woman respondent has two years of schooling and reports that she can read and write, though with difficulty. Nearly all her children attend school, at least sporadically. She usually boils drinking water, and her family eats a protein food such as fish or chicken once a day.

The economies of most sampled households are subsistence based, although with a small amount of market activity. The subsistence base has several aspects besides forest extraction—rice cultivation, garden production, and animal raising. Nearly half of all families cultivate rice in both paddies and swiddens, 29% rely on paddy cultivation only, and 22% rely on swidden cultivation only. Large gardens, often including vegetables, tubers, and fruit trees, are grown by 23% of the women, often by those whose families are already cultivating both paddy and swidden fields. Some of the produce from the larger gardens is sold. An average household owns three buffalo for paddy production and raises two pigs and a dozen chickens. Market activity appears sporadic rather than regular, with villagers either taking their produce to markets in the provincial or national capital or selling to traveling merchants. Animals, crops, or forest products are sold whenever the family needs cash.

Results

Forest Gathering

Women regularly enter the forest to gather items for use and for sale. A description of the gathering process depicts the common patterns of women's forest work, and an analysis of the items gathered indicates the contribution of forest products to the household economy. The forest work and forest product usage of women from villages having access to both old-growth forest and second-growth areas is somewhat different from the forest work and product usage of women having access only to second-growth areas.

Forest Gathering and Household Economies. Although all villagers depend on gathering to some extent, paddy-farming women in particular gather shoots and leaves only from the margins of their rice fields or from overgrown fallow fields near the village. In each village some families do not produce enough rice for their family's annual consumption. These families regularly supplement their rice with forest tubers and other forest products, though they may also work for neighbors to earn rice. During years of too much, too little, or poorly timed rain, most families must depend on the forest for subsistence. Forest tubers are largest in April and May, just when villagers may begin to experience rice shortages, and well before the October to December rice harvest.

All respondents together report gathering or hunting 141 different types of forest products: 37 types of plant food items, 68 types of medicinal products, 18 types of items for household use, and 18 types of animals. The most common items gathered or hunted (reported in at least 19 households) are listed in Table 2.

Composite lists combining all forest products gathered or hunted in each village reveal that each village gathers at least 28 forest products, with the longest reaching 53 items. The number of items does not seem to correlate with the type of forest available to a village. The composite lists for villages with access to old-growth forest are not noticeably different from the lists of villages without such access.

On the average, each household gathers eight different kinds of items from the forest. A number of families sell or barter forest products (38%), most commonly mushrooms or bamboo shoots; resin, rattan, and cardamom; squirrels, birds, rats, and freshwater shrimp. The gathering and sale of medicinal plants appears quite specialized and is usually done by an older man or woman. Only one respondent gathers medicinal plants regularly, selling nine different varieties. The respondents themselves were usually the main or one of the main household gatherers, and half of all respondents consider forest gathering as one of their central work roles.

In short, women regularly gather forest products, and forest products are an important aspect of the family economy in all villages. These products are a source of income for some families and a source of food and material for household items for nearly all families. This aspect of the rural economy becomes more important during years when other aspects of family and village economy are weaker, as during droughts or crop pest infestations, but the forest is a regular feature of the rural economy in all years.

The Process of Gathering. Gathering is an integral part of the weekly and often of the daily routine of village women, depending on the season. Twenty-five percent of the respondents report gathering every day, and 75% gather at least once a week. Women gather shoots and leaves in small groups or singly, moving steadily from plant to plant, cutting, trimming, discarding unusable material, dropping the usable shoot or leaf into a

Table 2
Commonly Obtained Forest Items

Products	Number of households
Plant food products:	
Bamboo shoots (*noh mai*)	105
Mushrooms (*het*)	92
Calamus shoots (*noh boun*)	52
Rattan shoots (*noh wai*)	49
Medicinal products:	
Sarsaparilla (*ya hua*)	44
Uvaria micrantha (*kamlang seuakhong*)	28
Other subsistence products:	
Rattan (*wai*)	39
Elephant grass (*nya kha*)	34
Thitga resin (*khi si*)	36
Bamboo for weaving (*mai hia*)	19
Forest animals:	
Fish/freshwater shrimp	34
Mouse/rat	45
Birds	39
Squirrel	25

basket, cutting a new plant, and so on. Root digging requires more time for each root. Obtaining the cores of banana and other stalks requires cutting a large plant to obtain a much smaller usable portion. Sometimes women in a large group sing or play on a bamboo flute fashioned as they walk, but they rarely stop walking or working. Most village women do not go great distances into the forest; they always come home by dark.

The majority of the women (63%) reported going with one or two other people on their most recent forest visit, although several women gathered alone, and a number went with larger groups. Respondents gathering in a group most often went with other women (68%), usually with neighbors (43%). Other gathering companions include mixed groups of relatives and neighbors (19%), husbands (14%), children (10%), non-household relatives (9%), and other household relatives (4%). Larger groups are more likely to spend all day in the forest, go into old-growth forest, and gather insects and small animals in addition to plant food, use items, and medicine. Smaller groups of people tend to spend three hours or less gathering. These groups usually visit fallow swidden fields or second-growth areas and gather only food and medicinal plants. Gatherers generally take only a machete and a basket or bag for carrying items gathered; a few also carry a fish net or basket.

Forest Type and Variation in Household Economies. The type of forest available to a rural household conditions the organization of its economy. Comparisons between respondents with access to both old-growth forest and second-growth areas and those with access only to the latter indicate the two groups use the forest somewhat differently.

Women with access to old-growth forest are more likely to report visiting the forest every day (41% versus 15%); women without such access are more likely to report that it had been more than two weeks since they last visited the forest (27% versus 16%). However, women without access to old-growth forest report selling more types of forest products than do other women ($t = -1.98$, $p < .05$). Other differences in forest activities between the two groups are negligible (Table 3). That is, village women from both kinds of villages report similar likelihood of eating forest products, using forest products during a rice shortage, and gathering three of four types of forest products (elements of the forest activities composite scale).

There are virtually no differences between the two groups of village women on any of the economic indicators, with the exception of product sales noted above: households without access to old-growth forest sell more kinds of forest items (Table 4). Women in these same households report 25% less rice available for their families than women with access to old growth, but this difference is not statistically significant.

Lower rice availability among this group of women is perhaps a result of their limited cultivation of paddy land. These women are much more likely than other villagers to grow rice only in swiddens. Fully 96% of villagers with access to old-growth forest report cultivating paddy fields, and 61% cultivate both swidden and paddy fields,[6] but only 64% of women without access to old-growth forest live in households that cultivate some paddy fields. Members of these households, though, are somewhat more likely to grow large gardens. Women in villages without access to old-growth forest report a greater variety of nonagricultural income sources than the other group of women, with more than one-fourth selling forest products; shopkeeping; selling prepared food or drink, crafts, or services; or clearing forest and planting tree seedlings (Table 5). Some of these women perceive that some commercial logging conflicts with village forest needs. In contrast, women with access to old-growth forest rely almost entirely on sales of animals or agricultural produce for their cash needs, although a few sell forest products.

When the categories of activity in Table 5 are combined so as to distinguish between traditional family farm product sales and participation in the nonagricultural market economy (Table 6), differences between village types become even clearer ($\chi^2 = 9.88$, $p < .01$).

In short, women with access to both old-growth forest and second-growth areas use

Table 3

Forest Activities: t-Tests of Differences Between Villages With and Without Access to Old-Growth Forest

Variable Name	Mean With	Mean Without	t-Test	Significance
Forest products sold (number of kinds)	0.78	1.56	−1.98	*
Types of products gathered	2.85	2.68	—	n.s.
Frequency of forest visits	5.48	4.88	1.70	n.s.
Forest activities	1.40	1.08	—	n.s.

n.s. = not significant.
*$p < .05$.

Table 4
Household Economy: *t*-Tests of Differences Between Households With and Without Access to Old-Growth Forest

Variable Name	Mean With	Mean Without	*t*-Test	Significance
Rice available (kg)	1422	1055	1.71	n.s.
Crops sold (scale 0–8)	1.52	1.64	—	n.s.
Forest products sold (number of kinds)	0.78	1.56	−1.98	*
Cash available (kip)	1757	2221	—	n.s.
Housing quality (scale 0–5)	1.61	1.66	—	n.s.
Large animals owned (number)	4.87	5.23	—	n.s.
Household goods (scale 0–8)	5.41	5.58	—	n.s.

n.s. = not significant.
*$p < .05$.

forest products to supplement their relatively more adequate subsistence rice crop. Small amounts of family surplus crop production and sale of animals seem to meet these women's minimal cash needs. Women with access only to second-growth areas, on the other hand, may have a somewhat more commercial view of the forest and forest products. Although these women visit the forest less frequently, they are twice as likely to sell what they gather. They describe a conflict between commercial logging and their forest gathering, two different commercial uses of a forest. These women are more likely to participate in other aspects of a market economy as well, through various selling and wage earning activities. Because their rice subsistence base is less adequate than that of women in the other group, they have developed alternative ways of supporting their families.

Without longitudinal data, it is hard to know if the subsistence base of women with access only to second-growth areas has always been inadequate or has only recently become so. It is possible that women with access to old-growth forest may have to adopt similar strategies as logging operations remove that source of subsistence. Interestingly,

Table 5
Income Sources: Cross Tabulation of Households With and Without Access to Old-Growth Forest

Income Sources	With	Without
Sell animals, rice, vegetables	29 (54%)	20 (30%)
Sell forest products	3 (6%)	7 (11%)
Sell cooked food, whiskey; shopkeeping	0 (0%)	4 (6%)
Craftwork, sewing, hairdressing	1 (2%)	2 (3%)
Clear forest, plant seedlings	0 (0%)	4 (6%)
None	21 (39%)	29 (44%)
Totals	54 (100%)	66 (100%)

Table 6

Income Sources: Cross Tabulation of Households With and Without Access to Old-Growth Forest

Income Sources	With	Without
Sell animals, rice, vegetables	29 (54%)	20 (30%)
Sell nonagricultural products/services/wage labor	4 (7%)	17 (26%)
None	21 (39%)	29 (44%)
Totals	54 (100%)	66 (100%)

$\chi^2 = 9.88, p < .01$.

the income sources reported in villages with access only to second-growth areas are mainly exploited by women (selling forest products, selling cooked food or whiskey, or shopkeeping), with the exception of clearing forest and planting seedlings, tasks open to both men and women. Any changes in the quality of forest available to a village, then, are likely to have effects on women's work. With further forest destruction, more women may market agricultural products or prepared items or may engage in wage labor, if the local market can support them.

The more subsistence-oriented women may experience difficulty entering the market economy, because some of them live in villages without roads. They may prefer to maintain their subsistence orientation. Further research on accessible villages with access to old-growth forests and remote villages with access only to second-growth areas would clarify this issue.

In any case, it is clear that women's forest work in both types of villages contributes regularly to the economies of their households: by supplementing household subsistence production, by contributing to family income, or both. This is one of several work roles that appear to support women's social influence.

Notes

1. See, for example Ehlers (1990), Stoler (1977), and Weiner (1986).

2. Moore identified several studies documenting this phenomenon, including her own work (1988, p. 61).

3. For information on the history of Laos and the region, see Ireson and Ireson (1989) and Steinberg (1987).

4. Copies of the original interview schedule and an English translation are available on request.

5. Besides the author, the research team consisted of a social science–trained official from the central government, four women interviewers (three former teachers and one technically trained forester), and the vice president of the provincial Women's Union. The central government official introduced the project to provincial authorities and gathered much of the supplementary province, district, and village data. Fieldwork took a total of six weeks, spread through the months of December 1988 to February 1989. Prior to this, Westerners had not been allowed to conduct field research in the Lao People's Democratic Republic. An independent research team was not permissible. Such a team would not have been granted travel papers or have been accepted by villagers. Our provincial Women's Union guide introduced the project to villagers at an explanatory meeting for all village women, but neither she nor other official Lao were present during individual interviews.

6. These figures are inflated by reports from the farmers of one newly established village who are still clearing and leveling their paddy fields, so most of these fields are still, in actuality, swidden fields. The villagers can't acknowledge this however, because they were given the land by provincial authorities so they would cease slash-and-burn cultivation. Even without this village, however, the differences are still striking.

References

Ayabe, T. 1959. *The village of Ban Pha Khao, Vientiane Province: A preliminary report.* Laos Project Paper No. 14. Los Angeles: Department of Anthropology, University of California.

Babb, F. 1986. Producers and reproducers: Andean marketwomen in the economy. In *Women and change in Latin America,* ed. J. Nash and H. Safa, pp. 53–64. South Hadley, MA: Bergin & Garvey.

Barker, R., and R. Herdt, with B. Rose. 1985. *The rice economy of Asia.* Washington, DC: Resources for the Future (distributed by Johns Hopkins University Press).

Belsky, J., and S. Siebert. 1983. Household responses to drought in two subsistence Leyte villages. *Philippine Quarterly of Culture and Society* 11:237–256.

Branfman, F. 1978. *The village of Deep Pond, Ban Xa Phang Meuk, Laos.* Amherst, MA: International Area Studies Programs, University of Massachusetts.

Charlton, S. E. 1984. *Women in Third World development.* Boulder, CO: Westview.

Condominas, G. 1970. The Lao. In *Laos: War and revolution,* ed. N. Adams and A. McCoy, pp. 9–24. New York: Harper and Row.

Ehlers, T. B. 1990. *Silent looms: Women and production in a Guatemalan town.* Boulder, CO: Westview.

Fortmann, L. 1986. Women in subsistence forestry: Cultural myths form a stumbling block. *Journal of Forestry* 84(7):39–42.

Fortmann, L., and D. Richeleau. 1985. Women and agroforestry: Four myths and three case studies. *Agroforestry Systems* 2:253–272.

Halpern, J. 1958. *Aspects of village life and culture change in Laos: Special report.* New York: United Nations Council on Economic and Cultural Affairs.

Hecht, S., A. B. Anderson, and P. May. 1988. The subsidy from nature: Shifting cultivation, successional palm forests, and rural development. *Human Organization* 47(1):25–35.

Hoskins, M. 1980. Community forestry depends on women. *Unasylva* 32(130):27–32.

Ireson, W. R., and C. Ireson. 1989. Laos: Marxism in a subsistence rural economy. *Bulletin of Concerned Asian Scholars* 21(2–4):51–72.

Kaufman, H. 1956. Village life in Vientiane Province. Vientiane: U.S. AID Mission to Laos. Unpublished report.

Keyes, C. 1975. Kin groups in a Thai-Lao community. In *Change and persistence in a Thai society,* ed. G. W. Skinner and A. T. Kirsch, pp. 278–297. Ithaca, NY: Cornell University Press.

Kunstadter, P., E. C. Chapman, and S. Sabhasri. 1978. *Farmers in the forest: Economic development and marginal agriculture in northern Thailand.* Honolulu: University of Hawaii Press.

Leacock, E. 1978. Women's status in egalitarian society: Implications for social evolution. *Current Anthropology* 19(2):247–275.

Liemar, L. M., and M. Price. 1987, November. Wild foods and women farmers: A time allocation study in northeast Thailand. Paper presented at Association for Women in Development Conference, Washington, DC.

Maynard, P., and P. Kraiboon, 1969. Evaluation study of the Muong Phieng Cluster Area. Report prepared for U.S. AID Mission to Laos. Bangkok: Stanford Research Institute.

Moore, H. 1988. *Feminism and anthropology.* Minneapolis: University of Minnesota Press.

Mueller, M. 1977. Women and men, power and powerlessness in Lesotho. In *Women and national development: The complexities of change,* ed. Wellesley Editorial Committee, pp. 154–166. Chicago: University of Chicago Press.

National Committee of Plan, Lao PDR. 1985. *1985 population census, preliminary results.* Vientiane, Laos: Author.

Orr, K. 1966. Behavioral research project for the Sedone Valley program, Southern Laos: Preliminary presentation of interviews and questionnaire data from Lao officials and villagers. Vientiane: U.S. AID Mission to Laos. Unpublished paper.

Potter, S. 1977. *Family life in a northern Thai village.* Berkeley: University of California Press.

Rosaldo, M. 1980. Use and abuse of anthropology: Reflections on feminism and cross-cultural understanding. *Signs* 5(3):389–417.

Sacks, K. 1974. Engels revisited: Women, the organization of production, and private property. In *Woman, culture and society,* ed. M. Rosaldo and L. Lamphere, pp. 207–222. Stanford: Stanford University Press.

Sacks, K. 1979. *Sisters and wives: The past and future of sexual equality.* Westport, CT: Greenwood.

Sanday, P. 1974. Female status in the public domain. In *Woman, culture and society,* ed. M. Rosaldo and L. Lamphere, pp. 189–206. Stanford: Stanford University Press.

Sanday, P. 1981. *Female power and male dominance: On the origins of sexual inequality.* Cambridge: Cambridge University Press.

Sen, G., and C. Grown. 1987. *Development, crises, and alternative visions: Third World women's perspectives.* New York: Monthly Review.

Steinberg, D., ed. 1987. *In search of Southeast Asia: A modern history.* Rev. ed. Honolulu: University of Hawaii Press.

Stoler, A. 1977. Class structure and female autonomy in rural Java. In *Women and national development: The complexities of change,* ed. Wellesley Editorial Committee, pp. 74–89. Chicago: University of Chicago Press.

Stuart-Fox, M. 1986. *Laos: Politics, economics and society.* London and Boulder: Frances Pinter and Lynne Rienner.

UNICEF. 1987. *An analysis of the situation of children and women in the Lao People's Democratic Republic.* Vientiane, Laos: Author.

Weiner, A. 1986. Forgotten wealth: Cloth and women's production in the Pacific. In *Women's work: Development and the division of labor by gender,* ed. E. Leacock and H. Safa, pp. 96–110. South Hadley, MA: Bergin & Garvey.

Wulff, R. 1972. *A comparative study of refugee and nonrefugee villages. Part 1: A survey of long established villages of the Vientiane Plain.* Vientiane: U.S. AID Mission to Laos.

Epilogue

Legal and illegal commercial logging expanded dramatically during the few years following this study, as I had predicted. During the same period, opportunities for wage employment increased in towns, especially Vientiane. The growing garment industry, for example, began to employ an increasing number of women, mostly young women. Observers report that some rural women have responded to these new opportunities by moving to town to seek employment. The negative effects of commercial logging on women's forest gathering are probably more widespread now than at the time of the study, and it is likely that fewer rural women now have access to old-growth forest. Rural women unwilling or unable to migrate to town, however, may be able to expand their income sources as urban economic growth filters into the countryside, as many women with access only to second-growth forest have done; but many may instead find alternative rural economic opportunities to compensate for the loss of their gathering forests.

Additional Reference

Ireson, C. 1996. *Field, Forest, and Family: Women's Work and Power in Rural Laos*. Boulder, CO: Westview Press.

Chapter 3

The Underground Proving Ground: Women and Men in an Appalachian Coal Mine

SUZANNE E. TALLICHET

Assistant Professor of Sociology
Morehead State University
Morehead, KY 40351
USA

Abstract *Coal mining has provided rural women with economic opportunities not otherwise available in more restricted rural labor markets. This study focuses on coal mining women's physical and social adaptations to work underground. Case study data were obtained from in-depth interviews with 10 coal mining women, archival research, and observations made at a large underground coal mine in central Appalachia. The data demonstrate the problematic nature of women's struggle to prove themselves as competent in performing traditionally male-identified work in a male-dominated work setting. Although over time men's overt work-related hostility and sexual harassment dissipate, their more subtle sexualization of work relations reveals the power of negative stereotypes regarding women's work capabilities and behavior.*

Keywords Coal mining, women, gender, women's nontraditional employment.

Introduction

Despite more limited employment opportunities compared with their urban counterparts, rural women have been rapidly increasing their labor force participation during the past few decades. However, unlike urban-based labor markets, nonmetro labor markets are comprised of relatively fewer industry types and offer predominantly occupations in blue-collar extractive industries such as agriculture, forestry, fishing, and mining, all of which heavily favor male employment (Tickamyer and Tickamyer, 1991). During the coal industry's "boom" period of the 1970s, employment in mining expanded rapidly, offering opportunities for rural workers (Tickamyer and Bokemeier, 1988), including women. Thus, during the mid- to late 1970s, several thousand pioneering women made history when they officially began mining coal (Hall, 1990).

Women's integration into underground coal mining has been constrained by the twin forces of capitalism, because it affects all miners, and patriarchy, because it affects women in particular. The present investigation deals with the issues of women's entry into coal mining and their physical as well as social adaptations to mining work at a large coal mine in central Appalachia. First, however, it is important to have some knowledge about the coal industry and the occupation itself. Therefore, this chapter begins by briefly reviewing the mining labor process and women's entry into the male culture of mining in general before turning to the case study and its analysis.

The Labor Process of Underground Coal Mining

Mines are a series of interconnecting and parallel passageways through which miners, their machinery, air, and coal are moved into and out of the mine. Usually, miners work about 2 miles or more from the mine entrance. Coal mines are noisy, dirty, and dangerous. Problems with roof supports, ventilation, lighting, drainage, access, coal extraction, and conveyance are always present. Operating highly powered equipment and using high-voltage electricity in tight working areas where footing is often unsure can result in twists and sprains of joints, broken bones, dismemberment, and hearing loss. Veteran miners face the possibility of developing coal miner's pneumoconiosis, or "black lung" disease.

Mining sections are defined geographically. A single crew works on each section in shifts. Although mine operations are essentially to extract coal, other tasks must be performed to maintain mine safety. Hence, sections and jobs in a coal mine are divided according to two basic work functions: production and maintenance. Functionally related jobs are classified into five grades (see Table 1). Skill and wage levels increase with the grade of the job. Relatively speaking, jobs in Grade 1 generally require fewer skills and more physical strength than jobs in higher grades, which require specific operative skills or certification. Work underground is highly interdependent. Grade 1 workers perform maintenance duties in support of those miners classified in higher ranking jobs, who either move or extract coal from the face. New miners, or "red hats," are usually assigned to the Grade 1 positions of either general inside labor ("GI") or beltman. After receiving their "mining papers," or miner's certificate, miners become "black hats," at which time they can bid on any posted job, provided they have the seniority and necessary skills.

Each section, whether it is a production or "down" section, is supervised by a section foreman or crew boss, who is a nonunion, salaried company employee. Although bosses stay in close contact with workers, by United Mine Workers of America (UMWA) contract they are forbidden to perform any work duties (UMWA, 1988). Bosses are given the authority by the coal company to make the day-to-day decisions regarding production activities, safety, and work assignments.

Historically, the most intense struggles between miners, represented by the UMWA, and the coal operators, known collectively as the Bituminous Coal Operators Association (BCOA), have occurred during periods when the demand for coal is either rising or declining. Thompson (1979) articulated this historical tension in terms of the dialectic relationship between capitalist accumulation and the relations of production. During the past century, in order to remain competitive as capitalist producers, the coal operators had to ensure increasing profits so they could continue to expand their operations. To do so, they needed to maintain greater control of the miners' work activities at the point of production to increase output and reduce their labor costs. By increasing the mechanization in the mines, the capitalist operators hastened the pace of production and increased the dependability of the output.

However, increasing mechanization and the establishment of a job hierarchy in the mines did more than simply reduce the miners' skill level. The removal of the skill barriers to entry-level mining jobs expanded the pool of potential mining labor and increased competition for mining employment. Another result of increasing mechanization was the proletarianization of the workforce, which served to raise the miners' consciousness as a laboring collective in opposition to the capitalist operators (Thompson, 1979). Above-ground miners were generally a gregarious lot. Mining communities,

Table 1
Coal Mining Jobs by Title Within Their Classification Ranks

Rank	Classification of Titles
Grade 5	A. Continuous Mining Machine Operator B. Electrician C. Mechanic D. Foreboss E. Longwall Machine Operator F. Welder, First Class G. Roof Bolter
Grade 4	A. Cutting Machine Operator B. Dispatcher C. Loading Machine Operator D. Machine Operator Helper E. General Inside Repairman and Welder F. Rock Driller G. Continuous Miner Helper H. Roof Bolter Helper I. Maintenance Trainee J. Electrician Trainee
Grade 3	A. Driller—Coal B. Shooter C. Precision Mason—Construction D. Faceman E. Dumper F. Shuttle Car Operator
Grade 2	A. Motorman B. Maintenance Trainee (6 mo.) C. Electrician Trainee (6 mo.) D. Electrician Helper E. Mechanic Helper
Grade 1	A. Beltman B. Bonder C. Brakeman D. Bratticeman E. General Inside Labor F. Mason G. Pumper H. Timberman I. Trackman J. Wireman K. Laborer

being small and isolated, furthered their common interests. Their increasing dependency on one another in the mines strengthened their solidarity.

The 1970s was an important period of conflict between the operators and miners that affected the terms and complexion of mining employment significantly (Simon, 1983). As a result of the energy crisis during the early part of that decade, both parties anticipated rapid growth within the industry. The miners had heightened expectations for winning numerous concessions in the 1974 contract and, to some extent, their expectations were fulfilled. Although the companies continued to look for ways to cut costs, they also agreed to increase miners' wages and benefits in order to attract new miners to help increase production—a move that made mining more attractive to nontraditional employees, such as women.

Women's Entry into the Male Culture of Mining

Outside the mines, miners have organized themselves politically and culturally in opposition to the coal operators' attempts to exploit them (Wardwell, Vaught, and Smith, 1985). Inside the mines, the twin forces of advancing technology and bureaucratic organization have made their work increasingly interdependent. Even so, work is still performed under threatening and anxiety-provoking conditions. Thus, the work underground strongly discourages work autonomy and results in correspondingly high levels of conformity. Under these conditions, workers come to value certain traits in one another as they collectively cope with the stressors in the workplace. A "good" miner is competent and tough. A competent miner works hard and observes safe work practices, and a tough miner never demonstrates fearful behavior even if he or she admittedly is afraid. In addition, miners with good reputations display a "team spirit" through cooperation and "give and take" jocularity among coworkers. Thus, they emphasize "getting along with others."

Having the above qualities enhances one's reputation among coworkers and supervisors, all of whom are locked into relational patterns of power and dependency. Workers are dependent on a boss for rewards, such as promotions, which stem from the boss's estimation of them as miners. Moreover, what a boss believes about individual workers can influence what workers are inclined to believe about each other. Conversely, a boss is dependent on workers to produce coal, which affects her or his own reputation as a company employee. Workers also have the ability to influence what a boss may come to believe about one of their coworkers.

Formally, the union promotes this solidarity. UMWA "brothers" and "sisters" are united in their collective militancy, as manifested in the union slogan: "An injury to one is an injury to all." Informally, ritualistic behaviors underground such as teasing, practical jokes, and horseplay serve to reduce tension about the dangers and incorporate individuals into tightly knit work groups. Ross (1974) found that miners are "open, friendly, helping but tough; hostile to the company but not lazy; with blunt, unvarnished feelings along with tolerance, always sharing and never cheap; everyone with a nickname, indicating individual acceptance in the group; a social solidarity recognizing individualism . . . " (p. 176). According to Althouse (1974), experienced miners, especially older ones, are immersed in what he called "the miner mystique—a sense of justice, toughness, manliness, respectability, pride, and above all, solidarity" (p. 16). Hence, most miners evaluate the "worth" of entry-level employees on the basis of their commitment to and stamina for working underground.

Traditionally, coal mining has been a "man's job" in which women had no place.

Although it is known that women worked underground in family-operated coal mines in Appalachia during the Depression and shortly after World War II, according to government records, women were not officially employed in underground coal mines until 1973 (President's Commission on Coal, 1980). During that year, women began entering coal mining jobs as the industry was prospering. Few women were hired, however, without pressure from government agencies. Into the late 1970s, although the women's rate of entry was steady, it was slow. According to advocates for women in mining, government agencies charged with enforcing equal employment statutes had failed to recognize the obvious discrimination in the industry (Hall, 1984).

During the late 1970s, a Tennessee-based women's advocacy group known as the Coal Employment Project (CEP) provided perhaps the greatest impetus toward women's entry and integration into the coal industry. In October of 1977 the CEP staff filed a lawsuit with the Office of Federal Contract Compliance Programs (OFCCP), eventually forcing 153 coal companies with federal contracts into paying thousands of dollars in backpay to women whom they had denied jobs and to begin hiring more women until women constituted approximately one-third of their total workforce (Hall, 1990). In 1978, women accounted for less than 5% of all new hires in the industry. By 1979 they were 11.4% of all new hires and their absolute numbers in the coal mining ranks began to rise rapidly (Reskin and Hartmann, 1986). By 1986 women constituted almost 2% of the total underground workforce (Butani and Bartholomew, 1988). However, the coal bust of the mid-1980s and the ensuing layoffs caused a relative decline in their numbers.

Theoretical Framework

Socialist feminist theorists use the dual processes of patriarchy and capitalism to explain women's oppression and inferior status in the family, the labor market, and society at large (Sokoloff, 1988). Although its proponents focus on the mutually reinforcing and sometimes conflicting relationships between these two forces, the key concept of patriarchy is seen as an autonomous force that, when combined with capitalism, results in the maintenance of male privilege and the sexual division of labor in the workplace. Hartmann (1976) defined patriarchy as "a set of social relations which has a material [and an ideological] base and in which there are hierarchical relations between men, and solidarity among them, which enable them to control women" (p. 138).

During the past century, neo-Marxist theorists have noted the changes capitalists have made in the work process to better control workers. The sequence of mechanization, task specialization, and closer supervision has brought workers under the capitalists' tighter control. The effects on the working class have been a systematic deskilling and further division among workers themselves (Gordon, Edwards, and Reich, 1982). Since these patriarchal relations are reproduced in the workplace, socialist feminism "emphasizes the role of men as capitalists in creating hierarchies in the production process in order to maintain their power" (Hartmann, 1976, p. 139). Therefore, men are united by their common vested interest in maintaining the status quo and are, therefore, dependent on one another to make these hierarchies work. Men at higher levels in the hierarchy "buy off" those at lower levels by offering them power over individuals who are even lower. Thus, workers, male and female, are exploited by capitalists, and women are exploited by men, resulting in women's "superexploitation."

However, out of the resolution between the forces of patriarchy and capitalism comes their renewed antagonism. For example, at times capitalists have used the threat of substituting male workers with lower wage female labor in order to increase their profits. In these cases, the unions, representing the patriarchal interests of males, have levied pressure on the capitalists to do otherwise (or to at least admit women so as to accommodate some basic beliefs in the patriarchal ideological system). The unions' role in the creation of internal labor markets, defining occupational hierarchies by establishing positions and corresponding wage rates as well as the rules for advancement, has been crucial to the realization of their power within the capitalists' industrial systems. Women's position in the workplace has been the result of the mutual accommodation between patriarchy and capitalism.[1]

In addition to the concept of patriarchy, another relatively recent theoretical formulation useful in understanding female coal miners is social closure theory. Social closure theory states that "a status group creates and preserves its identity and advantages by reserving certain opportunities for members of the group" using exclusionary and discriminatory practices (Tomaskovic-Devey, 1993, p. 61). Because women pose a threat to masculine-based privileges, men will tend to emphasize women's presumed incapability for doing "men's work."

The gendered status hierarchy is preserved through certain "social practices that create or exaggerate the social distance between status groups" (Reskin and Roos, 1987, p. 7). These practices dictate subordinates' behavior in the presence of dominant group members and shape the casual interaction between them. Such gendered status hierarchies are usually seen by both men and women as natural and, thus, appropriate, because they recreate gendered social relations occurring in the larger culture. Because women who do "men's jobs" are challenging the routinization of the presumably natural order of gendered relations, they are "at risk of gender assessment" (West and Zimmerman, 1987, p. 136). They are held accountable for engaging in gender-inappropriate behavior through other women's and men's (and their own) evaluations of their behavior based on "normative conceptions of appropriate attitudes and activities" for their gender category (West and Zimmerman, 1987, p. 139). Thus, women in male-dominated workplaces are required to prove their "essential femininity."

Kanter (1977a, 1977b) was among the first to document that a token woman's conspicuous presence leads to men's exaggeration of the differences between them. This is accomplished via men's "sexualization of the workplace" during which work relations between men and women are "sexualized" (Enarson, 1984; Swerdlow, 1989). Sexualizing the workplace and work relations consists of behaviors that express "the salience of sexual meanings in the presumably asexual domain of work" (Enarson, 1984, p. 88). As the literature on women in nontraditional blue-collar occupations has documented, most men engage in at least one of several forms of workplace sexualization, using sexual harassment, sexual bribery, gender-based jokes and comments, and profanity to make sex differences a salient aspect of work relations (Enarson, 1984; Gruber and Bjorn, 1982; Swerdlow, 1989). These behaviors, according to Enarson (1984), "constitute a continuum of abuse" and reflect "a cultural tradition which sexualizes, objectifies, and diminishes women" (p. 109).

Men's sexualization of work relations directly expresses the expectation that women should "act like women" by making their integration into a sexualized workplace contingent on their production of gender as they interact with men. Because men's sexualization of work relations identifies women primarily by their gender and not by their work role, it objectifies them. As Schur (1984) has pointed out, this "objectification" of female workers contributes to their stigmatization by men concerning their work-related inferiority. Because there are simply too few women present in a workplace dominated by men, women are usually unable to collectively counter men's expressions of the negative stereotypes upon which women's presumed inferiority is based (Kanter, 1977b). However, depending on their individual abilities to adapt to this set of social conditions, women in male-dominated workplaces are able to accommodate and simultaneously resist men's beliefs about their own superiority. These mechanisms and strategies employed by women working in an underground coal mine constitute the main focus of the present study.

Methodology

Primary data from a case study conducted at a single coal mining establishment in southern West Virginia were obtained using in-depth semistructured interviews, informal conversations, and on-site nonparticipant observation. Data collection in the field lasted approximately one month. Sampling is best characterized as a combination of snowball and purposive methods. From my earliest interviews with women, I obtained the names of others who, by virtue of their tenure, job rank, or other job-related experiences, such as discriminatory treatment, were selected for the study. All of these women consented to be interviewed. The in-depth semistructured interviews were conducted with ten women, and on numerous occasions shorter, 20-minute discussions were held with seven more women who were either unable or unwilling to speak at greater length. In addition, a company management official was interviewed, and conversations were held with a high-ranking union official and several male miners during my daily visits to the site.

Every effort was made to conduct interviews in quiet private settings, such as my motel room or in the women's homes, at times when the respondent would be at ease and feel free to provide information and opinions about sensitive topics. These in-depth interviews lasted about 1½ hours and were taped with the interviewee's consent. The company management official was interviewed in his office at similar length. The brief conversations with the additional women in the study occurred in the women's bathhouse. Brief conversations with the local union official and male miners occurred in the lamphouse.

Profile of the Case Study and Sample

Similar to other coal companies, the company used for this case study did not begin to hire women in appreciable numbers until it was forced to do so. In the fall of 1978, the company was sued for sexual discrimination in hiring and settled the charges against it by paying backwages to those women it had failed to hire and by adopting a new hiring ratio beginning in 1979. The management official explained that the company operated out of fear and, so, was forced to accept virtually any woman who applied. He expressed resentment at the government for infringing on his right to manage the workforce, adding that "management had to pay the price for social change." Indeed, the women who applied for jobs at the mine during that time were hired without delay. As relatively large numbers of women entered the mine, several changes in company policy were made. For example, the company reissued rules governing workers' conduct underground, strictly forbidding any form of harassment, horseplay, or profane and obscene language.

During the early 1980s the company's employment peaked at about 800 miners, more than 90 of whom were women. Since then, however, the pressures of industrial decline have forced even the largest of coal companies to lay off the least senior miners, many of whom are women. At the time of this study, in the fall of 1990, the company employed approximately a dozen assistant foremen or "bosses," all of whom were male, and 466 miners. According to the list provided by the company, there were 23 female miners, who constituted almost 5% of the underground workforce. Three pairs of women were working together on regularly assigned shifts; the others were working as token members on all-male crews. All of the miners were members of the UMWA.

At the time of the study, approximately 35% of all the miners in the case study were classified in Grade 1 jobs. Female miners were disproportionately represented among the laboring jobs (Grade 1) relative to men. Only 5 of the 23 women working at the mine held job classifications higher than Grade 1. Among the women in the sample, six out of ten were classified in Grade 1 jobs and the other four held jobs in Grades 2 through 5. The least experienced women had been mining for 9 years, the most experienced for 15 years. The ages of women in the sample ranged from 29 to 50. Most had a high school diploma, one had completed the tenth grade, and two others had attended but never graduated from college. At the time they were hired, seven of the ten were either single or divorced with children. The other three were married with at least one child. The youngest woman, a single mother, was black. The rest of the women in the sample were white.

From Red Hat to Miner: Women's Adaptations to Mining

During their early days underground, women's adaptation to working in the mine was both physical and social. Both women and men at the mine told me repeatedly: "Not everybody can be a miner, you know." Becoming a miner meant being physically capable and willing to adopt a specific orientation toward work. For the women this was particularly important because they were doing work deemed appropriate for males only. Moreover, beyond the instrumental challenges of working underground, the women were also hard pressed to form solid working relationships with male coworkers and bosses who had traditionally defined themselves by "what women are not" in the course of their everyday interactions with each other. As a result, the women miners often had to overcome coworkers' and foremen's work-related hostility and sexual harassment.

The "Brute Work" of Mining

Coal mining, considered one of the most dangerous occupations, requires stamina and strength regardless of whether a miner is doing heavy manual labor or operating heavy equipment. All new miners are assigned to the entry-level position of general inside labor (Grade 1) for a specified period usually lasting between 4 and 6 months. Hence, their tasks consist of what the women call "brute work," including some of the most physically demanding types of manual labor performed underground. Basically, brute work consists of rockdusting, hanging "rag" (ventilation curtain), setting timbers for roof support, shoveling coal that has gobbed off the beltline, moving the beltline structures and power cable, laying track, and keeping the mine free of debris.

Unlike miners who have operative jobs, general inside laborers are given their assignments daily by their section boss. Work assignments and, therefore, a general inside laborer's work location are made at the bosses' discretion. As one woman working on the belts said, "When you're general inside, they can make you do anything, like shovel a mud hole or hang rag. That's hard work." Two of the female miners who had started working at the operation at the same time commented on their first few weeks: "Remember [looking at her partner]? He [the boss] told us to get rock supports and timbers to use? Rough. It was rough for me [after the first few days]. Your body physically could not move, but you had to do it anyway. These jobs are something different and women aren't structurally built like men." She added, however, "They hired you here to work and that's what they expected you to do. They expected you to do what they'd tell you to do." Similarly, another woman talked about the difficulties of her early work experiences: "Like when you're hanging rag, that's the toughest job in the mines. You had to lift [and] drag like three boards and two timbers and lift them up and that old cloth stuff, the rag. You get real dirty and I'm short. It's just different stuff. All you got to depend on is the little light on your head. Seemed like nobody felt sorry for me, but I wasn't no man."

Although these women acknowledged that their own lesser strength or stature relative to men's was a limitation, similar to other women in nontraditional blue collar jobs (Deaux, 1984), they insisted that the discrepancy between the physical demands of their jobs and their own capabilities was one of the initial adjustments they had made long ago. One of the two women quoted above also declared, "It was just hard work. You know we can do it now." And the other commented, "It's still hard work. We've just adapted to the conditions. We've just gotten stronger and learned the ropes basically. [But back then] it was a whole new world."

In addition to the physical demands of mining, the women also learned to cope with the dangers. Although most miners admit being apprehensive about mine work, they refrain from showing their fears. The apparent paradox allows them to cope with the omnipresent threat of serious injury or death. Moreover, the women's demonstration of outward calm and restraint in the face of danger is a characteristic male miners associate with being masculine and doing a man's job. Few of the women mentioned being afraid of the mine and the possible dangers. Rather, as is typical of their male counterparts, one of the women commented on her approach: "I'm not scared. I have a fear of it because you know you have to, but it don't bug you all the time. You have a fear, you're conscious enough to know something can happen . . . [but] if you let it bother you or worry you, you wouldn't go back." And from another woman: "I used to be intimidated about all the big machinery, but I never worried about top falling on me or anything, never bothered me a bit. But that's wrong. You really need to be aware of it. But you get so used to it, it doesn't bother you."

Despite some of their initial difficulties with the work itself, at the time of the study most of the women expressed satisfaction with their jobs, most often mentioning the high wages and financial security. Some women were less enthusiastic about coal mining than others. One of the masons, who has held several different jobs during her 12 years at the mine, commented on her current job as follows: "[It's] another hard [kind of] work, a lot of lifting all the time, a lot of smashed fingers, broken fingers and broken thumbs and all that. It's got its good and bad points. It's a job and I make good money and that's it." But for others, although the higher wage was important, doing their jobs had certain intrinsic rewards, too. One woman said with pride, "I had to shovel gravel up [at the face] off onto the plow under the track. But I like my job. It's dirty, it's hard, it's cold and wet, but I like my job."

Several of the women mentioned that other women who had started working at the mine with them quit within weeks of being hired because they lacked the physical strength needed. Although a lack of strength and endurance affects job retention in coal mining, there was no indication that women's relative lack of physical strength during the initial adjustment period affected their prospects for advancing to a more skilled operative position. Rather, acceptance and recognition by male coworkers is more central to the issue of women's advancement in occupations, such as coal mining, that have strong male-identified traditions for work and social relations (Deaux, 1984). Most of the miners, female and male alike, reported that a miner's work reputation is important, not only for gaining respect and getting along with their coworkers, but also for gaining the kinds of opportunities necessary for advancement.

A miner's work reputation is usually established within the first few years of employment underground. Model coal miners are typically recognized by coworkers and bosses as being able and consistently willing to work, especially "brute" work. In order to establish a good reputation, "my advice to anyone going into mining," one woman said, "is to get the toughest job underground and go at it." But a good work reputation also depends on a good work record with few, if any, absences. Absenteeism was not only sanctioned by the company, but it created a hardship for a miner's crew. According to one woman, "You can be slow [on the job], but you have to be there." Another woman said that the combination of having a bad work record and making mistakes on the job is often grounds for dismissal and that a miner who has a bad work reputation risks losing the union's support.

Miners tend to feel a great deal of responsibility toward each other to get work done efficiently and quickly. Moreover, bosses' reputations with the company depend on their crews' willingness to pull together and work cooperatively. When work does not get done bosses look bad to the company and miners' resentment toward one another builds. Hence, when one miner slows down or fails to complete her or his assigned task, the others must take up the slack. Both the women and the men I spoke with had stories about recalcitrant coworkers. However, women found that they had to be equally as assertive with other men as the men were with each other when attempting to correct the situation. As one woman said, "There's some men like this one guy I used to bolt with. The boss told us one night to go get our pin supplies. Well, he was gonna sit on the back of the bolter and sleep. I kept carrying him and he never did come and help me. So I just made all the pins up and put them on my side. When we got ready to pin a place, he come over and I said if you take one of them pins I'll wrap it around your neck. I cussed a little bit and the boss got scared. You just have to put them in their place or they'll make it as rough on you as they can."

Crew members can influence what others, including the boss, think of each other based on a miner's work reputation. Two of the women who work together told me, "what we have to say about each other means a lot" regardless of gender. Just as a boss can refuse to take a worker on his crew, miners can affect his decision to do so. As one women stated, "If you're a lazy, good-for-nothing, they stick you somewhere where they can't depend on you. So the harder you work, the more they depend on you. So reputation is everything and once you get a lazy reputation, no matter how hard you work from that point on, you still have that reputation."

Most of the women in the sample agreed that establishing a good work reputation was harder for women than it was for men, although the extent to which they were willing to assert the existence of this double standard varied. One woman's awareness of the situation is demonstrated in the following: "There was a lot of women who didn't care and didn't do anything, but then there were a lot of men who was lazy. You couldn't get them hardly to move. They couldn't say much about the women, but they did. It's awful, but it's true. A boss would make it harder on that woman and they would [reprimand her] even if she did do a lot of work. It doesn't make any difference."

To the extent that women must work harder and have better work records in order to take advantage of available opportunities that could lead to promotion, they are disadvantaged relative to men.

Joining the Society of Miners

As tokens on work crews, the women posed a threat to male solidarity and those common bonds of masculinity vested in the culture and lore of coal mining. Threatened with the changes produced by the women's entrance, men reacted in ways to heighten the social boundaries between the women and themselves by exaggerating how the women were different. This was typically accomplished through work-related hostility and the sexualization of work relations in the form of sexual harassment, propositioning, and sexual bribery. This section discusses how the women adjusted to both sets of circumstances in the overall process of proving themselves as coal miners.

Work-Related Hostility. According to the women in the sample, many of their male coworkers made derisive comments about their presence, questioning the appropriateness of women mining coal and the women's ability to doing so. Two female miners related the following: "They would say, your husband works. What are you doing in here? You shouldn't be here. You're taking a man's job." Another woman said, "Even some of our union brothers [have said] I don't think women ought to be in here. They ought to get out here and let a good man have this job. They said we should be home cleaning house, raising kids, that that's no place for us, that that's a man's job."

Other male coworkers simply ignored the women. "They just avoid you. You couldn't even hardly talk to them or anything," one woman remarked. "Some will even tell you they don't like to work with women."

In turn, some women in the sample responded with justifications of their presence, such as: "[Male coworker said] why don't you go home and give this job to a man that needs it. I said, well, when I come up this holler to get my job they was begging for men to work and they didn't come and get it. It's mine. I'm keeping it." Another woman reported, "I even had a boss tell me he didn't like to work with women and he wanted to know why my dad let me come in the mines. I said, buddy, I was 28, divorced and single. I could do whatever I wanted whenever. And he said 'I just don't like to work with women.' And I said well, you just best get your dinner bucket and go the house [walk off the job] because I'm here to stay and I'll be here when you're gone." And she avoided some of these men: "There have been a few of them that's said, I really don't think women's got no place in the mines, but they're not being smart about it. They just tell you their feelings and when they do, I just kinda stay away from them. I think that's their right. But my right is here to work and I'll just do my work and not bother around them or anything." Unfortunately, whether a new female miner avoids or is ignored by her male coworkers makes little difference because the consequences are the same. The resulting social isolation makes a woman's socialization to the workplace and learning new work-related skills increasingly problematic.

All of the women said that male coworkers and bosses had complained that the women were incapable of performing the work required of them. One woman told of her early days on the job underground: "[Male coworker said] if you can't do the job, what'd they put you up here for, and just stuff like that. They didn't want you to [work], they don't want you to even try because you're crowding in on their turf." Two other women said that when they first started, some of the men told them that mining jobs were physically too difficult for them. Both of the women, miners now for almost a decade, felt that the men had substantially exaggerated their claims. They explained that this was male mythology designed to keep them from aspiring to become miners, not too much unlike the Irish folktale that women were bad luck in a coal mine.

Some coworkers and foremen sometimes used more explicit tactics to demonstrate women's incompetence in order to drive them out of the workplace. As one woman miner explained, "We were usually shoveling track, shoveling belt. And you had a lot of men that would want you to do all the hard dirty work while they sit. I heard one foreman say his sister-in-law was working there. He didn't want her there and he told me, we tried to run you off, but he said we couldn't." Another woman remarked that when she first began working underground, "I went through 8 or 9 bosses, all trying to break me, make me quit." And several other women reported that some foremen tried to mar their work reputations: "I had put up some ventilation [but] the curtain wouldn't reach the bottom. So I went off hunting another piece. [The foreman] came up and looked and I wasn't there. I went and got my ventilation and put it across the bottom. It was quitting time. [The foreman] didn't say nothing to me. Outside he told [the superintendent] that I didn't do my job right. I'd left the ventilation like that. I said I'm on my time, I don't want nothing outta this except us three to go back in that mine and go right over and look at that curtain. We did it. I demanded we do it. They saw that it was done." And she concluded, "You couldn't please [the foreman] no matter what you did or how hard you worked. He just had this thing against women coal miners." In sum, some of the male miners and bosses made it perfectly clear to the women that they refused to accept them as bona fide miners. As a result, the women felt that they had to prove their ability to perform some of the most strenuous tasks underground.

Previous studies have noted that "proving oneself" is a subcultural theme that is reflected in the Appalachian personality and that characterizes the approach many Appalachians take toward work as a means of self-sufficiency (Anglin, 1983). Moreover, Althouse (1974) found that new miners' job-related tensions stem from worries about their own technical competence and the extent to which they can rely on others. Similarly, the women in the sample reported that all new miners have felt the need to perform well by working hard, but that they felt more pressure to do so because they were women. As one woman said, "The women I have worked around [are] just as good workers as the men or better workers because they want to show people they can do it [even] if they do kill theirself in the meantime." And from another woman: "I think I worked hard and I did the jobs I was told so they respected me there. They didn't have to worry about, 'Well, we have a woman hanging rag today or we have a woman shoveling belt today so help out if you can, or we're really slow today because there's a woman hanging rag or running a roofbolter or whatever.' So I think that each of us has had to prove to ourselves also that we can do the job that we are in there to do."

However, as the women reported, some men have continued to make "proving oneself" problematic. Several male miners I spoke with said that when women began working at the mine, there was "trouble." One miner elaborated, saying that "the women wouldn't let nobody help them do nothing. They'd chew you right out and they've stayed here and become all independent." His attitude highlights the "double bind" or "catch 22" of the women's situations. On the one hand, receiving a man's assistance could be interpreted by others, both women and men, that a woman was either unwilling to or incapable of doing the work herself and, therefore, not deserving of her job. This perception could reinforce male miners' views about women's inability to do the work. On the other hand, those women who refused help, regardless of how tough bosses or coworkers made their work, were viewed as acting "independent," an inappropriate characteristic for females. Thus, the woman who is determined to prove herself risks offending male coworkers and losing their cooperation completely. The women reported that they usually reacted in the following manner: "You've got some men who will not, will almost refuse to help a woman, even though they'd help the men. [So] the men will help each other sometimes unless you ask for help. Sometimes you'll get people like that. I wouldn't ask for help." [Chuckles softly.]

Moreover, not only did the male coworkers presume that the women were incapable of doing "brute work," but, by both word and deed, the foremen communicated to them that they also were not suited for running machinery. Several women reported that foremen bypassed them in favor of men when assigning miners to jobs requiring operative skills. As one woman said, "They don't think women are smart enough to put something together. Which I can do. I've done a whole lot. And the boss goes right along with it." Indeed, one management official with whom I spoke at length said that men have more experience and, therefore, "a more mechanical approach" than women. He concluded that women having more menial jobs in the mines was a result of "the natural settling of their skills and their application."

The pressure for women to perform their jobs well by the men's standards persists because the women reported that they continue to respond to it in two distinct ways. Some adopted the following attitude about running machinery: "Sometimes a general inside labor job, it's not easy, but there's no pressure, there's no major head busting decisions to make, somebody else tells you what to do, somebody else takes the blame if it does not get done right. If you don't advance [by running machinery] you don't take a chance on being wrong or messing up. And when you make a mistake, they [male coworkers] really don't let you live it down." Others decided to take the challenge, such as the 14-year mining veteran who worked at the face cutting coal, who commented, "I think women have come a long way to prove to these men that we can do the job that they can do." However, she repeated the "proving" process when she assumed a new position operating machinery at the face. "Just like me when I went to the plow. I had to prove myself a jacksetter. I had to prove to the people that I worked with because it had been all men up to that point. I had to prove to the men I could do it, I had to prove to the boss I could do it."

Sexualization of Work Relations and the Workplace. Although the women had to prove they were capable of being coal miners, they did not have to do much to make their presence as women known. Male miners' initial responses were mixed. Some were supportive, but others responded to the women with various forms of sexual harassment. Half of the women in the sample reported they had been sexually harassed by male coworkers or foremen who used verbal innuendo and body language to convey sexual messages (Gruber and Bjorn, 1982). Two women reported that on occasion some of their male coworkers grabbed their own genitals in the women's presence and then pretended to have gotten "caught" urinating. Another woman recalled an incident of homosexual buffoonery with a particularly potent message, accentuating men's sexuality and solidarity: "They was pretending they was queers in front of me. It was like one was humping the other one, but they had their clothes on. And the boss said, 'You scared of us, ain't you?' I said, 'No, I'm not scared of you all.' And he said, 'Well, this is our little world down here and you don't belong.'"

Some male coworkers and foremen either directly solicited sexual favors from the women or repeatedly asked them for dates. When women first started working at the mine, one woman said that they were treated "like a piece of pussy." Another recalled that "a boss [once said] all the women made beds out of rockdust for the men. You know, like that's all we did was go in there to sleep with them?" Knowing that their male coworkers had these expectations, female miners consciously adopted certain social strategies for interacting with their male coworkers: "When I first came here I set myself up right away. I've made it known: Don't bother me, I'm here to work. I'm not here for romance, [but for] finance. Once you establish yourself, they know your boundaries."

Sexual propositioning by foremen posed a greater threat to women's work status than propositioning by male coworkers. Women in the sample recognized that when a woman failed to capitulate to a foreman's sexual demands, she usually faced the prospect of getting a more difficult work assignment. One woman who had been reassigned for her refusal to capitulate was told by a male coworker, "If you let these bosses pinch your titties, you'll get along. If you don't, you'll get the awfullest job that ever was." She said she preferred the "awful" job every time.

Another form of punishment, used by one foreman, was social derogation designed to humiliate the woman who refused his sexual requests: "One time [the foreman] told the guys behind my back that I had sucked his dick, is the way he put it. It came back to me about a week or so later. I went through pure misery for about a year because the boss lied to the crew that I worked with, telling them [other] stuff. I didn't even know why everybody all of a sudden quit speaking to me, giving me the cold shoulder." In front of her male coworkers, she retaliated: "I walked up to him and I said, 'When did I suck your god damned dick, down the jackline?' He goes, 'I don't know what you're talking about.' I said, 'You're a god damned liar. You told everyone of them and you didn't think that they'd find out I'm not doing the shit you said I was doing and come back and tell me things, did you?' Right there it proved to the guys [he was lying]."

Thus, men's sexualization of work relations underscored the women's sexuality at the expense of their work role performances and substantiated the cultural contradiction of a woman doing a man's job. Mining women felt a moral responsibility to themselves and to one another to avoid "loose" behavior. The sexual indulgences of other women were a reflection on each of them. As one woman explained: "[The boss] wanted to sleep with me. I wouldn't have anything to do with him. He thought if a woman worked for him she had to sleep with him because there was one woman working on the section [who was] sleeping with him. Everybody knew it. When it came my turn, I wouldn't sleep with him." Although the women recognized that the men's sexual harassment was usually unprovoked, some of them tended to place the responsibility for the men's actions almost entirely on women.[2] A woman who had received little or no sexual harassment explained that "the majority of the men up there are good to you if you let them. But they'll treat you how they see you act. See, men, they tend to watch women more, I believe it's just the male in them." Such a charge demonstrates the phenomenon known as "blaming the victim," characteristic of Kanter's (1977b) "exceptional woman," who as a token female plays the role of the "insider" by assuming the men's stereotypical orientation toward other females.

When the company issued a mandate against harassment, the superintendent told me it was necessary to "teach the men what harassment was." Although the rule seemed to have effectively eroded these incidents, the women complained that its enforcement put the onus of responsibility on them. Women themselves and not other men, such as foremen, were solely responsible for reporting harassment. Some women indicated their reluctance to report harassment because it created tension among crew members. Reporting harassment also violated the UMWA oath of solidarity, thus defeating the women's attempts to become socially integrated as unionized members of their crews. Although some of the women said they had never experienced any form of harassment, they insisted that they would readily report it if it occurred. However, others discussed having used the rule by directly confronting their harassers, but then were transferred to other work locations.

At the time of the study most of the women insisted that any kind of sexual harassment was largely a thing of the past, due, in part, to the enforcement of company rules. A few also explained that the saliency of sexual harassment was the result of media hype and not indicative of their current experiences. As one women said, "I think things have changed so much since the first woman come into the mines. She was harassed a lot [with emphasis]. But things have changed because they've accepted us." However, another said, "I think it's still going on, it's just more subtle now." Her comment indicated that although men's sexualization of work relations has changed form, it has not entirely disappeared.

The primary arena for women's socialization occurs within the social boundaries of the work crew on a section. As previous research on the social relations of mining has shown, a miner's primary identity is with the work unit or crew (Vaught and Smith, 1980). Looking back over her years at the mine, one woman commented on the adjustment process between herself and her all male crew members. When she first started working at the mine, she said, "I wasn't scared of the mine, I was scared of the men." But, she added, "The men I work with, they might talk about me behind my back, but in front of me, they got a lot of respect. They're family men and I guess we've growed onto each other we been there so long." Several other women who had similar experiences likened their crew membership to being in a family. One woman explained, "It's just like a family really, especially on sections. It's like you're just one big family. Everybody's working to help each other. If you don't, it makes your job hard. When you get on a section where people aren't like that, it makes your job hard." However, over time it had become clear to the women that their successful integration had done little to seriously disrupt men's sexualization of the workplace. So ultimately, the informal norms of the occupation continue to be male norms governing social behavior underground.

Over the course of their mining careers the women have been continually confronted with the conflicting expectations of being female and being miners. As a result they have been faced with two sets of prevailing norms: those governing female–male relationships and those governing peer relations in a masculine-identified workplace. Although some of the women reported conforming in varying degrees to the informal norms of their workplace, some women felt conflicted, as illustrated in the following account: "I guess if you're gonna be down there you get more and more like you're a man. It takes a lot out of you, like dresses and stuff. You wouldn't hardly see any woman [miner] in a dress outside the mines anywhere. There's nothing delicate about it, it just changes us all over."

Two types of men's behavior that contributed to workplace sexualization and helped maintain gender-based boundaries were sexual jokes, stories and profanity. Gutek (1985) concluded that sex, in the form of graffiti, jokes, comments, and metaphors for work, is a part of male-dominated workplaces regardless of women's presence. However, as women enter the work setting, they are obligated to set limits on some of the men's activities in order to avoid being degraded. Sometimes the male miners were careful about telling jokes in the women's presence. At other times the women found themselves in the position of having to "draw the line" on men's unacceptable behavior. On her crew, one woman said that although she generally "laughs stuff off," she was careful not to "get rowdy with them" because invariably the action would escalate. She commented that occasionally if they got carried away, she would "make them stop." Another woman attempted to curb the men's "sex talk": "They would start making sexual remarks about their girlfriends and women and I'd say, 'Hey, you shouldn't talk like that! What's the matter with you guys? You ought to be ashamed of yourself,' just to get them to watch what they say." Although she realized "you're not going to change people," she concluded, "all you can do is have them have respect for you."

Similar to other workers employed in dangerous occupations, coal miners are notorious for using profanity. The women said that men generally apologized if they thought a woman had overheard them using foul language. Their apologies strongly imply that there is a difference between men's and women's language. Language serves to maintain role boundaries. If profanity is not fit language for a woman to hear, then certainly she should avoid using it. The women varied considerably in their use of foul language and in their willingness to tolerate it from others. A few women did not swear and had no tolerance for it. However, most of the female miners admitted to using what constituted "men's language," but they said they were careful to conceal or curtail their profanity. For example, "There's a lot of stuff I will say. I used to not cuss too bad, but I'll cuss now. I'll say it under my breath. I don't think they've ever heard it. They'd die if they heard me say what I say to myself." Another said, "I cuss some when I get mad, but I always try to watch what I say because I'll lose that edge." That "edge," she explained, was the men's respect.

Conclusion

Female miners physically adjusted to doing hard manual labor underground, but their social adjustments were not made as easily or without compromise. They encountered sex bias and stereotyping of their inability to perform male-identified work and were viewed and often treated as sex objects. Men's beliefs and actions about women as workers and as sexual beings have been mutually reinforcing and have resulted in women's stigmatization and objectification.

Women are objectified when they are treated as sexual objects rather than individuals within their own right according to their own capabilities. Men attribute certain negative characteristics to women regarding their work performance, which results in the women's undervaluation based on the occupational standards of work as imposed by males in the work setting. Thus, as one woman succinctly put it: "The men look at our bodies and not at what we can do."

Until the men at the mine became familiar with the women they worked with, they were more apt to harass and, thereby, degrade them to the level of sex object. As noted by Swerdlow (1989), "men have a status stake in the sexualization of the workplace when the division of labor renders women equal to men" (p. 381). Or, as the case may be, men have a status stake in sexualizing the workplace and subordinating women's position in it when the division of labor and the way it is maintained provides the potential for rendering women equal with men. Moreover, although the more blatant objectification of women resulting from sexual harassment was regulated by company policy, more subtle forms of "sexualization" of the workplace remained, thus preserving male's sexual–social dominance underground.

Although most of the women conformed to the work norms expected of all miners, many also behaved in ways that contribute to the establishment and maintenance of gender-based boundaries as they were reset by men. That is, many women behaved in ways that maintained the status differential between the sexes by limiting their own visibility in the workplace. At the same time, the women also responded by continuing to prove themselves in the jobs to which they were originally assigned, such as beltman and general inside labor. Despite having earned good work reputations, the women still feel pressure to maintain their reputations. Some of the women who exceeded male work standards were held out as exceptions to the general rule of women's presumed inferiority. Thus, the rule about women's inferiority as a group was sustained. Still, many of the women expressed great satisfaction with their jobs and spoke of friendships with male coworkers, which provides them with the opportunity for successful integration as legitimate members of the underground workforce.

Notes

1. As conceptualized by socialist feminists, although the state often acts to support the material interests of capitalists and the ideological interests of patriarchy, it also serves to mediate their conflicts. In making their challenge, women's groups in support of female coal miners were instrumental in forming an alliance with government. The state responded with enactment and initial enforcement of federal antidiscrimination legislation that threatened employers with the loss of federal contracts and their profits. Again, the interests of the capitalists (as defined by the threatening actions of the state) were brought into direct conflict with the system of patriarchy (Sokoloff, 1988). As a result of the state's pressure, more women gained access to previously inaccessible types of male-dominated occupations, amid protests from male coal miners that women were taking "men's" jobs. Other previously held beliefs in the ideological system that reinforced women's exclusion were that women could not possibly do the work, so that the men would have to step in and do it for them, which would drive up the cost of coal. This could be viewed as an attempt by male miners to realign corporate interests with their own.

2. Moreover, the wives opposed female miners because of doubts about their fidelity, which reinforces the idea that the women are to be held responsible and may also partially explain the men's behavior toward their female coworkers.

References

Althouse, R. 1974. Work, safety, and lifestyle among southern Appalachian coal miners: A survey of the men of standard mines. *West Virginia University Bulletin Series*, 74, no. 11-9. Morgantown: West Virginia University.

Anglin, M. 1983. Experiences of in-migrants in Appalachia. In *Appalachia and America: Autonomy and Regional Dependence,* ed. A. Batteau, pp. 227–238. Lexington: University of Kentucky Press.

Butani, S. J., and A. M. Bartholomew. 1988. *Characterization of the 1986 Coal Mining Workforce.* Bureau of Mines Information Circular 9192. Washington, DC: U.S. Department of the Interior, Bureau of Mines.

Deaux, K. 1984. Blue-collar barriers. *American Behavioral Scientist*, 27:287–300.

Enarson, E. P. 1984. *Woods-working women: Sexual Integration in the U.S. Forest Service.* Birmingham: University of Alabama Press.

Gordon, D. M., R. Edwards, and M. Reich. 1982. *Segmented Work, Divided Workers.* Cambridge, England: Cambridge University Press.

Gruber, J. S., and L. Bjorn. 1982. Blue-collar blues: The sexual harassment of women autoworkers. *Work and Occupations,* 9:271–298.

Gutek, B. A. 1985. *Sex and the Workplace: The Impact of Sexual Behavior and Harassment on Women, Men and Organizations.* San Francisco: Jossey-Bass.

Hall, B. J. 1984. Coal mining women confront pattern of discrimination. *Mountain Life and Work* (Special Issue: Women in Nontraditional Jobs in Appalachia), 60(7):4–9.

Hall, B. J. 1990. Women coal miners can dig it, too! In *Communities in Economic Crisis: Appalachia and the South,* eds. J. Gaventa, B. E. Smith, and A. Willingham, pp. 53–60. Philadelphia: Temple University Press.

Hartmann, H. 1976. Capitalism, patriarchy, and job segregation by sex. In *Women and the Workplace: The Implications of Occupational Segregation,* eds. M. Blaxall and B. Reagan, pp. 137–169. Chicago: University of Chicago Press.

Kanter, R. M. 1977a. *Men and Women of the Corporation.* New York: Harper and Row.

Kanter, R. M. 1977b. Some effects of proportions on group life: Skewed sex ratios and responses to token women. *American Journal of Sociology,* 82:965–990.

President's Commission on Coal. 1980. *The American Coal Miner: A Report on Community and Living Conditions in the Coalfields.* Washington, DC: U.S. Government Printing Office.

Reskin, B. F., and H. Hartmann. 1986. *Women's Work, Men's Work: Sex Segregation on the Job.* Washington, D.C.: National Academy Press.

Reskin, B. F., and P. A. Roos. 1987. Sex segregation and status hierarchies. In *Ingredients for Women's Employment Policy,* eds. C. Bose and G. Spitze, pp. 1–21. Albany, NY: SUNY University Press.

Roos, P. A., and B. F. Reskin. 1984. Institutional factors affecting job access and mobility for women: A review of institutional explanations for occupational sex segregation. In *Sex Segregation in the Workplace: Trends, Explanations, and Remedies,* eds. B. F. Reskin, pp. 236–260. Washington, DC: National Academy Press.

Ross, M. H. 1974. Lifestyle of the coal miner: America's first hard hat. In *Humanizing the Workplace,* ed. R. P. Fairfield, pp. 171–180. New York: Prometheus Books.

Schur, E. M. 1984. *Labeling Women Deviant: Gender, Stigma, and Control.* Philadelphia: Temple University Press.

Simon, R. M. 1983. Hard times for organized labor in Appalachia. *Review of Radical Political Economics,* 15(3): 21–34.

Sokoloff, N. J. 1988. Contributions of Marxism and feminism to the sociology of women and work. In *Women Working: Theories and Facts in Perspective,* eds. A. H. Stromberg and S. Harkess, pp. 116–131. Mountain View, CA: Mayfield Publishing Company.

Swerdlow, M. 1989. Entering a nontraditional occupation: A case of rapid transit operatives. *Gender & Society,* 3:373–387.

Thompson, A. M., III. 1979. *Technology, Labor, and Industrial Structure of the U.S. Coal Industry: An Historical Perspective.* New York: Garland Publishing.

Tickamyer, A., and J. Bokemeier. 1988. Sex differences in labor market experiences. *Rural Sociology,* 53:166–189.

Tickamyer, A., and C. Tickamyer. 1991. Gender, family structure, and poverty in Central Appalachia. In *Appalachia: Social Context Past and Present,* 3d ed., eds. B. Ergood and B. Kuhre, pp. 307–315. Dubuque, IA: Kendall Hunt.

Tomaskovic-Devey, D. 1993. *Gender and Racial Inequality at Work.* Ithaca, NY: ILR Press.

United Mine Workers of America (UMWA). 1988. National Bituminous Coal Wage Agreement. Indianapolis, IN: Allied Printing.

Vaught, C., and D. L. Smith. 1980. Incorporation and mechanical solidarity in an underground mine. *Sociology of Work and Occupations,* 7:159–187.

Wardwell, M. L., C. Vaught, and D. L. Smith. 1985. Underground coal mining and the labor process. In *The Rural Workforce: Non-agricultural Occupations in America,* eds. C. D. Bryant, D. J. Shoemaker, J. K. Skipper, Jr., and W. E. Snizek, pp. 43–61. South Hadley, MA: Bergin & Garvey Publishers.

West, C., and D. H. Zimmerman. 1987. Doing gender. *Gender & Society,* 1:125–151.

Chapter 4

Gender, Culture, and the Sea: Contemporary Theoretical Approaches

DONA LEE DAVIS

Department of Social Behavior
University of South Dakota
Vermillion, SD 57069
USA

JANE NADEL-KLEIN

Anthropology Program
Trinity College
Hartford, CT 06106
USA

Abstract *Social science studies of fishing communities have tended to be highly focused on male activities and to regard women's work as domestic or as merely supplemental to that of men. This review article is intended to update the material presented in an earlier, more comprehensive essay on gender in the maritime literature. It examines some contemporary exceptions to this androcentric tendency, suggesting that understanding of local fisheries can be greatly enhanced by reexamining the role of gender in fishing communities and in fisheries production.*

Keywords Anthropology, ecology, economy, fisheries, gender, North Atlantic, Pacific, social science theory.

Introduction

In the early 1980s we began to prepare a review of the literature on women in fisheries for our edited volume, *To Work and to Weep: Women in Fishing Economies* (Davis and Nadel-Klein, 1988). As anthropologists who had recently conducted our own fieldwork studies on women in maritime settings (Davis in Newfoundland and Nadel-Klein in Scotland), we were anxious to provide a cross-cultural context for our work. In the course of surveying the existing literature, we found that much of general maritime studies was highly androcentric. Those scholars who did mention women frequently relegated them to a passing comment, paragraph, or discrete section on the household and/or family. Accounts focusing on women as major actors in fishing economies were relatively rare.

To use such scattered material effectively, we concentrated on organizing our essay into a descriptive review of the various roles that women could play within the world's diverse fishing communities. Our review was thus structured by the contents of the literature. We subcategorized the material to address each of the following issues: (1) the interrelatedness of environment, subsistence technology, and women's fishing activities;

(2) the roles of women in the production, processing, and distribution sectors of the fishery; (3) the fishing "way of life" and the various roles that women play in fisher families and the wider social realm of the local or occupational community; and (4) the less tangible, more expressive and ideological dimensions of women's experience of and contribution to the fishing effort. Our aim throughout the review was to encompass the entire range of women's involvement as active agents in the commercial and artisanal fisheries and to include a representative geographical coverage. Our critical focus, however, centered on the problems of androcentrism and of ethnographic presentation, rather than on gender theory per se.

In this review article, our aim is somewhat different. Here we turn our analytic attention to the relatively few and recent maritime studies that have dealt expressly with gender as a central theoretical construct. To do so, we sacrifice some geographic coverage as well as the full range of women's contributions to fishing economies. Instead of summarizing our earlier review,[1] we have opted to take a closer look at the relatively rare studies that encompass gender and its wider implications in terms of current trends in feminist theory. Finally, we suggest some directions that, in our judgment, studies of maritime economy ought to take.

Gender and Anthropology

Feminist theory has problematized the terms "sex" and "gender" (Caplan, 1987). Instead of assuming a total or necessary congruence between a person's apparent sex characteristics and the ascription of her or his social role, contemporary theorists ask how maleness and femaleness are culturally constructed and locally employed in organizing the division of labor, the allocation of social value, and the definition of the person. "Sex" normally refers to a category assigned a person on the basis of physical or phenotypic genital apparatus and conveys, at least to Westerners, an essential or irreducible "base." "Gender," on the other hand, entails the sociocultural construction and interpretation of masculinity and femininity. Gender understandings vary widely across cultures and often take forms that may be startling or surprising to readers accustomed to Western-based notions of what is "natural." Gender affects an individual's life and life expectations and social relations in complex ways. Anthropology is especially well placed to explore gender issues and cross-cultural variation and to expose the potential for Western bias.

In this section, we outline briefly some concepts and theories for understanding gender as a key component of any culture or society. Our analysis draws on two recent synopses of gender theory in anthropology. The first is Sandra Morgan's (1989) discussion of the value of examining gender for all areas of anthropology. The second is Henrietta Moore's (1988) book-length feminist critique of social anthropology. The clarity of presentation, identification, and critical assessment of conceptual frameworks and the comprehensive coverage of these authors have made their works valuable contributions to gender studies.

Both Morgan (1989) and Moore (1988) object to what has been called the "add women and stir" method of gender ethnography (a term coined by Boxer, 1982, p. 258). This add and stir metaphor refers to the characterization of gender in terms of simple identification and description of women's "contributions" to everyday social life. Although the "add women and stir" method succeeds in making women more visible in the ethnographic spectrum, it fails to make gender a central analytic construct and thus falls short of sophistication on more abstract theoretical levels. Moore's critique asserts that the study of gender should not be equated simply with "the study of women," arguing that this is a "remedial" rather than a "radical" approach (1988, p. 6). Instead, the analysis of gender and gender relations should be approached as a central structuring principle of human societies, their histories, ideologies, economies, and political systems. In a similar vein, Morgan (1989) advocates the treatment of gender as a fundamental aspect of the social relations of power, of individual and collective identity, and of the fabric of meaning and value in society.[2]

For the purposes of this review, we identify three different frameworks for gender analysis in the maritime ethnographic literature. The first approach includes those studies that focus on the lives of women as separate or distinct from the lives of men. Here, gender is articulated as the separation (either oppositional or complementary) of male and female domains of activity or spheres of influence. This has been widely characterized as the dichotomy between the public and domestic spheres of human organization, with women consigned fairly consistently to the domestic. Women's positions in a society are explained in terms of social and cultural constructs shaped by and expressed in the roles of motherhood, kinship, and marriage. In the maritime framework, domestic and public dichotomies may often be seen to overlap or reflect those of land and sea.

The second framework concerns the historical construction of the roots of power, powerlessness, and empowerment. Here gender is studied in terms of systemic models of inequality such as colonialism, global capitalism, race, and class. The aim is to cast relations between males and females in terms of Marxist/materialist models of production and reproduction. Since many fishing societies or communities are socially marginal and powerless,[3] this macrolevel approach is especially timely and relevant.

The first two frameworks deal with gender as a binary phenomenon. In contrast, the third approach aims to deconstruct such polarities. Assumptions about the binary nature of gender are challenged in light of critical reflection on our own androcentric and Eurocentric biases. Included in this section are studies that question as premature the search for universal hypotheses or global theorizing about gender as a source of stratification. These studies often advocate renewed attention to ethnographically intensive, microlevels of research, which show that there can be multiple concepts of power and value within a given cultural context, as well as across cultures. From this perspective, women and men can be portrayed as thinking social actors with multiple roles, statuses, and positions within the power structures and belief systems of particular societies. The maritime literature on Oceania and island Southeast Asia have made an important contribution to this body of thought. These studies expose the Western bias that treats gender as a discrete rather than a continuous phenomenon.

Putting the individual into the ethnographic process is clearly crucial here. Within anthropology, there has been increasing emphasis on looking at gender from an interpretive, reflexive, and highly personal perspective. Maritime studies in this frame are still rare but offer us interesting avenues for future analysis of gender.

These three theoretical, analytical, and methodological frameworks allow us to structure our review of the treatment of gender in fishing economies. Our focus here is on those studies where gender is an important and theoretically developed point of the analysis of fisheries systems. The discussion will be heavily skewed toward two areas: first, the North Atlantic and the industrial/commercial fisheries; second, studies of gender and exchange in Southeast Asian and Oceanic communities partly or heavily reliant upon fishing for subsistence.

Women on Land and Sea

In the early 1970s, the growing concern to make women visible led to presentations that often separated the study of women's lives and experiences from analysis of men's lives (Morgan, 1989). While a necessary and valuable innovation, this approach often left underexplored the relational context in which gender was socially constructed. To a significant extent, this relational context has been ignored in the maritime literature. Men, especially commercial fishermen, could be discussed with little or no reference to women. When women were considered at all, a land/women, sea/man dichotomy dominated the ethnographic accounts of maritime female and male work and family roles (see Davis and Nadel-Klein, 1988). Yet the land/sea dichotomy was usually somewhat mitigated by the widely recognized fact that even when women's lives were portrayed as land-bound, those lives (collectively and individually) remained dominated or shaped by women's various relationships to men and to the nature of men's work at sea. This land/sea, domestic/public approach to gender as a binary phenomenon against which to view the maritime configuration of gender roles continues to characterize the work of the majority of scholars in this area. However, a few researchers are beginning to take greater cognizance of the relational dimension.

Smith (1977) was one of the first to look for potential universals pertaining to women's position in maritime settings, where the marine econiche affects the structure or total configuration of maritime communities as sociocultural systems. Smith argues that maritime communities do vary from other groups. As Smith compares maritime to land-based communities, she proposes a distinctive gender role configuration. Maritime peoples often rely on the subsistence resources of both land and sea and thus are characterized by a greater dependency on women to control land-based food production. This control results in a greater role differentiation between males and females, as well as greater economic independence for women. When males are absent at sea along with other males, according to Smith, much of the daily life of the community is perforce controlled by nonfishers, such as women, through their kinship associations and networks. Women whose menfolk are at sea not only are responsible for childrearing, but also make important decisions and display considerable individualism. In this section, we review some sources that investigate the maritime construction of gender in terms of describing the land/sea, public/domestic dichotomy, and that focus on women's activities in terms of domestic, household, or family spheres of activity.

Drawing data largely from analyses of North Atlantic artisanal and industrial fisheries, where fishing could take men from home for long periods of time, many maritime studies have been stimulated by the sociological concept of extreme occupation (see Clark et al., 1985). These studies stress a contrastive, dichotomous view of the land as the domain of women and of the sea as the domain of men. A primary focus here is on the separation of workplace and family and its effect on the sexual division of labor and on the stressful nature of the fisher family life on shore.

A comparative perspective on the land/sea occupational dichotomy can be found in Binkley and Thiessen's (1988) study of the effect of the husband's employment on the marriage relationship and the psychosocial consequences of such work patterns on employees and their families. Binkley and Thiessen (1988, p. 39) describe how stressful life can be for wives of Nova Scotian offshore trawler men. For these women, stress comes from a double-role strategy, where successful adjustment to sea-time roles is often dysfunctional for family stability and the health and recreation of the labor force. While their men are at sea, women must become "reluctant matriarchs"; when their men are at home, women must turn into dutiful wives.

Gender conflicts based on disparate interests of husbands at sea and wives on shore are also described in a survey of wives of North Carolina fishers (Dixon et al., 1984), which indicated that wives were so unenthusiastic about their husbands' occupational choice that they encouraged their children to pursue other occupations. In North Carolina, women's lack of commitment to the fishery made it extremely difficult for older men to recruit the next generation of fishermen.

Some studies do focus on the complementary as well as contrastive aspects of gender relations in fishing families. An example is Levine's (1987) study of how wives in a small New Zealand fishing village act individually and collectively as "tranquil domestics," mediating the conflicts that crew formation and competitive fishing strategies breed among their fishing men. Nadel-Klein's (1988) ethnohistorical examination of Scottish fishwives as material and symbolic mediators between land and sea also emphasizes the idea of complementarity. Women of the east coast Scottish villages were widely recognized as indispensable to the fishery (Nadel-Klein, 1988).

Thus far we have referred to studies where women stay on shore and men work out at sea. It is striking to note that even when women are described as entering the fishery as fishers, the domestic/public dichotomy dominates ethnographic descriptions. These women are portrayed as exceptions within the male domain of the sea. They are still women doing men's jobs—women in nontraditional occupations. Although seafaring women (DePauw, 1982) have been largely neglected, they are becoming a more popular topic of research in recent years. Here we will mention a couple of examples.

Kaplan (1988) compares women who fish in the modern commercial fisheries to women who enter other blue-collar occupations, such as mining or truck driving, which are traditionally reserved for men. Kaplan interviewed women fishers in Massachusetts to assess their attitudes toward work and to determine what factors lay behind their choices to become fishers and the strategies they use to remain in the fishery. Kaplan concludes that her informants like the pay and challenges of their work, have supportive husbands, and come from stable families. She offers a social structural analysis, suggesting that the low percentage (1–2%) of women in fishing rests not in the nature of the work, but in the lack of anticipatory socialization for young girls. All women fishers interviewed by Kaplan seemed to have fallen into the job by circumstance.

A different note is sounded in the recent volume by Allison, Jacobs, and Porter (1989), who stress the flexibility of transition between land and sea roles for women in the Pacific Northwest. Many of the women in the Pacific Northwest fisheries have had considerable encouragement from family or community. They play many roles in the fisheries, and do not appear to suffer much from doubt about the gender-appropriateness of their chosen occupation. Allison et al. present a series of life histories that show how individual women have a range of fishery-related experiences in the course of a lifetime. Some have been, at different times, at home on land (as "fishermen's wives" or fish processors) as well as at home at sea (as fishers).

Gender, Power, Production, and the State

The largest and fastest-growing body of literature on gender and fishing tends to focus on the macro level of political economy. The works reviewed in this section represent the growing body of literature dealing with gender in terms of women's work and women's relationship to the modern capitalist state. The women's work literature deals with the relationship between the sexual division of labor and the status of women in society. It explores the connection between the organization of gender relations in the household and the entry of women into the wage labor force, as well as how women's unpaid work at home is related to reproducing the capitalist labor force (Moore, 1988).

Women's power, the valuation of women's labor, and the position of women in the nation state, as well as in the world capitalist system, have been central to a series of studies of gender and the fisheries in the North Atlantic. Here we will start by looking specifically at the political economy approach of Porter (1987) and of Connelly and MacDonald (1983) in Atlantic Canada.

In an article significantly titled "Peripheral women. . .," Porter (1987) points to ways of integrating gender-based Marxist-feminist concepts such as patriarchy, subordination, production, reproduction, and the sexual division of labor, with debates over the nature of Atlantic Canada as a peripheral economic region. Porter takes the Kirby report (1982) on the state of the North Atlantic fisheries to task for its sexist bias or "sin of omission" (Porter, 1987, p. 49). Not only did the report overlook the issue of gender but it failed to locate male and female either in the family or in the community, wrongly took for granted that the male was head of the household, and totally overlooked women's contributions to family economy, particularly through fishplant labor. Porter (1987, p. 53) also argues that sins of omission can be compounded by particular kinds of "omission." She challenges assumptions that as Atlantic Canada developed from a peasant society into a capitalist one, men were drawn into wage labor while women remained in noncapitalist modes of production. In the Newfoundland inshore fishery, for example, Porter argues that men, who continue to operate in the harvesting sector, sell their fish to the fishplants and are the commodity producers. On the other hand, it is the women, as fishplant laborers working in the processing sector of the economy, who are full-fledged industrial wage earners.

Porter also shows how on the local level, because male absence intermediates, ceaseless negotiation, countervailing strategies, and different locales of power characterize gender relations in maritime communities. These limit the ability of males to co-opt the power structures of the larger society in which they live (i.e., patriarchy). Yet modernization, industrialization, and delocalization of the fishery can undercut women's traditional sources of power. Porter's (1987) work echoes Moore's (1988) critique of the gender literature for assuming female subordination, for lacking historical or ethnographic detail, and for treating women as a homogeneous category. Moreover, Porter's argument adds a new twist to assumptions about the sexual division of labor and the nature of capitalist development.

Connelly and MacDonald's (1983) examination of the relationships among class, gender, the sexual division of labor, and women's differential experience of development on Nova Scotia's maritime periphery is informed by a combined micro-macro frame of analysis. By presenting a comparative analysis of women's work in neighboring Nova Scotian fishing and logging communities, they show how women's wage work is directly related to developments in the fishing industry. In the fishing community, women worked at the local fishplant to achieve a higher standard of living. As the local fishery entered a boom period and fishing husbands' incomes increased, women left work at the fishplant. Yet fishplant wages remained low because women from a nearby logging community (then in a bust period) readily accepted the work. Underdevelopment and its characteristic boom and bust cycles appear to affect different groups of women in different ways. Ultimately, however, Connelly and MacDonald argue that the entire region suffers the same net effect of perpetual low wages. The Nova Scotia study also contributes to the debate concerning the relation between women's domestic and wage work and the family wage. Connelly and MacDonald address the dual issues of women as (1) a reserve army of extradomestic labor that can be called out and returned home as economic conditions change, and (2) an unpaid domestic labor force that keeps the total family wages at less than the true costs of family subsistence. They describe women's wages as a form of secondary income that perpetuates poverty and underdevelopment among a large underclass of dispossessed farmers and poor inshore fishermen who survive on the fringes of capitalist labor markets.

These community studies are presented in terms of the wider perspectives of the history of women's work in the Canadian economy and the nature of underdevelopment in the Nova Scotian economy and in Atlantic Canada in general. Because the Atlantic Canadian fisheries have been chronically underdeveloped, opportunities for women are limited by low labor force participation rates, high rates of unemployment, and seasonal work. Connelly and MacDonald (1983) show how people in communities respond to changing conditions, not just as individual workers but as members of households. Strategies that may be adaptive for one household may not always be adaptive for another. They conclude that in a state where the fishing industry is more developed and central to the overall economy (for example, Iceland and Norway), the situation for women with respect to domestic and wage labor may be very different.

Women's resistance is depicted in several studies from the other side of the North Atlantic. For example, Buchan (1977), Nadel-Klein (1988), and Thompson, Wailey, and Lummis (1983) have noted the political activism of the Scottish "gutting quines." These were women from east coast fishing villages who worked to clean and pack herring in processing stations from Yarmouth to Shetland, following the transhumant herring fleet on seasonal contracts. With miserable working conditions and low pay, they openly questioned the capitalist relations of production, and even occasionally went out on strike.

Thompson (1985) advocates the importance of good community studies that relate local findings vertically to province, region, and state in his review of the roots of power between the sexes in various maritime settings. He includes both historical and cultural construction approaches within a political economy framework in his examination of women's lives in a selection of British fisheries. Thompson's work gives ethnographic note to the phenomena of power, identity, and the fabric of meaning, all offered by Morgan (1989) as interrelated and crucial to understanding the nature of gender. He shows how the status of women in a particular fishing context cannot be presumed or automatically read from their relations to the means of production, since within the same frames of production gender ideologies as well as women's roles and relative power and autonomy can be quite variable.

Thompson argues in the following manner. Although fishing is commonly thought of as a man's business, fishing has a special interest for understanding the position of women both past and present. As men go to sea, they become particularly dependent on the work of women on shore. First, women have direct productive roles in the fishery (e.g., the Scottish herring girls, or gutting quines). Second, women create the next generation in both a physical and moral sense (the bearing and rearing of children). Third, women take on added responsibilities for family and community when their men are at sea. Men's dependence gives women more responsibility but also, according to Thompson, the *possibility* of more power both at home and in the community.

However, Thompson notes that although the sexual division of labor can be quite uniform across fishing communities, the sexual division of power is not. Although male absence is an important element in the fishing family social milieu, it alone does not explain situations in which women are ascendant. Thompson argues that women's ascendancy is the result of the conjunction of two dimensions of the sexual division of labor, which may or may not run parallel. When female control of space (household) is combined with economic responsibility for preparation and sale of the catch and the control of property (either in boats, land, or independent business) and when women have socially legitimate roles (based on formal laws, property laws, family need, or moral convention) governing their control of property, their base of power is enhanced. Yet even then, power relations between the sexes are still subject to negotiation over time in particular contexts by individual men and women. Moreover, other factors can intervene to reduce women's power. One such factor is the context and conditions of men's work, such as how and when men are paid. For example, when men are bitter or when men sense the degree of their own exploitation by merchants or employers, their experiences may harden them and drive them toward compensation, self-indulgence, and an assertion of their own male authority when they come home.

Thus Thompson offers a list of social forces that qualify women's power in fishing communities. These are spatial (male absence), economic (women's production), social (recognized entitlement to and control of property), situational (individual proclivities), and emotional (the nature of the husband's relationship with his employer). The list ends with culture as the ultimate qualifier or mediator. With his penchant for listing and sublisting, Thompson expresses four ways in which cultural traditions or local influences can mediate gender power relations. First, the culture may have a traditional family form that encourages strong women, or, relatedly and second, a distinctive family pattern. Third, the separation of sexes can allow women to practice or maintain traditions different from those of men. For example, women may practice mild punishment of children and men may believe in more severe forms of punishment. Fourth, extensive regional traditions such as a strong patriarchal heritage or Calvinism can mediate gender power relations. Finally, nonlocal influences such as the status of fish stocks, changes in technology and markets, or international conflicts over fishing rights can have profound effects on all of the local patterns referred to above. Thus, Thompson makes it clear that women's power is contingent on a number of special features and is ultimately rather precarious.

Critiquing Gender

This final level of analysis looks at women as persons. In recent years, anthropology has seen a reorientation, away from traditional ethnographic methods and theoretical models of the world and toward more microlevel particularistic description and theories concerning thinking social actors and the strategies they employ in day-to-day living. The studies reviewed in this section focus on the experiencing self or persona and how this is culturally constructed. Here we focus on the analysis of how individual and collective identities are formed and maintained through interpersonal relationships and situational contexts as they shape behavior on a daily basis. Relevant here are Davis's work on emotional expression among Newfoundland women (1983a, 1983b, 1986, 1989) and Porter's (1990) very personal and moving account of her historical research project on the diaries of seafaring wives of whalers in 18th and 19th century New Bedford, MA.

Davis's studies of a Newfoundland inshore fishing village employ ethnographic detail to capture the processes of homogeneity and heterogeneity as they are shaped in the local community by individual interactions on a day-by-day, play-by-play basis.

Davis's original intent was to study women's experiences of menopause in a small, isolated fishing village. She found she had to deconstruct the meaning of menopause, confront feminist and medical models of stress, and question the applicability of accepted measures of women's status in order to describe the process by which identity and meaning are actively negotiated on family, household, and community levels. She recounts, "I began to see that there was a danger in overemphasizing the relationship between women and their bodies and ignoring the relationship of women to men and the relationship of both sexes to the basic economic adaptation of their culture" (Davis, 1983b).

Davis found that although women accepted their land-based work roles as part of the division of labor by sex, their major means of articulating the maritime heritage they shared with men was on a more emotional, affective level and was expressed through the vocabulary of nerves and worry. In Newfoundland, worry is inextricable from matters of everyday life, of which the fishing and fishing heritage are important components. Nerves, as an idiom, captured present, past, and future. Yet the vocabulary of nerves had multiple and complex meanings, was context and situation specific, could be both highly personal and impersonal, and was either well or ill defined. Nerves unified women and divided them. By following individual women through their daily social interactions, Davis's work on nerves and worry shows how the rules and strategies women use individually and collectively shape their maritime identities. In her articles on social change in the Newfoundland fishery, Davis (1983a) confronts maritime models of the Newfoundland patriarchal extended family and shows how insider views of social change differ from those of the academic or sophisticated critic of social policy. She highlights the importance of understanding insider/outsider dichotomies, rather than female/male or land/sea dichotomies, as a key to understanding "outport" life. In her articles, Davis lets the women speak for themselves and examines the process by which their words are mined for relevance to academic publication and theory building (Davis, 1986).

An excellent example of reflexivity in this more personal approach is found in Porter's (1990) study of the diaries of wives of whaling captains in New Bedford. In this unabashedly personal and moving account, Porter explores her own feelings as a woman sailing captain and a feminist in relation to the women she had felt she had so much in common with at the outset of her research. She began to feel both closer to and more distant from her subjects as the reading and analysis of their diaries progressed. Her strivings to become more sympathetic with women she had hoped would be radical, feminist, and firmly rooted in and supportive of the gender and class standards of their day is insightful and illustrates this reflexive perspective.

Finally, in this section, and most importantly for this essay, we address the large body of literature on gender in cultures with gender ideologies and subsistence strategies quite different from our own. Although women's fishing complements and overlaps men's fishing in many cultures, many maritime scholars have overlooked the growing body of literature in this area. This oversight is probably due to several reasons, including an androcentric and industrial bias that associates fishing with men, with complex and capital-intensive technologies, and with the drama of deep-sea fishing in dangerous oceans like the North Atlantic.

While our review thus far has encompassed work largely from the industrial/artisanal North Atlantic, our third level of analysis leads us to the Pacific, where there have been fewer studies focusing on fishing communities as such, but far more studies that seek to analyze the significance of gender and the construction of the person in fishing-dependent societies.

McDowell (1984) questions the public/domestic, sea/land dichotomy using the concept of complementarity among Bun men and women in Papua New Guinea. Although McDowell's analysis starts with a refutation of Western notions of the division of labor in fishing, her argument soon moves into a phenomenal realm of analyzing the interrelatedness of gender beliefs and complex local structures of cross-cousin relations as well as an abstract sense of self. According to McDowell, fishing is neither gender specific nor gender stratified. Bun men and women both fish. Women fish by wading in streams and swamps with individual nets. Both men and women fish with hooks and lines, but only men fish with poison. During peak fishing time for women, men sit with the children. McDowell goes on to discuss the ethnographic complexities of cross-cousin marriage and world view as they act, like fishing, to intersect and equalize male and female realms of experience and meaning. McDowell concludes that in the case of the Bun, gender cannot be reduced to such simple constructs of analysis as the assumption of hierarchy in gender relations or the clear-cut universal and oppositional existence of public and domestic domains.

This conclusion is also reached by Firth (1984), who compares women's contributions to the fishing effort in terms of the range and intricacy of women's involvement in two distinct fishing economies—the Tikopia of the Solomon Islands, where fishing is for home consumption and part of a mixed subsistence economy, and the coastal Malays of Kelantan Malaysia, where a peasant market economy predominates. Like McDowell for the Bun, Firth describes the ideological ramifications of overlapping competition and complementarity between male and female fishing patterns among the Tikopia. Although both men and women fish on the reefs, the high-prestige sea fishing is the man's world. Yet this does not mean that women are absent from that world, for female as well as male deities control the fish and the canoes. In many ritual situations relating to male fishing activities, female spirits are involved. As Firth sees it, the role of women secularly excluded from the high-status sea fishing reappears as compensation or revenge at another level, that of spirit control. Thus, in his interpretation, Firth portrays the sexuality of women and the female nature as dangerous to men. To neutralize the potential danger of women, men keep them from sea fishing, yet the pervasiveness of female activity is too powerful to ignore, so the principle of female control or intervention is allowed on the more abstract or spiritual level.

F. Errington and Gewertz (1987) also reject an individualized, economic, and Western-based model of how persons are valued and gendered in their analysis of the Chambri. Chambri women are responsible for fishing. Their ability to produce fish is seen as part of their nurturing, reproductive nature. Yet their important contribution to subsistence in itself gives them neither inferior nor superior status as women in relation to men. For the Chambri, dominance is an issue within genders, rather than between men and women, as they compete for power in entirely different, unranked symbolic arenas.

These three analyses alert us to the ways in which Western culture places great emphasis on occupation and economic organization to conceptualize, actualize, and rank gender and personal status. However, while North Atlantic maritime gender models rest on notions of fishing and maleness as inherently connected, non-Western models may not use fishing in this way at all. First, fishing itself may not be seen as a unitary and gendered occupation. For the Melanesian Bun, for example, the specific technique of fishing, rather than the goal of fish capture, appears to be the engendered category: Fishing with poison is different from fishing with nets. Second, gender may be marked without becoming a significant source of collective stratification. As S. Errington (1990) points out for island Southeast Asia, personal prestige may not derive from work and economic resources, but from the control of manners, language, and ritual, as well as from nongendered sources of rank, such as kinship and generation.

These non-Western studies of gender make us aware that the appropriate question for maritime ethnography is not whether men are to the sea as women are to land. Rather, one must ask in what ways fishing forms part of the symbolism of gender and thus informs people's decisions about who can work where, how, and when. It may be the case that this huge and important body of literature has often been overlooked by those in maritime studies precisely because the treatment of fishing is seldom a central focus, but is embedded in the wider context of ideology and social relations.

Conclusion

Our first two frameworks point to several difficulties in the maritime gender literature. First is the need for more detailed ethnography. With so few studies that have made gender a central concern, theoretical generalization becomes problematic. The second difficulty is that many studies have been dominated by a model of male absence that implies women are merely passive "reactors" to a social deficit of men. A third difficulty is the economic reductionism that still tends to equate status with control of economic resources. Fourth is the constraint of Western-derived, binary oppositional frames of analysis that ignore the relational components of gender construction and assume female subordination. Such oppositions often do not fit the data even in a North Atlantic context.

It is important for future maritime research to pursue a gender, rather than an "add women and stir," approach. Despite the scarcity of existing studies, it is abundantly clear that women play vital roles in fisheries worldwide. It is also clear that women are entering fisheries in many places, such as the Pacific Northwest. Cultural conceptions of how fishing becomes an engendered process need much more analysis. These conceptions relate to issues of economic competition, changing technology, reorganization of households, the relationship of communities to the nation-state, and ultimately to cultural survival.

Notes

1. *To Work and to Weep* is a collection of articles on gender and women's roles in fishing economies within societies ranging from Massachusetts to Malaysia.
2. See also Lamphere (1987), Strathern (1987), and Yanagisako and Collier (1987).
3. Many now face the problem of sheer survival. For a recent discussion of fishing community problems in the modern world, see Cordell (1989).

References

Allison, Charlene, Sue-Ellen Jacobs, and Mary A. Porter. 1989. *Winds of Change: Women in Northwest Commercial Fishing*. Seattle: University of Washington Press.
Binkley, M., and Thiessen, V. 1988. Ten days a "grass widow"—Forty-eight hours a wife: Sexual division of labor in trawlermen's households. *Culture*, 8(2):39–50.

Boxer, Marily. 1982. "For and about women": The theory and practice of women's studies in the United States. In *Feminist Theory: A Critique of Ideology,* ed. Nancy Keohane, Michelle Rosaldo, and B. Gelpi, pp. 237–271. Brighton: Harvester Press.

Buchan, Margaret. 1977. The social organization of fisher-girls. Unpublished manuscript, Aberdeen, Scotland.

Caplan, Pat, ed. 1987. *The Cultural Construction of Sexuality.* London: Tavistock.

Clark, D., K. McCann, K. Morrice, and R. Taylor. 1985. Work and marriage in the offshore oil industry. *International Journal of Social Economics,* 12(2):22–36.

Connelly, Patricia and Martha MacDonald. 1983. Women's work: Domestic and wage labour in a Nova Scotia community. *Studies in Political Economy,* 10:45–72.

Cordell, John, ed. 1989. *A Sea of Small Boats.* Cambridge, MA: Cultural Survival.

Davis, Dona. 1983a. The family and social change in a Newfoundland outport. *Culture,* 3(1):19–32.

———. 1983b. Woman the worrier: Confronting archetypes of stress. *Women's Studies,* 10(2):135–146.

———. 1986. Occupational community and fishermen's wives in a Newfoundland fishing village. *Anthropological Quarterly,* 39(3):129–142.

———. 1989. The Newfoundland change of life: Insights into the medicalization of menopause. *Journal of Cross-Cultural Gerontology,* 3(4):1–24.

Davis, Dona and Jane Nadel-Klein. 1988. Terra cognita? A literature review. In *To Work and to Weep: Women in Fishing Economies,* ed. Jane Nadel-Klein and Dona Lee Davis, pp. 18–50. St. John's, Newfoundland, Canada: Institute of Social and Economic Research, Memorial University of Newfoundland.

DePauw, Linda. 1982. *Seafaring Women.* Boston: Houghton Mifflin.

Dixon, Richard, Roger Lowery, James Sabella, and M. Hepburn. 1984. Fishermen's wives: A case study of a Middle Atlantic fishing community. *Sex Roles,* 10(1,2):33–52.

Errington, Frederick, and Deborah Gewertz. 1987. *Cultural Alternatives and a Feminist Anthropology: An Analysis of Culturally Constructed Gender Interests in Papua New Guinea.* Cambridge: Cambridge University Press.

Errington, Shelley. 1990. Recasting sex, gender, and power: A theoretical and regional overview. In *Power and Difference: Gender in Island Southeast Asia,* ed. Jane Atkinson and Shelley Errington. Stanford, CA: Stanford University Press.

Firth, Raymond. 1984. Roles of women and men in a sea fishing economy: Tikopia compared with Kelantan. In *The Fishing Culture of the World: Studies in Ethology, Cultural Ecology, and Folklore,* ed. Bela Gunda, vol. II, pp. 1145–1168. Budapest: Akademiai Kiado.

Kaplan, Ilene. 1988. Women who go to sea: Working in the commercial fishing industry. *Journal of Contemporary Ethnography,* 16(4):491–314.

Kirby, Michael J. 1982. *Navigating Troubled Waters: A New Policy for the Atlantic Fisheries.* Ottawa: Canadian Government Publishing Centre.

Lamphere, Louise. 1987. Feminism and anthropology: The struggle to reshape our thinking about gender. In *The Impact of Feminist Research in the Academy,* ed. Christie S. Farnham, pp. 11–33. Bloomington: University of Indiana Press.

Levine, Marlene. 1987. Exclusion and participation: The role of women in a New Zealand fishing village. In *Stewart Island: Anthropological Perspectives on a New Zealand Fishing Community,* ed. Hal Levine and Marlene Levine, pp. 33–49. Wellington: Victoria University Occasional Papers in Anthropology, no. 1.

McDowell, Nancy. 1984. Complementarity: The relationship between female and male in the East Sepik village of Bun, Papua New Guinea. In *Rethinking Women's Roles,* ed. Denise O'Brien and Sharon Tiffany, pp. 32–52. Berkeley: University of California Press.

Moore, Henrietta. 1988. *Feminism and Anthropology.* Minneapolis: University of Minnesota Press.

Morgan, Sandra. 1989. Gender and anthropology: Introductory essay. In *Gender and Anthropol-*

ogy: Critical Reviews for Research and Teaching, ed. Sandra Morgan, pp. 21–40. Washington, DC: American Anthropological Association.

Nadel-Klein, Jane. 1988. A fisher laddie needs a fisher lassie: Endogamy and work in a Scottish fishing village. In *To Work and to Weep: Women in Fishing Economies,* ed. Jane Nadel-Klein and Dona Lee Davis, pp. 190–120. St. John's, Newfoundland, Canada: Institute of Social and Economic Research, Memorial University of Newfoundland.

Porter, Marilyn. 1987. Peripheral women: Toward a feminist analysis of the Atlantic region. *Studies in Political Economy,* 23:41–73.

———. 1990. *Not Drowning but Waving: Reading Nineteenth Century Women's Diaries* (No. 28 Monograph Series, Study on Sexual Politics), ed. Elizabeth Stanley. Manchester, UK: University of Manchester Press.

Smith, M. Estellie. 1977. Comments on the heurisic utility of maritime anthropology. *The Maritime Anthropologist,* 1(1):2–8.

Strathern, Marilyn. 1987. An awkward relationship: The case of feminism and anthropology. *Signs: the Journal of Women in Culture and Society,* 12(2):276–292.

Thompson, Paul. 1985. Women in the fishing: The roots of power between the sexes. *Comparative Studies in Society and History,* 27(1):3–32.

Thompson, Paul, Tony Wailey, and Trevor Lummis. 1983. *Living the Fishing.* (History Workshop Series.) London: Routledge and Kegan Paul.

Yanagisako, Sylvia, and Jane Collier. 1987. A unified analysis of gender and kinship. In *Gender and Kinship,* ed. Jane Collier and Sylvia Yanagisako, pp. 86–118. Stanford, CA: Stanford University Press.

Epilogue

In the five years since this article was written, anthropologists have continued to explore and rethink the implications of gender theory for the study of culture. Scholars have increasingly conceptualized gender as an aspect of how persons are constructed and identities are contested. Moreover, the insight that "gender" itself may be a nonunified category within cultures has led to a greater focus on how the concept continues to be dynamically created, negotiated, and experienced in individual lives, rather than being received simply as a given. Thus, this addendum is prompted by theoretical concerns as well as the need to acknowledge new ethnographic contributions. Those who look today at how women in maritime societies relate to men, to each other, and to marine resources must more than ever acknowledge the extent to which gender has become an analytically problematic aspect of social identity.

The bulk of recent studies continue to come from North Atlantic communities, where "crisis" is the watchword of the day. From Norway to Newfoundland, ecosystemic breakdowns (attributed primarily to overfishing and pollution) have resulted in the virtual collapse of many stocks. How the loss of fishing opportunities affects women's life chances and empowerment, and how women's articulations with their communities become altered as a result, is becoming an increasingly popular topic (though more with anthropologists and sociologists than with policy makers, who continue to target most of their efforts toward men's livelihoods).

Some of these studies have an applied, even an activist focus, especially as questions of occupational continuity and even community survival become urgent in localities where no other work is available. An example of this activist approach is found in the work of Neis (1988), who argues that to understand the workings of patriarchy in Newfoundland emphasizes the need to be more explicit about how class and gender mutually inform each other and how gendered social policies affect the lives of women in outport communities.

Pettersen (1996) explicitly considered the impact of the crisis on households in the Lofoten district of Norway, where women have increasingly become significant or even dominant income providers. Economic stress has forced many households to adjust their occupational strategies, but often at some cost to domestic harmony and even to individual health. Pettersen's goal is to integrate "social and cultural concepts of sustainability . . . with those relating to the biological resource" (p. 247).

Norwegian anthropologist Munk-Madsen (1996) has also addressed psychological issues for female members of fishing couples in the small-scale fishery of northern Norway. She takes an unusual but much-needed look at how male–female fishing partners—often, but not always, husband and wife—negotiate the sometimes competing demands of power, sex, and love.

In writing about women's position in Icelandic fishing communities, Skaptadottir (1996) began with the premise that gender is a culturally contested domain, that "Culture is neither a restrictive blueprint for thought nor a prison for action but a vocabulary of argument and action" (p. 89). For women in these communities, identity is not fragmented between work and home, as is so often the case in Western societies. Rather, according to Skaptadottir, "their identity as fish processors cannot be clearly separated from their identity as members of their fishing community or as members of families . . ." (p. 102).

Davis's (1993, 1995) recent study of a single Newfoundland community dramatically changed by the resource crisis emphasizes the negative impact of loss of livelihood on gender relations. It also describes how a multiplicity of gender roles has come to supersede the earlier traditional ones that characterized the community in the 1970s.

The intersection of gender with occupational and community identity continues to be an important thread in the literature. Nadel-Klein (1991, 1993) examined how tourism and the newly burgeoning "heritage industry" represent and reinterpret women's roles in Scottish villages, where fishing is no longer viable, but where fisher identity continues to be highly important for local people's self-esteem.

Exceptions to this North Atlantic focus are relatively rare, but all the more valuable because they provide some much-needed basis for comparison. The following two studies, one from Africa and the other from Indonesia, emphasize opportunity and growth rather than crisis and decline. Overa (1996) explored the economic centrality of women's participation to development in the Ghanian canoe fisheries. Women there take active entrepreneurial roles in shore-side processing and trading activities and most recently have begun investing in canoes and motors. Volkman (1994) sees the Mandar women of Indonesia, who have recently turned to fish-trading, as embarking on a potentially risky, though lucrative, course, because government attempts to "modernize" and thus take control over fish-selling may force women once again back to their previous occupation as weavers.

Additional References

Davis, D. L. 1993. When men become women. *Sex Roles*, 29(7/8):1–18.

Davis, D. L. 1995. Women in an uncertain age. In *Their Lives and Times: Women in Newfoundland and Labrador*, eds. B. Neis, M. Porter, and C. McGrath, pp. 279–295. St. John's: Killick Press.

Munk-Madsen, E. 1996. *Wife the deckhand; husband the skipper*. Paper presented at the research seminar: Global Resource Crises—Marginalized Men and Women? University of Tromso.

Nadel-Klein, J. 1991. Reweaving the fringe. *American Ethnologist*, 18(3):500–517.

Nadel-Klein, J. 1993. *Hopping down the heritage trail*. Paper presented at the American Anthropological Association.

Neis, B. 1988. Doin' time on the protest line. In *A Question of Survival*, ed. P. Sinclair, pp. 133–156. St. John's: Memorial University of Newfoundland Institute of Social and Economic Research.

Neis, B. (In press). From "shipped girls" to "brides of the state"? *Canadian Journal of Regional Science*, 16(2):185–211.

Overa, R. 1996. *Social dynamics and economic change: Female entrepreneurship in the canoe fisheries of Ghana*. Paper presented at the research seminar: Global Resource Crises—Marginalized Men and Women? University of Tromso.

Pettersen, L. T. 1996. Crisis management and household strategies in Lofoten. *Sociologica Ruralis*, 36(2):236–248.

Skaptadottir, U.-D. 1996. Housework and wage work. In *Images of Contemporary Iceland: Everyday Lives and Global Contexts*, eds. G. Palsson and E. P. Furrenberger, pp. 87–105. Iowa City: University of Iowa Press.

Volkman, T. A. 1994. Our garden is the sea. *American Ethnologist*, 21(3):564–585.

Chapter 5

Women's Status in Peasant-Level Fishing

JAMES L. NORR
KATHLEEN F. NORR

University of Illinois at Chicago
Chicago, IL 60680
USA

Abstract *The women of Minakuppam, a small hamlet of ocean-going fishermen located just outside the city of Madras in Tamilnadu, India, are more active and less limited in their daily social activities and have more power than women in most Indian farming villages. This contrast is extended with evidence on women's status in fishing and agricultural communities in other predominantly agrarian societies. Several crucial features of political economy account for women's status in these communities. Participation of women in economic production and low status for fishing occupations in societal stratification are the major factors accounting for women's distinctive activities and relationships. These features of the political economy affect women's control and use of resources and their opportunities for social support from kin and other women.*

Keywords Agriculture, economy, fishing, gender, inequality, production, stratification, techniques.

Introduction

We use a comparative perspective to analyze the activities and social relationships of women in a south Indian fishing community. Women in this ocean fishing community are less dependent and have more power than women in most communities located in this peasant agrarian society. Two distinguishing features of Indian fishing communities—their adaptation to oceanic natural resources and their low status in the society at large—suggest the major sources of these gender differences.

Comparative reviews of fishing work and community social patterns emphasize fishing's technical and environmental features (Acheson, 1981; Anderson and Wadel, 1972; Smith, 1977). Nonroutine techniques employed in a risky and uncertain environment consistently structure fishermen's work and status relations as flexible and egalitarian. This conclusion is strongly supported in our comparative analyses of ocean fishing in societies at various levels of complexity (J. Norr and K. Norr, 1977, 1978; K. Norr and J. Norr, 1974). These material relationships of fishermen, by their effect on men's relations and the model they provide, may indirectly influence the position and interaction repertoires of women. In addition, maritime adaptation may directly determine women's status. Access to independent income from fish selling, long periods of male absence, and high male death rates may influence the position of women in fishing communities. Thompson (1985) stresses these spatial and economic factors in accounting for women's power in fishing economies.

One of the many contributions of the Nadel-Klein and Davis (1988) collection of 11 ethnographies of women in fishing economies is to establish both the commonalities and the wide variation in women's roles in fisheries. The similarities mostly derive from the economic and political marginality of fishing. In their evaluative review of these case studies, Nadel-Klein and Davis suggest low status of fishing occupations and ideological supports for societal stratification as two potentially significant factors in accounting for differences in gender inequality. Low status for fishermen would limit their control over the behavior of women. At the same time, societal elites are not threatened by greater independence and more public activities by low-status women.

Our comparative examination adds evidence to support Nadel-Klein and Davis's claims of the importance and significance of women both in their roles as economic producers in fisheries and in their household and family roles. In addition, these data, when compared with other crucial cases, provide a unique opportunity to specify a model of gender relations that includes the influence of technical and environmental aspects of marine adaptation to natural resources along with the influence of ideological and political aspects of societal stratification.

The Community

Minakuppam is a small hamlet of ocean-going fishermen, located just outside the city of Madras in Tamilnadu, India. The village is being transformed from a peasant-level fishing village to a lower-class suburban slum (J. Norr and K. Norr, 1982).[1] All the fishermen are members of the same low caste of *Pattanachettiars*. This caste is lower in status than agricultural castes but above those scheduled castes whose traditional tasks include handling dead animals and services at polluting events such as births and funerals.

In 1965, the village's 150 homes, mostly thatch-roofed huts, were clustered together between dense palm groves behind and a broad sandy beach and the blue waters of the Bay of Bengal in front. The surrounding area was barren, treeless sand, making the village appear very isolated and desolate. Two other fishing hamlets were half a mile distant along the beach in opposite directions. Just a few hundred yards from the village was a paved road, used by city residents who come to the beach to swim and by villagers who walked 3 miles to a bustling suburb with many shops and with buses running into the city of Madras. Despite its isolated appearance, Minakuppam residents relied on the city as a market for their fish, as a place to buy supplies, and for entertainment. In addition, the city and visitors to the beach provided full- and part-time job opportunities for some villagers.

By 1980, urban expansion radically altered both Minakuppam and its surroundings. Suburban middle-class housing covered the formerly empty land north and northwest of the village. Immediately west of the village a large cluster of mud huts inhabited by scheduled castes grew up. Regular bus service to the city is now less than 100 yards from the community. Since 1965, a large Catholic church and an impressive Lakshimi temple compound have been constructed adjacent to the village. Both draw large numbers of devotees and pilgrims. Various small stalls, some of them run by fishermen, sell religious articles and refreshments near both the church and the temple. The beach's popularity has diminished considerably as it has lost its isolated and picturesque character. In 1965, a significant number of fishermen earned regular income providing services for recreational visitors to the beach. Offsetting the loss of this income is the greatly increased demand for arrack, illicit home-brewed rice beer. There has been a growth in the number of fishing-caste households, mainly due to natural increase. An influx of non-fisher-caste residents now comprises one-third of the village's registered voters. They rent living space from fishermen who have moved out of the village.

All these changes within and outside the fishing community have radically altered what was formerly a major factor in the life of the community—its relative isolation from other castes. In 1965, Minakuppam was a single-caste community. While fishermen had many important economic and noneconomic links with urban residents, they had relatively little day-to-day contact with non-fisher-caste people. Contacts with non-fishermen were almost entirely initiated by the fishermen and usually took place outside the village. Today the community is still predominantly the residence of traditionally occupied fishermen, but it is increasingly lived in and open to a wide variety of urbanites.

Women's Economic Activities

Peasant-level ocean fishing dominates the economic and social life of the village. Of 190 male workers in the village in 1965, about three-quarters were totally dependent on fishing, and only 29 men did not fish at all. Fishing-related activities provide part-time employment for many women and older men. The daily and seasonal variations in fishing activities and success govern the rhythms of village daily life.

While the men are away fishing, the beach is deserted, and the village is quiet as women go about their household tasks and children play in small groups or go to school. When the boats return, everyone gathers on the beach. Wives come to care for their husband's catch; the widows of the village and other women who want to earn a little money come to buy small amounts of fish they can sell in the market; men from outside the village come to buy fish on a larger scale. In addition, many other villagers, including children, old men, and those who did not go fishing, are drawn to the beach to see who was lucky and perhaps to receive a few fish from a friend or relative.

Fishing-Related Activities

The owner's wife helps unload the boat. Selling the fish is now her responsibility, although the husband often stands by and gives advice. The fish auction occurs on the beach as soon as a boat is unloaded. Everyone crowds around to watch the drama; in the end the highest bidder pays cash to the fisherman's wife and takes his fish. Then the noisy process starts all over again, ending only when the last boat has returned to shore.

Although women are excluded from fishing, they take an active part in fish marketing. Women have traditional duties and rights in ocean fishing that begin when the fish arrive on shore. Men from outside the village sell fish, but the fishermen do not. Every man who fishes has a female relative who looks after his catch: his wife if he is married, his mother or elder sister if he is young, a daughter or daughter-in-law if he is elderly and widowed. The women come down to the boat and help sort and unload the catch. Each crew member receives a few fish for the evening meal, and these are often doled out by the boat owner's wife. It is the owner's wife who is in charge of selling the fish. Other wives as well as crewmen give a great deal of commentary and advice during the sales. The crew's shares are sometimes paid by the owner; on other occasions the owner's wife gives the money to the crew's wives directly. In any case, there is a strong tradition among the fishermen that the earnings from fishing should be immediately turned over to the senior woman of the household. The senior woman of the household usually manages the household throughout India in both low- and high-caste families. However, in agricultural households the resources women manage are actual food and supplies, while any cash income is controlled by the senior male head of the household. It is likely that this fishing tradition of giving cash income to the senior woman may reflect her traditional role as the seller of fish; traditionally she would be given the fish to sell or dry and now receives the income instead.

An elderly village woman who sells fish runs the fish auction and is elected by the fishermen for her honesty and bargaining skills. While holding office, she usually does not sell fish herself. This office is changed every year or two, so that most of the village's elderly women fish-sellers eventually are auctioneers.

Women perform a number of minor tasks related to fishing. Net making and repair are tedious chores. Men who have fished with nets in the morning work together repairing the net in the afternoon. If there are large tears, the wives, as well as older children, will help out. In addition, many women weave sections of new net. The pay for this work is low, but in poorer households it can be an important contribution. Older widows who have no son to support them often help pull in the beach seines, along with older men and young boys. This work is not very regular, but provides enough fish for a meal and a small cash payment. Neither of these activities would enable a woman to be self-supporting.

Selling fish is the primary traditional economic activity of fisher-caste women. Women can begin fish selling with very little capital, and a small profit is virtually certain. If all the fish cannot be sold, they can be dried or salted and sold later. In 1965 there were about 10 older or widowed women in the village who supported themselves and their unmarried children through daily fish sales. A number of women whose families are poor also sell fish regularly. All the village women know how to sell fish, and most have done so at least occasionally. Fish selling is hard work and takes a woman away from the village for at least 3 hours, often longer. Women with nursing infants, large families without other females to watch the children, or those whose husbands make an adequate income seldom sell fish regularly. However, even a relatively well-off owner's wife may sell fish occasionally if dealers don't offer a good price, if the family is temporarily experiencing a run of bad luck, or if abundant catches make fish selling attractive.

Nonfishing Economic Activities

Some women with extra money lend it to others at high interest rates, thus earning small sums. A few women run small stores selling candy, betel leaf, and other miscellaneous items. Others make and sell illicit arrack, mostly to fishermen within the village. In 1965, only one woman in the village had a modern job. She was a schoolteacher. However, she had run away with someone from another village and was ostracized by the other women. In the past, economic activities outside fishing were rare and offered little opportunity for profit.

Between 1965 and 1980, economic opportunities outside fishing expanded dramatically for both men and women. In 1980, one old woman was working as an *ayah,* and one younger woman was working in a hospital. The two new religious institutions on the north and south sides of Minakuppam provide service jobs for a number of villagers, both men and women, mostly as stall sellers of religious items. Several women run stalls independently, and others work in stalls run by their husbands. Urban jobs for women will probably increase as girls stay in school longer but will continue to be rare for some time to come.

The most dramatic change affecting women has been tremendous expansion of arrack sales. Fishermen traditionally make and drink arrack for special occasions. In 1965, arrack-making was a minor activity for a few old women who sold mainly to fishermen. At that time it was not the regular occupation of any men. The increased numbers of nonfishing residents both in and close to the village have made illicit liquor sales highly lucrative. However, the increased sales have also exacerbated the conflict between those who profit from and those who disapprove of these activities. At least some of the village residents characterize arrack sellers as shiftless, lazy, and morally dubious. Because alcoholic beverages are prohibited in Tamilnadu State, arrack sales make Minakuppam vulnerable to periodic police raids. Villagers also complain about the noise and drunken behavior of the customers, who are mainly nonfishermen. What was once an unimportant and exclusively female-dominated activity is now greatly expanded, controversial, and dominated by young men.

Family Life and Daily Activities

Many activities in Minakuppam center around the household. The household is the unit of economic consumption. Children are reared and taught within the household. The household is also the center of the closest interpersonal relationships—exchanges that are often frustrating and antagonistic but that also provide warmth, security, and meaning in life.

As in most societies, women maintain the household. The amount and type of work to be done depend on the household structure. Most of the village women devote the majority of their day to meeting their family's needs. The fishing women have relatively light housekeeping burdens, primarily because they are relatively poor. Homes are small, sparsely furnished, and easy to straighten and sweep.

The traditional pattern among fishing families is to prepare only one hot meal a day, in the late afternoon or evening. Cold leftovers are consumed in the morning by fishermen at sea and by the rest of the family at home. No noon meal is cooked. Snacks may be purchased for the children, and the men may enjoy a cup of tea if the family can afford it. In 1965, some of the more affluent families cooked rice twice a day. By 1980 both a second cooked meal and more purchased snacks and tea had become common. Meals are quite simple, rarely more than rice and one other dish, usually a fish curry with vegetables or *dahl* added or substituted occasionally.

Water must be carried, but there are several good wells or pumps right in the village, and water is always abundant. Laundry is time-consuming, but most families have only a few garments per member. There is little wood or dung available in the village, and most women purchase firewood. Only the very poor go out and forage for fuel. Children old enough to crawl are usually watched by an older sister, or a brother if no sister over 5 years old is available; adult women are free for more strenuous or productive chores.

How busy a woman is in doing household chores depends mainly on the number of young children and available helpers. Widows who live alone or have one or two children at home have relatively little housework, which is fortunate, because they must spend most of their time earning a living. Women with many young children and no other adult women or older children work the hardest at home. Much of their day is taken up bathing, nursing, and feeding their babies. Women with older children, a cowife, mother, or mother-in-law have a somewhat easier time. In households with two adult women, it is quite common for one to sell fish regularly.

Leisure and Friends

Most women, like the fishermen, rise very early and do their hardest work in the morning. Many women nap with their children in the heat of the early afternoon. Late afternoon is a time for relaxation. Women gather together in pleasant, shady areas near their homes. Usually lighter tasks such as spice grinding, net weaving, and combing, oiling, and braiding hair for themselves and their daughters are also accomplished during these social gatherings. The little children play nearby, under the supervision of the eldest child, and small babies sleep or play in their mothers' arms.

The women who sell fish regularly form a somewhat separate friendship group. They are off selling fish in the afternoon, but they gather together on the beach each morning while waiting for the fish to come in. There they discuss both business and personal matters.

Friendship groups of women are shaped primarily by ecological factors, and a group may contain women of widely differing ages who live near each other. However, the group does not include men. In a society in which women are inferior to men and interaction between the sexes is severely limited, rigid protocol governs the relationships between men and women. The easy familiarity of friendship is only possible within one's own sex.

Women have little recreation outside these social hours. Men friends often go together to a tea shop, to a movie, or on a similar excursion, and such outings increased greatly between 1965 and 1980. Occasionally a group of young girls goes to a movie together, but adult women almost never went to a tea shop together in 1965, and this was still relatively unusual for women in 1980.

Women do, however, participate in family-based leisure activities. In 1965, these included celebration of festivals at home, going to annual temple festivals in Minakuppam and nearby communities, weddings, and funeral feasts. By 1980, these activities had expanded to include occasional city excursions such as movies or sightseeing for the more affluent.

Women whose relatives live outside the village go to visit them occasionally. These visits are frequent during the early years of marriage. Women may go alone, with some or all of their children, or with both children and husbands. These visits home are usually a vacation for the woman who is visiting.

Marriage and Kinship Networks

Most women look forward to motherhood, and childlessness is a great misfortune. In a society where children are the primary resource for one's old age, no one wishes to be childless. If a woman does not bear children or her children die, her husband may take a second wife, or the couple may rear one or more children that a poorer relative is having trouble supporting. The most frequent cause of deviation from the usual life course is early widowhood. Fishing is a hazardous occupation, and accidents sometimes leave a widow with dependent children. These women may return to their own parents, or support themselves through fish sales. Less often, their husband's family may provide support, and levirate remarriages are not unknown. If she is widowed at an early age, it is acceptable for a widow to live with another man.

While the most important kinship unit is the patrilineage, kinship ties through women are also very important among fishermen, like most other south Indian castes. Each man in the same patrilineage has a different, partially unique set of kinship ties through his mother, wife, sisters, and daughters. The preference for cross-cousin marriages means that individuals are often related in several ways.

Like most castes in south India, fishermen allow brides to be married into their own village, and certain close relatives are permitted and even encouraged to marry. The fishermen see definite advantages in arranging marriages with brides nearby. It is easier for the families of both the boy and the girl to find out about one another. When the bride and groom live close to one another, the marriage is easier to arrange and the cost of transportation is reduced. Families like to have their married daughters nearby so that visits can be frequent. It is not surprising that the fishermen prefer to marry their sons to girls from their own village, or, failing that, to girls from nearby villages. Nearly 40% of the wives in 1965 were also from Minakuppam, and about three-quarters of the wives came from within 10 miles of the village. Because of these marriage patterns, many women have women relatives in the village to whom they can turn for support in times of trouble or in domestic quarrels.

Marriage to close relatives has some of the same advantages as marriage within the village. The risks and costs of arranging a marriage are reduced. Instead of giving a girl to a stranger, she is married to someone whom the family already knows. About 45% of the married women in the village were related to their husbands in some way before marriage; in most cases, they were classificatory cross-cousins or classificatory maternal uncle and niece.

Women's Place: Behavioral Standards in the Fishing Village

Women of the fishing caste, like low-caste women in most Indian villages, do not follow the behavioral model of high-caste Hindu women. They are boisterous and even aggressive, rather than shy and retiring; they have formal as well as informal sources of authority and power within the home; and they often play an important economic role in the family. Women move freely about the village and go to nearby markets and the city of Madras. Friendship groups are important, and most women spend much of their leisure time with their friends outside the family. Women are the ones who manage the household's money on a day-to-day basis.

The fishing community, like many low-caste groups, is looked down upon by higher castes for moral laxity. Compared to high-caste Hindus, many moral violations occur, and fishing women suffer less for these lapses. The fishermen and other low-caste groups have different standards for male–female relationships, just as they are also nonvegetarian. At the same time, higher-caste Hindus label these different standards as unacceptable and stigmatize lower-status groups for their behavior. There are a number of now-respectable women in the village who ran away with their lovers or conceived before marriage. In a few cases, a marriage breaks up due to adultery or incompatibility, and the village *panchayat* can declare a valid divorce. Such cases cause considerable scandal for a while, especially if the wife committed adultery, but eventually all the parties settle down and are accepted into village life. Occasionally, the lovers may be forced to leave the village, but they can easily migrate to another fishing community. When a widow is relatively young, a second relationship is not considered disgraceful. No formal marriage occurs; the couple just begins living together.

Certain aspects of the upper-caste Hindu view of the relationship between men and women do prevail in the fishing village. Husbands are supposed to dominate their wives. On formal and ritual occasions the husband comes first. Most women give a certain amount of formal deference to their husbands. They do not usually eat until their husbands have finished. A man's public pronouncements are not usually contradicted by his wife. The belief that women are inferior to men is generally accepted by both sexes. Young girls should be watched and secluded, at least as far as it is practical to do so.

Among fishermen, the special economic role of wives also has an impact on husband–wife relations. The wife usually manages the family's money and gives her husband and the older children daily spending money. The husband has to ask her for any additional money he wants for personal use or family investments. However, husbands generally make decisions about large amounts of money. When there is fishing equipment or other resources for the husband to manage, his authority within the family often increases.

When a village man has an urban job instead of fishing, it affects the traditional economic role of his wife. The urban worker has no fish for his wife to sell; he himself receives his paycheck and he also takes over money management. In at least one case, the husband deliberately deceives his wife about the amount of money he earns and spends nearly half of it on himself. With the reduction of her economic power, the city worker's wife also seems to become more subservient toward her husband. The wives of urban workers tend to be much less vocal when their husbands are present, even about purely household matters. Wives of urban workers are much less likely to have a strong friendship group in the village with whom they interact. They are more likely to stay at home during their leisure time and to send their children to market rather than go themselves. They are the only women in the village who show some tendency toward the behavior patterns of higher-prestige women and look down on the other women of the village because they are "too free."

Comparative Analysis

In many respects, activities and relationships of women in the fishing village of Minakuppam are similar to those of women living in agricultural villages in India and in most other industrializing agrarian societies. The same general level of technology in both peasant fishing and peasant agriculture and similar relations with both urban centers and advanced industrialized nations make for similar levels of wealth, standards of living, and consumption patterns. The women of Minakuppam are also basically similar to most other Indian women in their ties to religious, political, educational, and urban structures; in their celebration of family ceremonies and festivals; in their fulfillment of the economic, political, and ceremonial obligations of kinship; and in their increased power and status as they become older.

However, within the context of a peasant level society and its general male dominance, women in Minakuppam are relatively independent and unsubservient in their everyday behaviors and relations with men. To understand what accounts for this distinctive pattern of activities and relationships, we need to look to those features of economic production and societal stratification that affect women's control and use of resources and their opportunities for social support from kin and other women.

Fishing economic production differs significantly from most other economic activities, especially the prevailing work of agriculture in agrarian societies and of most manufacturing and service industries in modern societies (Norr and Norr, 1974, 1976, 1978). The distinctive technical and environmental constraints of fishing include exposure to physical risks, uncertainty, separation of work from residential community, difficulty in maintaining clear-cut control of productive factors, and the need for teamwork, skill, and reciprocal coordination. As a consequence of these constraints, work organization in fishing is nonroutine, rational, and difficult to bureaucratize. Most often there is recruitment based on skill and compatibility, equality, minimal administrative hierarchy and authority, workers' involvement in decision making, flexible working arrangements, more even distribution of the means of production, and larger returns to workers. Fishing encourages autonomous, unsubordinated workers.

Ocean fishing's production and work relationships can influence women's status in at least three ways. First, the special nature of the fishermen's occupational community with its nonhierarchical emphasis provides a model that women may follow. Second, fishermen's long separation from the community and frequent accidental death can make women more self-reliant and decrease their dependence on men. Third, and probably most significant, fishing can provide women with access to economic resources and an independent role in production. Fish selling and net making and repair make women in fishing never completely dependent economically on men. They have a familiar way to earn money on their own. Moreover, their traditional role as the marketers of fish means that fisher-caste wives often handle the family's daily income and have a great involvement in family spending and saving. Their market activities regularly expose at least some village women to urban activities and to non-fisher-caste people outside the village.

In addition to the effects of fishing economic production, Indian social organization also greatly influences the activities and relationships of the women of Minakuppam. There are substantial regional and caste variations in the extent to which Indian men dominate women. These differences in occupational status and cultural supports for societal stratification also contribute to the distribution of economic resources and opportunities for social support that result in the women of Minakuppam having greater independence, power, and authority. To understand how fishing and factors of social organization interact to contribute to these distinctive behavior patterns for women, we compare the fishing village women first to women in Indian agriculture and then to women in fishing villages in other peasant-level, predominantly agrarian societies.

Comparisons with Indian Agriculture

Hindu culture and social organization pervade the fishing community, and the activities and values of village women manifest these influences. The "ideal woman" of Indian society is the patient, selfless mother, modest and always deferential to men and elders. This ideal is recognized throughout India, but actual behavior in different regions and social classes does not always follow the model.[2]

Women in the south have greater independence than women in the north. They move about outside the home more freely, talk more frequently and freely, and are more independent of male authority. The historical legacy of Muslim conquerors and their *purdah* ethic are undoubtedly important in accounting for the restriction of women in the north. Differences in marriage patterns also contribute to greater freedom of movement for southern women. In the south, marriage within the village among close relatives increases the importance of kinship ties outside the patrilineage. The geographic closeness of the wife's relatives and kinship ties to her husband's patrilineage before marriage combine to give her more security and make her less dependent on her husband's patrilineage in marriage. A girl's transition to wifehood is a much less radical change in the south. After marriage, a girl in the south tends to be closer to her own family and visits them more often. She has relatives other than her husband's to call on for aid and support.

In addition to the marked variation in marriage patterns by region in India, marriage customs also differ by caste status. Higher castes tend to have a dowry system while lower castes usually pay a bride-price. Among higher castes, and especially among Brahmins, there is no legal dissolution of marriage bonds through divorce or legal separation. Women can have only one husband and are expected to remain chaste before marriage and faithful to one husband for their whole life, both during marriage and after separation or his death. Among the lowest castes, there are recognized procedures for divorce, and widow remarriage is permitted. Although there is also a double standard of sexual behavior among lower castes, the lapses of women are not usually punished severely, and indeed are often tolerated quite openly for older women. The upper castes pay more attention to ancestry and the patrilineal line than do lower castes. Upper castes can often trace their patrilineal descent over many generations and have large, well-recognized patrilineal clans (Dube, 1955; Gough, 1956). All of these differences tend to make low-caste women in the south more independent than high-caste women.[3]

In many south Indian villages, the lowest castes are residentially separate from higher castes. Relative isolation of low castes permits them to follow their own behavioral standards with less interference or competing world views from their more powerful and prestigious neighbors. Ocean fishing villages along most of the Tamilnadu coast consist of members of the same low caste of *Pattanachettiars*. Minakuppam and other fishing villages, unlike most agricultural villages, contain only members of a single low-status caste. Fishing is considered a dirty, polluting occupation, and other castes look down upon the fishermen. As in the rest of India, low-caste status is associated with greater freedom for women and less emphasis on male superiority. The village is similar to a scheduled caste hamlet in an agricultural village in that its members are all relatively poor and are bound together by common caste sentiment and customs, kinship bonds, and a community temple and festival. However, it is very different from a scheduled caste hamlet in its economic independence and general lack of informal contacts and formal ties with higher-caste persons. This isolation was even more pronounced in the fishing village in 1965; at least a mile of empty land lay between the fishing community and its higher caste neighbors. Unlike agricultural low castes, neither men nor women in the fishing villages work as economic dependents of higher-caste individuals. This further reduces the influence of upper-caste behavioral standards on the women of the fishing community.

Fishing Villages in Other Predominantly Agrarian Societies

Within the context of Indian society, it is difficult to determine the importance of fishing's low social status relative to its economic impact. Comparisons of fishing and agriculture in other complex peasant societies can help.[4] We were able to locate ethnographic reports to examine women's roles and status in 10 other fishing communities located in five peasant-level, predominantly agrarian societies: Japan (Glacken, 1955; Norbeck, 1954), Taiwan (Diamond, 1969), Peninsular Malaysia (Raymond Firth, 1966; Rosemary Firth, 1966; Fraser, 1960, 1966), Brazil (Forman, 1970; Kottak, 1966), and the Caribbean islands of St. Kitts (Aronoff, 1967) and Jamaica (Davenport, 1954).

The roles, status, and power of women in fishing communities compared to women in agriculture vary substantially among these societies. Agricultural women are most dependent and submissive in Japan and Taiwan. In both, the family system is patrilineal and patrilocal, and men ideally dominate women. In Taiwan, the ideal family is the lineal-collateral joint family, while in Japan the lineal-joint or stem family pattern is ideal. In Japan the eldest son is usually the main heir, while in Taiwan brothers usually inherit equally. Ideally, women are chaste, obedient, and self-effacing, but there is widespread deviation from the norms. These norms appear to have penetrated the lower socioeconomic levels of Japanese society more deeply than they have in Taiwan or in India. But overall, as in India, women in Taiwanese and Japanese fishing villages are more independent, active, and powerful than women in surrounding agricultural villages.

Women in agriculture have a great deal more independence in Malaysia and the southern Malay-speaking portions of southwestern Thailand. Both areas are Moslem. Both men and women can divorce and remarry fairly easily. Women own property, move about freely, and show relatively little formal deference to men. Family structure is patrilineal, but married couples can reside near either the wife's or the husband's family. In Rusembilan, Thailand, and Perupok, Malaysia, there are few indications that women in fishing communities are any more aggressive and independent than women in agricultural villages.

Women in the Brazilian and Caribbean fishing villages also do not appear to be more active and independent than women in agriculture. In fact, women in these fishing communities, especially those in Farquhar Beach, Jamaica, seem to be less independent than women either in the other fishing villages or in surrounding agricultural villages. Davenport (1954) attributes women's great dependence on their husbands to the lack of economic opportunities for women and the lack of supportive relatives nearby. In Dieppe Bay, St. Kitts, the wives of fishermen seem to be somewhat less independent than the wives of cane cutters, and fishermen seem to play a more dominant and central role in the family (Aronoff, 1967). Kottak's (1966) and Forman's (1970) reports of the relative lack of opportunities and dependence of women in Brazilian fishing communities is supported by Robben's (1988) data for wives of fishermen in a Bahia community who use canoes, a peasant-level technique, in contrast to modern motor boats.

A *machismo* cultural pattern prevails throughout much of Latin America including Brazil and the Caribbean, especially among the agrarian upper classes. Among the middle and upper classes, premarital chastity, a religious marriage, and seclusion as much as possible within the home, both before and after marriage, are the prevailing patterns of female behavior. The lower classes frequently do not fully accept the upper-class ideals and have quite different patterns. Common-law unions are far more frequent than religious marriage. Frequent breakups of informal unions result in a relatively large number of households headed by a woman with no male, or with only a casual lover present. The least stable unions and the highest proportion of matrifocal households exist among plantation workers and urban slum dwellers. Hazel Du Bois (1964) compares four settings in Puerto Rico and finds that a higher proportion of matrifocal families seems to be related to economic insecurity more than to poverty per se. Women have a great deal of sexual freedom, although there is still a double standard. They are economically active and move about freely. Despite the objective independence of women, there seems to be a great deal of underlying hostility and lack of trust in relationships between men and women. Lower-class women in these societies, in both fishing and agricultural communities, are relatively independent and nonsubmissive because of distrust between men and women and because of unstable unions, as well as because of the absence of men and ease of separation.

Discussion

Our comparative analysis of peasant-level, predominantly agrarian societies suggests that fishing as a particular technological adaptation to natural resources does not have a single effect on the degree of independence of women. This is also Thompson's (1985) conclusion in his review of fishing communities primarily located in industrial societies.

In Minakuppam, men have no opportunities to engage in farming or to invest in agricultural lands as an alternative to fishing, women have greater freedom and equality than the wives of most farmers, and there are no mechanisms for generating personalized ties with nonfishermen outside the village. All three of these differences also appear in Japan and Taiwan. In India, Japan, and Taiwan—but not in Brazil, the Caribbean, or Peninsular Malaysia—women in fishing communities have more independence and power than women in agricultural villages.

One important factor that seems to be related to the position of fishermen's wives is their economic opportunities. In K'un Shen, Taiwan, and Rusembilan, Thailand, women customarily sell their husband's catch and so collect their family's income, as do the women of Minakuppam. The women of Perupok, Malaysia, also earn independent income from petty trading and fish selling and sometimes own their own plots of land or coconut groves (Rosemary Firth, 1966). In Takashima, Japan, the fishermen's wives make a separate noncash economic contribution to the family because they farm while their husbands fish. In each of the Brazilian and Caribbean villages, jobs for women are relatively rare, and most do not make enough to be self-supporting. It is significant that in Dieppe Bay, St. Kitts, the wives of higher status fishermen are less independent than wives of cane cutters because their husbands tended to earn somewhat more (Aronoff, 1967).[5]

Another factor related to the role of women in fishing and agricultural villages is the relative status of fishermen compared to most agriculturalists. In Japan, Taiwan, and India, fishermen are definitely lower in status than farmers. In Malaysia and South Thailand the two occupational groups are about equal. In the areas near the fishing villages in Brazil and in the Caribbean, agrarian workers are lower in status than fishermen. Most of these agriculturalists work on large sugar plantations, own no land and sometimes not even a house site, and suffer a long period of unemployment or underemployment. In Negril, for example, the men of the fishing village regard working on the sugar plantations as beneath them no matter how poor they are.

In all of the societies examined, women have more independence in the lower strata of society. When fishermen are lower in status, their wives tend to be more independent than wives of farmers. When fishermen are equal to farmers in status, their wives do not behave very differently. Where fishermen are higher in status than agricultural workers, women in the fishing community have less power. And where there are significant class differences among fishermen, the wives of higher-status fishermen are less independent than the wives of lower status fishermen. These findings support those theories of gender stratification that emphasize women's contribution to production and women's participation in public activities as integrating women into a broader network of relationships that extends beyond the isolation of the domestic unit (Huber and Spitze, 1988).

Acknowledgments

Our account of women's activities and relationships in Minakuppam is based on detailed observation for 11 months in 1965–1966 and a 1-month revisit in 1980. Kathleen Fordham Norr carried out the fieldwork in 1965–1966. The contributions of A. Gopala Krishna as co-investigator and translator are gratefully acknowledged. The 1980 fieldwork took place when James L. Norr was an Indo-American Fellow, and we appreciate the support of the Indo-U.S. Subcommission on Education and Culture. Dona Lee Davis, Shelley Feldman, and Jane Nadel-Klein in addition to the issue editors kindly provided very helpful comments on earlier versions of this report. A preliminary version was presented at the meeting of the American Sociological Association in San Francisco in August 1989.

Notes

1. For the most part, we emphasize the situation before urban expansion for comparisons with other peasant-level fishing communities in agrarian societies, but we also include a number of comparisons between 1965 and 1980 to help evaluate the effects of social change. We focus on development questions in Norr and Norr (1974, 1982); other details on the village are found in K. Norr (1972, 1975, 1976).

2. These comparisons are based on 19 ethnographies of villages throughout India with plow agriculture but without extensive use of fossil-fuel machines. (For details and citations, see K. Norr, 1972.)

3. Gough (1956) explains the caste differences in women's roles as a complex result of both prestige and economic status. For the highest castes, the need to maintain their prestige is usually very strong. In addition, they are usually relatively well off economically and can afford to forgo having their women work outside the house. Brahmin men also usually prefer to avoid manual labor. The prestige factor is strong enough so that even quite poor Brahmins in rural villages do not allow their wives to do agricultural work.

4. Peasant-level ocean fishing uses large nets and produces large quantities of fish. It thus requires the support of a peasant-level social system and economy with a well-developed system of agriculture and regional marketing. The markets provide raw materials for fishermen and can dispose of large amounts of fish. Fishing in modern societies uses motor power for boats and hauling nets and requires an even more complex supporting economy and social system. Likewise, modern agriculture's use of motorized implements is technologically more complex than peasant sedentary agriculture based on animal, wind, or water power. These distinctions and their implications for work organization are discussed in Norr and Norr (1974, 1977, 1978). We restrict our comparisons to case studies in which both fishing and agriculture are at a peasant level of technology and in which it is possible to evaluate the status of women in fishing relative to women in agriculture. For good discussions of the uses and methods of comparative case analyses, see Feagin, Orum, and Sjoberg (1991) and Ragin (1987).

5. Additional evidence of the importance of access to economic resources for women and low societal status for fishermen is found in Portugal (Cole, 1988) and Peru (de Grys, 1988). Both cases are similar to the Brazilian and the Caribbean comparisons in the importance of a *machismo* culture. But in both cases, fishermen are lower in status, women have access to economic resources, and women in the fishing community are more active and more independent than women in surrounding agricultural villages. It may be that the dominant enterprise in agriculture is the

crucial variable (Stinchcombe, 1961). Plantations are dominant for our Brazilian and Caribbean cases; family-size tenancies and family small-holdings, with a more egalitarian class structure, dominate in both the Portuguese and the Peruvian cases. Where plantations are the dominant enterprise, agricultural lower classes have fewer legal privileges and less access to technical knowledge.

References

Acheson, James. 1981. Anthropology of fishing. *Annual Review of Anthropology,* 10:275–316.

Andersen, Raoul, and Cato Wadel, eds. 1972. *North Atlantic Fishermen: Anthropological Essays on Modern Fishing.* St. John's, Newfoundland, Canada: Institute of Social and Economic Research, Memorial University of Newfoundland.

Aronoff, Joel. 1967. *Psychological Needs and Cultural Systems.* Princeton, NJ: Van Nostrand.

Cole, Sally. 1988. The sexual division of labor and social change in a Portuguese fishery. In *To Work and to Weep: Women in Fishing Economies,* ed. Jane Nadel-Klein and Dona Lee Davis, pp. 169–189. St. John's, Newfoundland, Canada: Institute of Social and Economic Research, Memorial University of Newfoundland.

Davenport, William. 1954. A comparative study of two Jamaican fishing villages. Unpublished Ph.D. dissertation, Yale University, New Haven, CT.

de Grys, Mary Schweitzer. 1988. Does absence make the heart grow fonder or only the influence stronger? Women in a Peruvian fishing village. In *To Work and to Weep: Women in Fishing Economies,* ed. Jane Nadel-Klein and Dona Lee Davis, pp. 91–105. St. John's, Newfoundland, Canada: Institute of Social and Economic Research, Memorial University of Newfoundland.

Diamond, Norma. 1969. *K'un Shen, a Taiwan Village.* New York: Holt, Rinehart and Winston.

Du Bois, Hazel. 1964. Matrifocality and courtship in four Puerto Rican communities. Paper presented at the annual meeting of the National Council on Family Relations in Miami, Florida.

Dube, S. C. 1955. *Indian Village.* London: Routledge and Kegan Paul.

Feagin, Joe R., Anthony M. Orum, and Gideon Sjoberg. 1991. *A Case for the Case Study.* Chapel Hill: University of North Carolina Press.

Firth, Raymond. 1966. *Malay Fishermen: Their Peasant Economy.* Hamden, CT: Anchor Books.

Firth, Rosemary. 1966. *Housekeeping among Peasants.* London: Athlone.

Forman, Shepard. 1970. *The Raft Fishermen.* Bloomington: Indiana University Press.

Fraser, Thomas, Jr. 1960. *Rusembilan.* Ithaca, NY: Cornell University Press.

———. 1966. *Fishermen of South Thailand, the Malay Villagers.* New York: Holt, Rinehart and Winston.

Glacken, Clarence. 1955. *The Great Loochoo.* Berkeley: University of California Press.

Gough, Kathleen. 1956. Brahmin kinship in a Tamil village. *American Anthropologist,* 63:826–853.

Huber, Joan, and Glenna Spitze. 1988. Trends in family sociology. In *Handbook of Sociology,* ed. Neil J. Smelser, pp. 425–448. Beverly Hills, CA: Sage.

Kottak, Conrad Phillip. 1966. The structure of equality in a Brazilian fishing community. Unpublished Ph.D. dissertation, Columbia University, New York.

Nadel-Klein, Jane, and Dona Lee Davis. 1988. Introduction: Gender in the maritime arena. In *To Work and to Weep: Women in Fishing Economies,* ed. Jane Nadel-Klein and Dona Lee Davis, pp. 1–17. St. John's, Newfoundland, Canada: Institute of Social and Economic Research, Memorial University of Newfoundland.

Norbeck, Edward. 1954. *Takashima, a Japanese Fishing Community.* Salt Lake City: University of Utah Press.

Norr, James L., and Kathleen F. Norr. 1977. Societal complexity or production techniques: Another look at Udy's data on the structure of work organizations. *American Journal of Sociology,* 82:845–853.

———. 1978. Work organization in modern fishing. *Human Organization,* 37:163–171.

———. 1982. Impact of urban growth: Change in a south India fishing community from 1965 to 1980. *Ethnology,* 21:111–123.
Norr, Kathleen F. 1972. A south Indian fishing village in comparative perspective. Ph.D. dissertation, University of Michigan, Ann Arbor.
———. 1975. The organization of coastal fishing in Tamilnadu. *Ethnology* 14:357–371.
———. 1976. Factions and kinship: The case of a south Indian village. *Asian Survey,* 16:1139–1150.
Norr, Kathleen F., and James L. Norr. 1974. Environmental and technical factors influencing power in work organizations: Ocean fishing in peasant societies. *Sociology of Work and Occupations,* 1:219–251.
Ragin, Charles C. 1987. *The Comparative Method.* Berkeley: University of California Press.
Robben, Antonius, C. G. M. 1988. Conflicting gender conceptions in a pluriform fishing economy: A hermeneutic perspective on conjugal relationships in Brazil. In *To Work and to Weep: Women in Fishing Economies,* ed. Jane Nadel-Klein and Dona Lee Davis, St. John's, Newfoundland, Canada: Institute of Social and Economic Research, Memorial University of Newfoundland.
Smith, M. Estellie. 1977. *Those Who Live from the Sea: A Study in Maritime Anthropology.* St. Paul, MN: West.
Stinchcombe, Arthur L. 1961. Agricultural enterprise and rural class relations. *American Journal of Sociology,* 67:165–176.
Thompson, Paul. 1985. Women in the fishing: The roots of power between the sexes. *Comparative Studies in Society and History,* 27:3–32.

Epilogue

Since we last visited Minakuppam, 15 years have passed, years marked by worldwide changes. Globalization of economies has accelerated trends in Minakuppam: technological modernization of fishing and production for an international market; rapid transition from an isolated, homogeneous community into a socially diverse peri-urban community; overfishing; and movement of many individual fisher families to other fishing hamlets with better harbor facilities and into nonfishing jobs throughout the city. Some men and women no longer identify themselves primarily as fishermen but have been absorbed into a broader urban identity that incorporates caste, occupation, income, and education.

When we wrote this article, there was widespread interest in the status of women in different times and places and a search for the underlying structural cause of gender inequality. Explanations proposed included those centered around the structural requirements of biological reproduction, differences between private and public spheres in determining male versus female status, and the degree of access to independent economic means. Although interest in gender asymmetry and gender status continues, the search for a single structural cause has been disappointing. Structural factors can help make sense of specific situations, but no single structural cause for gender inequality has emerged from cross-cultural comparisons.

In hindsight, it seems clear that this search for a simple solution was hampered because it asked the wrong question. Gender inequality is robust and persistent in societies precisely because it is reinforced in a variety of structures, organizations, and beliefs. Thus, even if the specific structures that reinforce inequality in a specific context are identified, this does not necessarily mean that the absence of or change in these structures will result in a decrease in gender inequality. In reflecting on the lack of fruitful insight from this line of inquiry, many current feminist scholars have turned away from structures toward culture and values in the search for explanations. Moreover, interest has shifted from the search for a single explanation to an understanding of the diverse and specific contexts within which women live and gender inequality exists (Rosaldo, 1980).

The lack of a single structural explanation does not mean that structural factors are unimportant. Women's inequality is multidimensional and includes restrictions on movement, speech, and self-expression, reproductive freedom, mate selection, and education; greater exposure to physical, sexual, and emotional abuse and violence within and outside the home; heavy labor, restrictions on economic activities, and lack of control of economic resources; lack of status, respect, or honor; and absence or disparagement of women's cultural expressions. Not only are these dimensions different, they are often mutually exclusive, so that gains in one area lead to a reduction of equality in another. In many agrarian societies, there has been a crystallization of women's inequality into two distinct patterns related to family status and class. Women in upper-class, landowning families enjoy higher status and respect; greater protection from violence, especially violence outside the home; more dependable family support for themselves and their children; less heavy labor; and often more education. But they also have less freedom of movement and expression, less partner choice or sexual freedom, and less economic independence. Low-status women have the "freedom" to work hard outside the home and often have independent (though meager) economic resources. Their movements and speech are less restricted and they have more partner choice. But they suffer low status and little respect; less certain family protection—at least partly due to the pressures of poverty; and greater exposure to potential violence and exploitation outside the home.

As outsiders, social scientists have tended to view lower-class women as less oppressed by gender inequality than higher-status women in the same society. In part, this reflects the fact that lower-class women have less gender inequality in the more readily observed areas of public behavior and economic activities. Additionally, the greater public "freedom" of lower-status women may be more consonant with the values of the observer, especially observers who are educated working women from Western societies who place high value on independence and public participation. However, among those living within agricultural societies there is strong social consensus that the lot of the upper-status woman is preferable to that of a lower-status woman. Thus, it is not surprising that upward economic mobility often leads women and their families to adopt the "proper" behavior of upper-status women. As women reduce gender inequality in some area, they may simultaneously increase their relative inequality in others. Similarly, in industrial and postindustrial societies, gains in women's education and employment outside the home have been accompanied by crushing new workloads due to lack of change in the household division of labor. Concurrent increases in divorce and nonmarital childbearing have decreased stability of family support for many women. Because these dimensions of gender inequality are so qualitatively different, there is no simple answer to the questions of which group of women experiences greater gender inequality and whether gender inequality is decreasing or increasing overall.

Additional Reference

Rosaldo, M. Z. 1980. The use and abuse of anthropology: Reflections on feminism and cross-cultural understanding. *Journal of Women in Culture and Society*, 5(3):389–417.

PART TWO

PROPERTY RIGHTS: ACCESS TO LAND AND WATER

Photograph courtesy of Carolyn Sachs.

Chapter 6

Women and Irrigation in Highland Peru

BARBARA DEUTSCH LYNCH

Department of Natural Resources
Cornell University
Ithaca, NY 14853
USA

Abstract *Women remain invisible in Peruvian highland irrigation despite their increasing participation in other aspects of political life. Despite norms that define irrigation as men's work, women irrigate crops and contribute labor to system surveillance and maintenance. Women who work in irrigation tend to be members of poor, smallholder households and single heads of households. Their labor contributions are not matched by participation in system management, because, with the bureaucratic transition in highland irrigation, the critical task is no longer controlling water per se, but gaining and controlling access to external resources and government assistance. Interactions with the male-dominated irrigation bureaucracy are dominated by the paternalistic politics of normalcy, rather than the open, confrontational politics of crisis. With token exceptions, women are informally excluded from these interactions or co-opted into peripheral roles as users of domestic water.*

Keywords: women, gender, irrigation, water management, labor, bureaucratic transition, participation.

Introduction

Women's roles in Peru have changed substantially in the past 15 years. Women's issues have captured national attention; women's groups are in the forefront of grassroots democratization that has occurred in the past decade, and women have played prominent roles in rural social movements (Radcliffe, 1989). Yet, curiously, given the importance of water in highland Peru and the increasing involvement of women in irrigation, women's roles in water management have received little attention from rural women's organizations, government officials, or students of Andean irrigation. The first part of this article examines the effects of class, gender ideologies, and integration into the national economy and polity on (1) women's participation is system operation and maintenance and (2) the degree to which they share in control over resource allocation. The second

Research on the social organization of irrigation in Cajamarca was conducted as part of Plan Piloto, an AID-sponsored program to improve water management and use of lands irrigated as a result of the Plan MERIS irrigation project. The author acknowledges the insights into San Marcos household economies and agricultural production provided by Plan Piloto staff members Jorge Orillo, José Luis Villarán, Rodolfo Flores Chanduví, Ruth Aguilar Cobián, and Carlos Nonone. Plan MERIS extension agents Nelly Saavedra and Juana Paz shared with me their experiences as organizers of rural women's groups. The continuing support of Jose Hermosa Jerí, then head of the Plan MERIS for the Cajamarca region, was invaluable, as were the many fruitful issue discussions that he stimulated. Paul Gelles and Nancy Peluso have provided valuable comments on initial drafts of this paper.

part assesses the effects of government intervention in local irrigation development on women's participation in system management.

The fieldwork upon which much of the following article is based was carried out in two Cajamarca government-assisted irrigation systems, San Marcos and Santa Rita, as part of a U.S. Agency for International Development (USAID)-funded evaluation of Plan MERIS I, a project consisting of about a dozen small- and medium-scale construction and rehabilitation projects in the Peruvian Sierra.[1] The objective of the study was to learn what factors at the household and community levels affected farmer participation in irrigation system development, operation, and maintenance and, ultimately, system performance. Data on water use, agricultural and nonagricultural household economic activities, and participation in irrigation tasks; the history of irrigation development in the two project areas; and assessment of management problems were collected by means of extended informal interviews with men and women in the fields and in their homes. Informal interviews with local elected irrigation officials and appointed representatives of the Ministry of Agriculture were used to obtain information on project history and government–community relations. The discussion on women's work in agriculture relies heavily on data gathered by Deere (1978), also in the department of Cajamarca.

Cajamarca is a highlighted department in northern Peru whose population is largely Spanish-speaking and *mestizo*. The San Marcos and Santa Rita project areas adjoin district and departmental capitals and enjoy reasonably good road access to the coast. They are integrated into the national economy, albeit on unfavorable terms. The project areas are not necessarily representative of Andean societies, but they are undergoing changes common throughout the Sierra: (1) a transition to an open, cash economy, (2) replacement of indigenous irrigation organization and infrastructure by systems modified and regulated to some extent by irrigation bureaucracies, and (3) increased participation of women in irrigation-related tasks as a result of male outmigration and occupational diversification.

Women's participation in irrigation varies widely in the Andes from one area to another; it has also changed as the dynamics of community–state interactions have evolved. Allocation of labor to irrigation varies with the nature of the physical landscape, the centrality of irrigated agriculture to income generation in the local economy, and local cultural norms with regard to women's visibility and participation. In drier zones, lands cannot be cultivated without irrigation. In parts of the highlands that receive more rainfall, irrigation allows for flexibility in the timing of agricultural and herding activities in different production zones (Golte, 1980; Mayer, 1985), but it does not always permit additional cropping cycles, nor does it increase smallholder income from agriculture sufficiently to eliminate the need for other sources of income. Thus, women's roles vis-à-vis water are closely tied to the agricultural calendar.

Where they are integrated into the national economy, families often find it hard to meet subsistence needs from local agriculture (see, for example, Collins, 1989); seasonal migration and search for employment outside of agriculture are common. Both men and women in rural areas may work outside agriculture, but women are more likely to remain at home and perform irrigation tasks as heads of household. (De jure heads of households include widows and single women; de facto heads of household are women whose partners have abandoned them, have migrated for the season, or whose major employment obligations lie at some distance from the irrigated zone.)

Women's Work and Structural Subordination

My point of departure for looking at recent changes in women's participation is Bourque and Warren's discussion of Mayobamba irrigation in the context of other agricultural activities (1981), the only specific, albeit brief, discussion of gender-based division of labor in irrigation. However, their analysis focuses almost exclusively on irrigation in the field, rather than on operation and maintenance of the irrigation system. Bourque and Warren argued that in the Central Sierra, "women are structurally subordinated to men by patriarchal mechanisms that link access to key institutions to sex role stereotyping and the sexual division of labor." They see irrigation as one of a set of key tasks that are assigned to men; these tasks "serve as gateways to critical resources," including land, water, transportation, and cash (p. 123). Women are prevented from performing key tasks so that they depend on men for resource allocation. Thus, to the extent that women are structurally excluded from key institutions, their power is limited.

A second mechanism for women's subordination is the definition of work as that which men do. Bourque and Warren (1981, p. 119) distinguished between doing work and defining what women do as work:

> In talking about work such as irrigation, which integrates tasks from the domestic and agricultural spheres, men and women perceive work in different ways. Women who prepare meals for workers and bring burros as transportation between the fields and town, perceive these tasks as part of an integrated whole called "irrigation." In fact, women speak directly of being involved in irrigation when they carry out tasks that complement men's work in the fields. Men, on the other hand, more narrowly define irrigation as the work they perform in actually opening the channels with shovels so that water flows into the fields.

Thus, in a patriarchal community in the Central Highlands, men maintained power by excluding household women from a task—controlling water at the farm gate—that involved control over access to a critical resource. Their power and status was reinforced by their narrow definition of irrigation to exclude the roles that women play. The following section examines the extent and nature of women's participation in irrigation and the degree to which this participation has reduced women's subordination to men in terms of control over the resource.

Women and Irrigation

Technical decisions about irrigation are made in a social context and reflect the priorities of those making decisions. These priorities are shaped by beliefs and moral and aesthetic values as well as by economic considerations. The moral economy of water management in the Andes has the following tenets: (1) use rights to irrigation water are vested in land, rather than in individual users (Sherbondy, 1985); (2) landowners must participate in system upkeep to exercise existing rights to water; (3) when water is scarce, irrigators adhere to a rotation schedule that allocates the entire flow of a given canal to each user along its length for a fixed time period; and (4) during periods of extreme scarcity, domestic use of water has priority, followed by irrigation of a very small parcel for subsistence or corn cultivation.

Because water rights adhere to parcels rather than to individuals, obligations and decision making vis-à-vis the system are vested in the household rather than the individual. In principle, those who contribute their labor to operation and maintenance of small-scale irrigation systems and who make management decisions are a single set of individuals cooperating in a common endeavor. In practice, not all users contribute their labor to system maintenance, and not all those who use water and contribute their labor make management decisions. Gains and losses attributable to water management decisions are distributed unevenly and, in general, benefit the decision makers. To the extent that decision making is vested in the household as represented by the head, generally male, the interests of other household members in the water resource may be overlooked.

Definitions of women's work in Andean rural households, not to mention women's actual participation in agricultural work and management, depend upon the household's wealth and on the availability of family or hired labor.[2] Most women who participate directly in irrigation and system maintenance in Cajamarca are de facto or de jure heads of household (wives of migrants or the infirm, widows, or single women). If household income is insufficient to hire labor and household men are not available, women will participate in canal cleaning and construction in order to exercise their rights to water. Thus, it is a particular subset of rural women—smallholder heads of household—who are most likely to participate in irrigation activities on the farm or at the system level. When asked to describe participation in system construction and maintenance activities, 67 of the 172 irrigators interviewed in Santa Rita addressed women's participation; of these, 34, or slightly more than half, acknowledge that women participate. Fourteen responded that widows participate by sending a *peon* or paying a fee; 20 acknowledged women's work in construction and meal preparation.

Women's uses of water are diverse, owing to the range of their economic activities in irrigated zones. In San Marcos and Santa Rita, where care of domestic animals from cows to guinea pigs is often women's work, women use irrigation water for livestock as well as for crops. They gather hedgerow products whose yields increase with frequent water applications. These include tara (*Cesalpina tintorea*) pods that are exported to the European Common Market and are used to produce tannin; cochineal scale insets, also exported to the Common Market and used locally as a textile dye; and prickly pears. Where a domestic water source is lacking, women haul household water from irrigation ditches. They wash clothes and raw wool in canals and leach vegetables to remove toxins.

Thus, from a woman's perspective, we need to ask the following questions: (1) Do the physical infrastructure and social arrangements that characterize the systems satisfy these women's needs as consumers of water? (2) Do women participate in irrigation management decisions? (3) What kinds of contributions are women obliged to make to keep the system functioning? (4) Is women's control over irrigation infrastructure, water allocation, and water delivery commensurate with the contributions they make to system development, operation, and maintenance?

Women and Irrigated Agriculture

Delivery of water from the farm gate to the root zone of the crop—on-farm water management—is considered men's work in the Andes. A study of traditional forms of household labor in highland Peru concludes that although planting, cultivation, and harvesting are done by both men and women, "irrigation is a male responsibility" (CIAT, 1984, p. 37). In Cuzco, according to Radcliffe (1986),

> The exclusion of women from the use of the plough and from irrigation is strongly reinforced by taboos and myths illustrating the danger involved if women were to participate in these labours, as well as the belief that women are not strong enough to do so. . . . Irrigation potentially threatens women's fertility due to the contact with water.

These norms crumble quickly in the face of expediency, but they do mask women's work.[3]

The women most likely to play active roles in crop irrigation are members of smallholder households (less than 5 ha) where agriculture is not highly remunerative and whose household men are engaged in productive activities away from the field. According to Deere, land fragmentation in Cajamarca produced a decline in the contribution of agriculture to household income and a corresponding increase in women's participation. "Women's greatest agricultural participation, relative to men, is found among the poorest strata of the peasantry, those without sufficient access to land to produce their full subsistence requirements, and among those households where the man works full time in wage labor" (1978, pp. 311-312).

Comparing landless, smallholder, and middle-class and wealthy peasant households, Deere found that women in landless households had the highest rates of participation in agricultural tasks and decision making, and that participation generally varied inversely with farm size. Women in 61% of her sample of 93 Cajamarca households engaged in irrigation activities. In 42% of the sample households, men and women participated with equal productivity. Deere did not break down irrigation participation by size of land holdings, but it is likely that this task would have followed the general pattern of women's participation.

Because the labor of all women is valued less than that of men,[4] and because irrigation is traditionally defined as a male task, women in Cajamarca are not hired to perform on-farm irrigation tasks for others either as day laborers, sharecroppers, or participants in reciprocal labor arrangements. Figueroa (1984) reported a similar pattern in eight southern Sierra communities; in all but one, women were hired only for harvest work and were paid in kind rather than cash. Thus, even if they play a major role in agricultural decision making, rural women who lack irrigated land do not irrigate crops. Where households are wealthy enough to hire labor, women are equally unlikely to irrigate. Thus, a disproportionate number of women who irrigate fields come from poor households that own irrigated parcels.

Deere also suggested that where agriculture is a declining activity, men and women share more equally in agricultural work and decision making (Deere, 1978; Deere and León de Leal, 1982). The ethnographic data from San Marcos indicate that where men are employed as drivers, masons, or workers, they tend to do agricultural chores on weekends, leaving to women the job of irrigating fields during the week. In Santa Rita, an irrigation system whose tail end serves a suburb, the phenomenon of the weekend farmer was even more pronounced. Women who participated most in crop irrigation were heads of households or the partners of men who engaged in agriculture only on weekends.

Women's work in on-farm irrigation is not restricted to delivering water to crops. Water theft is ubiquitous; irrigation entails not only taking water in turn and moving it across the field, but preventing others from taking the supply. Bunker and Seligmann found women playing a crucial role in surveillance and conflict management in a small-scale system in the department of Cuzco:

For the most part, women control the canals while their husbands and other male relatives irrigate, and it is not uncommon to meet a grandmother walking above and below the canal with a large stick in her hand and a ferocious look. Men explain that women play this role because men have to respect one another and not fight. But, as all the men are in agreement that the women should control the ditch, the women fight among themselves like cats and dogs. . . . Full negotiations take place at the side of the canal, but even when it seems that they have reached an accord, as soon as the petitioner goes his way, the other opens the gate again. Thus, much work time is lost. When the conflict cannot be resolved peaceably, both go to the president of the community or to the Justice of the Peace, or they are taken to the police post, where they stay in jail for twenty-four hours (Bunker and Seligmann, 1986, pp. 161–162, my translation).

Another important water-related task is provision and transport of meals to men working in the field. Bourque and Warren (1981) observed that Mayobamba women regard provision and transport of meals to irrigation workers as an irrigation task, but men do not. Deere and León de Leal (1982) found that Cajamarca women tended not to include these activities as part of agricultural production. However, male and female irrigators in Santa Rita refer to meal provision as a form of participation in irrigation.

Women and System Construction

Andean topography requires irrigation infrastructure that is elaborate in its conception, but that can be continuously maintained and reconstructed at frequent intervals with minimal recourse to external funds. Highland systems are small in terms of area served and the volume of water conveyed, but main canals are long and vulnerable to landslides; system upkeep is arduous relative to the returns from water applications.

The chief source of labor for system construction in the highlands is and has been unpaid community labor. This is equally true for locally funded and for externally assisted efforts. Labor contributions may be voluntary or they may be obligations associated with membership in the system, water rights, or entitlement to use water.[5] Methods for assigning donated labor to a construction project generally follow some principle of equity, if not equality. However, the labor burden falls most heavily upon those smallholders who contribute their labor as sharecroppers and maintain canals for wealthier landowners as wage workers in addition to fulfilling the labor requirements associated with their own plots.

Women rarely participated directly in system construction in Cajamarca. In San Marcos and Santa Rita, their contributions consisted mainly of meal preparation for workers who lived far from the work site. During construction of the Santa Rita system, women provided parched and boiled maize, water, and chicha (corn beer) for workers. A male informant reported that women also participated in construction when the works were small, but when it was a matter of a whole day's work, women would send a *peon* or pay a fee. Women's cash payments in lieu of labor are used to pay for alcoholic beverages as well as for building materials.

In contrast, during the course of a 1983 field visit, I observed work crews in three Puno communities with very high levels of female participation. In one community, internationally funded food assistance was offered as an incentive for irrigators to participate in construction of a small reservoir. More than half of the labor force was female, and a number of toddlers accompanied their mothers to the work site. In the course of the visit, I learned that the workers were not local landowners. The landowners hired day labor and paid workers partly in cash and partly in foods—dry milk, oil, and so forth. The package amounted to less than the minimum wage; as a consequence, attendance lagged, and progress on the tank was slower than anticipated. In this instance women participated in system construction as wage workers, but at a rate which, because it was below the minimum standard set by the government, was unattractive to men.

Women and System Operation and Maintenance

The norm in Santa Rita and San Marcos is that men participate in canal cleaning and system maintenance. Widows and women with absent partners are expected to send a family member or hired hand to fulfill their labor obligations. Alternatively, they may pay a fee equivalent to the minimum daily wage to cover food and beverages for the workers. The wife of a Santa Rita migrant worker outlined the full range of options, stating that she usually sends a *peon* to substitute for her husband, who is generally in the *jalka* (high, rainfed agricultural zone). Once, when she had no *peon,* she helped haul rocks; at other times she paid a fee—a bottle of brandy or cane alcohol.

Other women irrigators in the Santa Rita project area described their contributions to maintenance as follows:

> Only men work; the women who do not send their *peon* are charged 1,000 soles that serve to buy cane alcohol for the workers to keep them from catching cold.

> Men do the participating, the widows get a *peon* or pay the fee; they only go to help at times with moving water through the canals. The fee is invested in the workers, buying *chicha,* bread, fish. If there is an excess, the treasurer can buy some things that are lacking for the canals, gates, and turnouts.

Although it is preferable to send a *peon* or make a cash contribution to system maintenance, women heads of households in Cajamarca who cannot afford to hire labor contribute their own. The divergence between norm and reality is revealed in the following statements by Santa Rita irrigators:

> Participation is by men, although women also cart stone and sand.

> Work is done by men, although sometimes the women help to load stones and clods.

> Only men work; at times the widows attend and help to load rocks.

> Work is done by men and some women who go when there is no one else in their family who goes. They gather rocks and branches in sacks.

One Santa Rita informant reported that four widows who had no money to pay workers or to pay their fee were excused from their contribution, but, in general, it is the female heads of households who are too poor to pay the quota or provide male substitutes who violate norms by participating in the community labor force. Poor, landless women may use irrigation water for domestic purposes, but because labor obligations are tied to water rights, which in turn belong to parcels of land, these women are not asked to contribute labor to system maintenance.

Women in Water Management

Water management implies the acquisition, allocation, and distribution of water and mobilization of the labor, cash, and material resources needed to maintain and repair infrastructure. Women are still largely excluded from these activities. Small-scale systems regulated by the state are governed by a subset of water consumers: peasant community officials, cooperative leaders, or elected representatives of a water-user organization.[6] It is expected that irrigators in government-assisted systems will form committees, some of which incorporate elements of indigenous organization.

In general, a committee comprised of all inscribed irrigators along a given canal or sector elects a board of directors to take responsibility for day-to-day system management. These officials tend to be men with status in the community—wealthier landowners who hold other community offices and/or individuals with experience and ties outside the community.[7] San Marcos had no women irrigation officials in 1985. One woman appears on the list of irrigator committee officials in Santa Rita for the same period. She was treasurer of a subcommittee for one sector of a lateral canal. Oral histories also indicate that the widow of a *hacendado* participated in the early phases of Santa Rita system planning. In interviewing irrigator organization leaders in two irrigation systems in the Department of Puno, I encountered no women. It is argued that women exercise informal leadership roles in irrigation development (Walter, 1983), but women in these roles are generally at a disadvantage when they deal with the bureaucracy.

Class, Integration, and Participation in Irrigation Tasks

System maintenance is defined as men's work in Cajamarca, but women are neither excluded from participation in moving water from canal to farm field nor barred from participation in local irrigation organizations. Whether or not women labor in the field moving water from turnout to furrow or in cleaning canals depends on both class and the role of irrigated agriculture in the household economy. Women who lack access to agricultural land are not hired as day labor to do either crop irrigation or system maintenance. Women who can afford to employ day labor are similarly invisible, although they may participate in irrigation organizations. Women owners of irrigated land who cannot hire labor engage in reciprocal labor arrangements and may irrigate crops and participate in system maintenance.

The participation of women from smallholder families is more prevalent where women are widowed or where household economies dictate that male members are absent from the community either for extended periods as migrants or on a daily basis as they pursue off-farm economic activities. The likelihood that household men will be employed away from the farm is greater in communities that are thoroughly integrated into the national economy and in those where agriculture is declining in importance as a source of income. Thus, class alone does not determine gender roles. Equally important are two closely related variables, contribution of irrigated agriculture to the household economy and integration of the community into the national cash economy. The effects of class and integration on participation of women in system management, however, are less pronounced; gender appears to be the primary constraint.

The fact that class and economic variables have become increasingly important determinants of women's participation in crop irrigation and system maintenance but continue to be relatively unimportant determinants of participation in system management indicates that the symbolic significance of water control at the farm gate for men's domination of women has declined. But elimination of control over water at the farm gate as a male prerogative does not imply the elimination of gateway tasks in irrigation. I would argue that the interface between the household and the community—the farm gate—was less important in Santa Rita and San Marcos in the 1980s than it was in Mayobamba in the 1970s and that its importance has declined relative to that of the interface between community and state. The locus of the critical gateway has shifted from management of the flow of water from canal to field to management of the flow of resources from the state to the irrigation system. This is in essence the bureaucratic transition.

Women and the Bureaucratic Transition in Highland Irrigation

Elsewhere (Lynch, 1988), I have identified three discrete traditions in Andean irrigation: the indigenous or Andean, the Hispanic, and the bureaucratic. The last tradition consists of the institutions, norms, and values associated with state and agency-built or agency-managed irrigation systems. Some features of this tradition are characteristically Peruvian; others bear strong resemblance to features of state- or international donor–financed systems throughout the world, a resemblance not unrelated to sources of project funding. I define the bureaucratic transition as the process of technological, social, and political change in local irrigation systems resulting from government intervention.

The bureaucratic transition was a response to stimuli for change from both communities and governments beginning in the 1960s. Low prices for agricultural products, mounting pressure on irrigated land due to land fragmentation and resettlement in irrigated zones, labor scarcity resulting from occupational diversification and migration, and a diminution in the authority of traditional irrigation leaders have progressively eroded local water management capacity. Irrigation institutions that had worked in the past were unable to respond to recent changes in command areas: new cropping patterns, agricultural intensification, more irrigators, and the resultant increased total demand for water relative to supply.

Under these conditions, rules were increasingly violated, and, as the number of irrigators on a canal increased, existing rotation systems could no longer contain conflict (Mitchell, 1976). Challenges to local irrigation authorities became commonplace, as did delicate negotiations with neighboring landowners and communities for the right to transport water across their property. Individuals and communities increasingly turned to government agencies to legitimate their claims or settle conflicts.

Conflict can be mitigated by rewriting the rules to reflect new realities, by changing the physical infrastructure to facilitate enforcement and enhance reliability and adequacy of the water supply, and by maintaining it in excellent working order. Where diminishing returns to agricultural production within the irrigated zone have necessitated occupational diversification and male outmigration, labor-intensive solutions to improving water efficiency have not been feasible. Capital-intensive solutions—canal lining, drip irrigation, and gate installation—may be beyond the repayment capacity of the community. Thus, communities have sought external assistance in allocating water and in making infrastructural changes that would facilitate regulation of water deliveries. To effect these changes, communities compete for the ever scarcer material and human resources of the state. Successful competition requires access to state agencies, the ability to make a claim on state resources through political loyalty or identification with a target group for state assistance, and the legitimization of ad hoc proconstruction committees as legal entities representing farmers in the command area.

Local demand for government intervention coincided with growing government interest in rural development. Reformist governments in Peru since 1963 have seen smallscale irrigation as a cost-effective way to increase national agricultural production and to build regime support through highly visible, yet widely dispersed public works. Unlike the massive coastal schemes that have dominated public investment in irrigation, highland projects are relatively cheap. Their cost effectiveness is enhanced in the eyes of government promoters by the potential for shifting the burdens of operation and maintenance to local irrigators. The visibility of irrigation projects and the divisibility of their benefits make them attractive to bureaucrats whose tenure depends on regime stability, to the political mobilization arms of governments and political parties, and to departmental development corporations.

The 1969 Peruvian agrarian reform and rural programs of the Velasco government vastly increased the direct presence of the state in the highlands. This change did not happen overnight, but, before the agrarian reform, the interactions of many rural families with the state were mediated by *hacendados,* their administrators, and local notables. The reform was followed by an invasion of young, urban, often inexperienced change agents and project promoters whose task was to mobilize peasant support for the "revolution from above." The result of this conjuncture of local need and government activity was the imposition of bureaucratic culture onto irrigation institutions in the Sierra.

How has this transition affected women's control over water and irrigation systems? Their participation in water management depends on the extent to which they can exercise power within local government institutions that are increasingly shaped by a bureaucratic state. This in turn depends on the extent to which women have a voice in these new institutions, the style of politics that dominates competition for the state's irrigation resources, and, where resources are forthcoming, the position of women within the culture of the irrigation bureaucracy.

Women in Local Government

Regarding questions of gender and power in Mayobamba, Warren and Bourque (1985) noted the structural exclusion of women from local political assemblies where decisions are made about community leadership, organization, and resource mobilization and distribution. In Mayobamba, it is the single mothers and widows who are most frustrated by this exclusion, because they are obliged to contribute labor to communal projects but have no voice in the distribution of tasks. The authors concluded that "the Mayobamba assembly of male heads of household is a major forum where the dominant idiom of community affairs is reproduced" (p. 266).

In contrast, women from a nearby market town had access to public forums but felt unable to express their needs or initiate projects whose principal beneficiaries would be women except in the context of women's groups.[8] This muting of women's voices extends to the irrigation agencies. Women employed in the irrigation bureaucracy tend to be extension specialists or sociologists who work with women's groups on women's issues. Rarely, if ever, do they play central roles in system design, construction, or water allocation.

Despite these impediments to participation in local government institutions, Peruvian women have been politically active in any number of spheres. In irrigation, however, their participation in management continues to lag behind their participation in the workforce. As a political arena, irrigation has several features that make it less accessible to women than others such as food and health policy. First, societal norms in the highlands that define water management as men's work, as well as sexist stereotypes present in the larger society, make women seem invisible to government officials. Because irrigation institutions at all levels are male-dominated, irrigation is not a sector where the state has encouraged movement of women into a political sphere beyond the community. Nor has the state tried to use irrigation projects to mobilize peasant women. Finally, irrigation issues are seldom enveloped in ideological controversy and do not involve the opening of new political spaces.

If men constitute the overwhelming majority of irrigation leaders, it is at least in part because outside agencies, particularly government agencies, are more willing to recognize men as community spokespersons.[9] Thus, in ordinary circumstances, Andean men are more likely than women to engage in interactions with government officials. This pattern is reinforced by the greater participation of men in the cash economy (Collins, 1986).[10] The sphere of Andean peasant women has been largely limited to the community and the household.

This does not imply that the state has excluded women entirely from political participation. By directing political mobilization efforts at rural women, the state since the Velasco government has created a political space for peasant women and opportunities for them to move out beyond the confines of the community in pressing their claims (Radcliffe, 1986). But, in defining rural women as a political category, the state has also circumscribed the areas in which this movement could take place.

Irrigation and the Politics of Normal Times

Even in the crisis-ridden Peru of the mid-1980s, the overwhelming number of political interactions in the Sierra were routine claims on a share of an ever-shrinking pool of resources or petitions to prefects to legitimize local activities. Water in Peru is tantamount to a motherhood issue; it would be unthinkable for any political faction within the system to oppose small-scale irrigation development.[11] Irrigation is the stuff of normalcy politics. It relies on efficient exploitation and manipulation of the political resources at hand, which in the case of highland agricultural communities means reliance on assistance from better-educated, wealthier, and politically better connected individuals in the community—the large landowner or merchant, the schoolteacher, or administrative officials at the district and hamlet levels. Intermediaries in the behind-the-scenes brokering that dominates irrigation politics include schoolteachers, lawyers, prefects, nongovernment organization staff, well-placed kin in departmental and national capitals, and officials of the irrigation and agricultural bureaucracies. Thus, the politics of the irrigation in highland Peru are characterized by the normalcy politics of maneuver rather than the politics of confrontation. Irrigation politics are personalistic, dyadic, and mediated by men.

With the bureaucratic transition, the frequency of these political interactions has increased as has their relative importance to system management as a whole. They are important not only for water acquisition but for tasks that were formerly internally managed: water allocation and financing small works and system rehabilitation. Because the style of politics that dominates these interactions is manipulative, rather than confrontational, the interactions are dominated by men. Where women have entered the regional and national political arenas during periods of normalcy politics, it has been in roles complementary to traditional male roles (Radcliffe, 1989).

Male Biases in the Irrigation Bureaucracy. Male domination of normalcy politics is reinforced by the male biases within the irrigation bureaucracy. One bias is the designation of the household as the smallest unit of policy attention. The Ministry of Agriculture has brought to Sierra irrigation a Spanish colonial concept of family and household and has defined this abstraction as the "the unit of production." Post-agrarian reform rural development policies proceed from the assumptions that (1) all parcels that make up a household agricultural production unit are cultivated by the household as a unit, (2) production decisions are made by the head of household, and (3) the head of household is a male, except where women are widowed or permanently abandoned by their partners. In this scheme, women become a residual category. Therefore, participation in rural development activities is largely confined to heads of households who are male (Collins, 1986; Deere and León de Leal, 1982; Radcliffe, 1989). Surveys are administered to heads of households, unless they are specifically targeted at women.

A second bias inherent in the bureaucratic approach to small-scale irrigation is the perception of irrigation as a technical rather than a social process. Works are identified with men and social considerations with both men and women. The staffs of irrigation agencies in Peru include civil engineers, agricultural engineers (*ingenieros agronomos*), economists, sociologists, and extension specialists, but status hierarchies within the bureaucracies reflect the emphasis on works and male-dominated activities: construction takes precedence over water management[12]; large works, particularly those that must be managed by civil engineers, are more important than small works.[13] Third, the role of women within the irrigation bureaucracies is largely restricted to statistical analysis and to organizing and teaching women. Women do not make it very far up in the hierarchy, and they have little input into system design and construction.

These biases notwithstanding, the irrigation bureaucracies, responding to the political mobilization programs of the state, do acknowledge the presence of women in the irrigator community. What is more, government intervention in small-scale irrigation in the highlands has yielded some benefits for women. Perhaps most important is the introduction and enforcement of rotation systems that obviate or at least reduce the need for political clout and brute force in order to receive the water allocated to their parcels. Irrigation projects reflect some awareness of women as domestic consumers of water and often include specific programs directed at women. In Santa Rita and San Marcos, Plan MERIS designed several infrastructural features to make women's chores easier—pits for leaching the toxins from *tarwi* and small tanks near populated centers where women could draw buckets of water and wash clothes. Where token women cultivators took part in scheduled Plan MERIS program activities in San Marcos, project staff, both male and female, fully supported their participation. But generally the irrigation agencies attempt to co-opt women rather than encourage them to play central management roles, and what little attention women do receive may be due largely to the dictates of donor agencies or Peruvian interests outside of the irrigation bureaucracy.

Conclusions and Implications

Since Bourque and Warren (1981) first addressed gender issues in Peruvian highland irrigation, substantial changes have occurred in the roles of rural women in farm households, in the viability of peasant agricultural production—and, as a consequence, the importance of water as a resource—and in the nature of small-scale irrigation. These changes, which began in the 1970s but accelerated in the 1980s, have produced some contradictions in women's relationships to water as a resource.

Women in many parts of the Andes now enjoy more control over water in the farm field than they did when Bourque and Warren studied Mayobamba women in agriculture. However, this gain is offset by other changes. First, although women exercise more control over water on the farm, it is as a result of the declining ability of agricultural production to meet family subsistence needs. With male outmigration and increasing dependence on men's wage labor to generate cash income, more women are participating in irrigation in communities where irrigation has traditionally been defined as a male task. Women already heavily burdened with domestic chores and responsibility for the care of livestock find themselves saddled with men's tasks in agriculture and system maintenance as well. Women with young children are particularly hard-pressed in Andean rural societies (Collins, 1989; Deere, 1978).

Second, the declining ability of agricultural production to meet subsistence needs due to land fragmentation and national price policies has had devastating consequences for highland cultivators. Thus, increased female participation coincides with increasing costs of water management coupled with declining returns for good water management.

Third, women's increased control over water at the farm gate does not mean that the gateway tasks in water management (and with them, access to resources) are now controlled by women. The scarce resource for water management is money for repair and maintenance of infrastructure, not water per se. The gateway in the government-assisted system has simply moved: Control over access to resources no longer occurs at the interface between the community and the household (between the canal and the farm field), but at the interface between the community, as represented by its formal irrigation organization, and the funding agencies. Women irrigators interact only as exceptions with these agencies, and then only as heads of household. Paternalism is particularly pronounced in interactions among male bureaucrats and peasant women, although it also typifies daily relationships between agricultural technicians and their male clients.

Is it necessary for women to participate directly at the interface between the local system and the agency in order to meet their needs as consumers of irrigation water? The Plan MERIS experience in Cajamarca indicates that the irrigation bureaucracy can incorporate women's concerns into projects in highly visible ways; domestic water and *tarwi* washing troughs are examples, as are the women's groups. However, this attention is in areas peripheral to system operation and management. By providing mechanisms for women's direct involvement in extension activities and for the development of women's groups, Plan MERIS has fostered consciousness-raising at a more general level, but no attempt has been made to address the particular problems of women as irrigators and cultivators.

In sum, whether as a result of the bureaucratic transition or economic changes in the highlands, women are doing more irrigation and exercising more control over water at the farm gate. Yet they continue to play only token roles in system management. Rather than excluding women from irrigation roles, the irrigation agencies have sought to co-opt women and to marginalize them in their domestic role. Despite government attempts to co-opt and marginalize, poor urban women in Peru have successfully turned other reformist government programs to their own ends. Why then are women not taking advantage of these small openings to further their own agendas with regard to water? It is because the very normalcy of the politics of irrigation development make it exceedingly difficult to break male-dominated, patron-client patterns of external resource mobilization.

A consequence of women's organization and of increased participation in irrigation activities is what Babb (1989, p. 40), describing highland market women, referred to as a "combination of a sense of potential power and a recognition of relative powerlessness." Growing consciousness about problems is offset by pessimism about achieving solutions. Levels of consciousness owe much to the grassroots efforts of women extension workers to turn women's groups from devices for co-option into vehicles for empowerment. Pessimism about reorienting irrigation activities to serve the needs of rural women owes much to continuing male control over access to a shrinking pot of nonlocal resources that are becoming ever more necessary for the management of small as well as large systems.

Notes

1. For comparative purposes, I also refer to my own observations in the Department of Puno and to the following case studies: Mayobamba (Bourque and Warren, 1981); Huanoquite (Bunker and Seligmann, 1986); San Pedro de Casta (Gelles, 1986).

2. Not all smallholder households derive equivalent proportions of income from agriculture in their own fields, and landholding size does not necessarily indicate wealth or social class. For a useful discussion of class relations and land ownership see Montoya (1982).

3. Collins (1989), looking at family division of labor in a Puno community marked by seasonal migration, found that, despite norms that dictate a sexual division of agricultural tasks, all agricultural tasks, even breaking fallow ground with a plow, have been performed at times by women.

4. In the highlands surrounding Cajamarca, temporary harvest laborers are paid a daily income in kind—a fixed quantity of grain (*almud*). Women receive three-fourths of this set amount of grain for a day's work; children receive half what men get.

5. In highland Peru, there is a history of conscripted labor for local public works. Notions of community self-help have often been invoked to legitimate conscription, and it is a mistake to assume that all local labor contributions to irrigation development and maintenance are freely given (Lynch, 1986; Mallon, 1983).

6. In principle, the Peruvian water law applies to all community irrigation systems. In fact, application of the law is patchy in the Sierra. It is generally restricted to communities that have received government assistance.

7. Increasingly, the latter group is assuming leadership positions in community organizations. This group is generally younger and better educated; it is actively engaged in commercial activity, transport, and livestock raising, activities that maximize interaction with the outside world (Montoya, 1982).

8. Lapasini (1984, p. 24) made similar observations in Cuzco: "Outside situations of conflict, the participation of rural women is favoured neither in the community nor among public service officials who are generally representative of urban culture and society." Her women informants reported that cooperative directors opposed their participation on the grounds that "women create confusion" and that women are not allowed to speak in meetings.

9. The relative absence of women in irrigation leadership roles at the local level may also be due to the fact that women have fewer years of formal education than do rural men and have lower literacy rates (Cassava de Valdes and Portugal Bernedo, 1985).

10. Although rural women do participate in the cash economy as market sellers, Babb (1989) found that they acted as a critical opposition within their union as their numbers in leadership positions declined in the 1970s.

11. The position of Shining Path on this issue is not clear. While they oppose all agricultural development efforts as band-aid solutions to Peru's problem of regional inequality, I do not know whether they have done any irrigation development in liberated zones.

12. The irrigation bureaucracy in Peru is in a state of perpetual evolution. In the mid-1980s, government bodies directly engaged in irrigation development and water management included the prime minister's office, responsible for developing large-scale projects through semiautonomous authorities created to manage individual large projects; the Instituto Nacional de Ampliación de la Frontera Agricola (INAF), an irrigation development institute or special agency of the Ministry of Food and Agriculture, and its (largely donor funded) development units, Pequeños y Medianos Irrigaciones (PEPMI), Plan MERIS, and Linea Global; the Dirección de Aguas, Suelos, y Irrigaciones, a line agency of the Ministry of Food and Agriculture charged with regulating water use; and the Ministry of Economy and Finance (MEF), which exercised direct and powerful controls over the disbursement of funds for irrigation development. In addition, irrigation activities were carried out by the Departmental Development Corporations and by political mobilization organizations. Even in the highlands, most irrigation development is internationally funded, and the shape of the irrigation bureaucracy reflects the demands of the international donors, the World Bank, and the InterAmerican Development Bank.

13. Coastal systems that serve agribusiness enterprises are emphasized at the expense of the generally small, local market or subsistence-oriented highland systems. State investments in highland systems have never exceeded 9% of the total irrigation budget. From 1968 to 1976, they averaged 3.1% (from Maletta and Foronda, 1980, pp. 204-205).

References

Babb, F. E. 1989. *Between field and cooking pot: The political economy of marketwomen in Peru.* Austin: University of Texas Press.

Bourque, S. C., and K. B. Warren, 1981. *Women of the Andes: Patriarchy and social change in two Peruvian towns.* Ann Arbor: University of Michigan Press.

Bunker, S., and L. Seligmann. 1986. Organización social y vision ecologica de un sistema de riego andino. *Allpanchis* 18(27):149-178.

Cassava de Valdes, M., and J. Portugal Bernedo. 1985. *La mujer en cifras. Estudio comparativo sobre la situacion de la mujer en los censos 1972-1981: Poblacion, empleo y education.* Lima: Centro de Investigacion y Promocion Popular (CENDIPP).

CIAT (Centro Interamericano de Administración del Trabajo). 1984. *Formas tradicionales de trabajo campesino en los Andes Peruanos.* Lima: OIT (ILO) Documento de Trabajo CIAT/DT/80/17.

Collins, J. L. 1986. The household and relations of production in southern Peru. *Comparative Studies in Society and History* 28:651-671.

Collins, J. L. 1989. *Unseasonal migrations.* Princeton, NJ: Princeton University Press.

Deere, C. D. 1978. The development of capitalism in agriculture and the division of labor by sex: A study of the northern Peruvian Sierra. Ph.D. diss., University of California, Berkeley.

Deere, C. D., and M. León de Leal. 1982. *Women in Andean agriculture: Peasant production and rural wage employment in Colombia and Peru. Women, work, and development 4.* Geneva: International Labor Office.

Figueroa, A. 1984. *Capitalist development and the peasant economy in Peru.* Cambridge: Cambridge University Press.

Gelles, P. 1986. Sociedades hidraulicas en los Andes: Algunas perspectivas desde Huarochiri. *Allpanchis (Cusco)* 18(27):99-148.

Golte, J. 1980. *La racionalidad de la Organización Andina.* Lima: Instituto de Estudios Peruanos.

Lapasini, G. 1984. Why rural women don't speak up. *CERES* 17(1):20-24.

Lynch, B. D. 1988. The bureaucratic transition: Peruvian government intervention in Sierra small-scale irrigation. Ph.D. diss., Cornell University, Ithaca, NY.

Lynch, B. D. 1986. Mobilization of local labor for small-scale irrigation development in the Andes. *Water Management Review* 1:14–15.

Lynch, B. D., J. L. Villarán, and R. Flores. 1987. Irrigacíon en San Marcos. Transición a la tradición burocratica. *Allpanchis* 18(28):9–46.

Maletta, H., and J. Foronda. 1980. *La acumulacíon de capital en la agricultura Peruana.* Lima: Centro de Investigación, Universidad del Pacifico.

Mallon, F. 1983. *The defense of community in Peru's central highlands.* Princeton, NJ: Princeton University Press.

Mayer, E. 1985. Production zones. Chap. 5 in *Andean ecology and civilization: An interdisciplinary perspective on Andean ecological complementarity,* ed. S. Masuda et al. Tokyo: University of Tokyo Press.

Mitchell, W. P. 1976. Irrigation and community in the central Peruvian highlands. *American Anthropologist* 78:25–44.

Montoya, R. 1982. Class relations in the Andean countryside. *Latin American Perspectives* 9:62–78.

Radcliffe, S. A. 1989. Between persons and women: Organized peasant women in Peru. Paper prepared for XV International Congress, Latin American Studies Association, San Juan, Puerto Rico.

Radcliffe, S. A. 1986. Gender relations, peasant livelihood strategies, and migration: A case study from Cuzco. *Bulletin of Latin American Research* 5:29–47.

Sherbondy, J. E. 1985. Water and power in Inca Cuzco. Paper presented at Annual Meeting of the American Anthropological Association.

Walter, M. F. 1983, April 20. Small scale irrigation in the Sierra. Water Management Synthesis II Project trip report, Cornell University, Ithaca, NY.

Warren, B. K., and S. C. Bourque. 1985. Gender, power, and communication: Women's responses to political muting in the Andes. Chap. 8 in *Women living change,* ed. S. C. Bourque and D. R. Divine. Philadelphia: Temple University Press.

Epilogue

Peru has changed appreciably in recent years. How have these changes affected Sierra irrigation and women as players in the micropolitics of water management? In brief, Shining Path insurgency and counterinsurgency activity abated by 1993. Authoritarian President Alberto Fujimori's structural adjustment programs and the apparent demise of Shining Path led to record economic growth and an air of optimism in Peruvian society. These domestic changes occurred against a backdrop of curtailed international assistance for agricultural development programs.

To assess the impacts of these changes on Sierra women, one needs to look at three phenomena: demographic patterns, rural development programs, and bureaucratic downsizing. During the 1980s, many *campesinos* died in the crossfire between the Shining Path and the Peruvian military and many more left the Sierra for Lima, abandoning their fields. Now the central highlands are being repopulated by ex-urbanites (Starn, 1994), who in some cases are returning to fields that had not been cultivated for a decade (Starn, 1994) and engaging in the construction of houses, schools, and water systems. The countryside is still riven by conflict and fraught with uncertainty, but within this environment, Anean social institutions and patterns of community organization have survived more or less intact.

Repopulation affects the dynamics of local politics. The politics of normalcy continue to prevail, but outsiders (*forasteros*) share certain characteristics that give them an advantage in the personalistic, coalitional politics associated with state-dominated irrigation arenas: education, command of Spanish, and participation in broader social networks (Guillet, 1992). Although Guillet neglects gender, he concurs with my findings that with the increasing state presence in highland irrigation, water management has become a process of community–agency negotiation (see also Oré, 1989). In these negotiations, local men who enjoy the advantage of old-boy networks and education predominate, because they have better access to male-dominated prefectural and agricultural bureaucracies.

Because counterinsurgency and political goals have taken precedence over economic considerations in the Sierra, the region has been shielded from the full force of the Fujishock. Export-oriented and privatization policies have focused on coastal agriculture, but Sierra agricultural programs have been undertaken for political objectives. Wholesale privatization of community lands has not taken place, and the current government continues to promote small infrastructure programs. By 1989, repressive counterinsurgency had given way to a carrot-stick strategy linking nationalism to rural development programs. When his 1992 "autogolpe" vastly increased the power of the presidency, Fujimori could fly to highland communities in the presidential helicopter and inaugurate projects with funds from the presidential purse (Manrique, 1995). Rural electrification and road building are increasingly important, but small irrigation remains a staple of the project repertoire for highland Peru.

Support for rural projects comes from national rather than international sources. This poses two problems for female irrigators. First, the fact that these projects are driven by their public relations potential rather than by local demand is symptomatic of the increasing distance between citizens and power and a decline in the significance of representative institutions (Córdoba, 1994). Second, withdrawal of the international community from rural infrastructure programs decreases the likelihood that gender and/or women in development (WID) considerations will be addressed.

It has been argued (Jahan, 1995; Jiggins, 1994; Lynch, 1993) that until women are well represented at all bureaucratic levels rural women's agendas are likely to remain invisible. What effect will Peru's shift toward authoritarian neoliberalism have on women's roles in the agricultural bureaucracy? The data are not yet in. However, reduction in the size and power of traditional line agencies is likely to have a differential effect on women, who were already marginalized in remote areas and ancillary disciplines and concentrated in the lower ranks. The privileging of technical and economic over social discourse in neoliberal thought is likely to further marginalize women within the irrigation bureaucracy (see Ferguson, 1990), and as international agencies play a smaller role in highland irrigation, both WID and gender-based concerns are less likely to find their way into agency mandates.

In sum, although the neoliberal policies of the Fujimori regime have not resulted in major changes in land tenure or in state intervention in Sierra irrigation, they are not likely to enhance women's visibility at the local level. At the local level, with demographic shifts, old bonds of communal cohesion are giving way to new networks of social and economic power. These networks remain male-dominated, however. Women irrigators are still unlikely to have a major impact on local decisions, nor will their ability to direct resources to meet their own needs increase substantially. The decoupling of irrigation project development from international assistance will probably reinforce the tendency toward male domination of irrigation micropolitics. Finally, downsizing of the irrigation bureaucracy, coupled with the economistic ideology of neo–liberalism, will mean fewer women in the bureaucracy at all levels. In sum, the politics of reconstruction, structural adjustment, and authoritarianism do not appear to be creating new social spaces for Sierra women as water users or agricultural producers.

Additional References

Córdoba, M. S. 1994. Reconstruir los sueños de los peruanos. *Quehacer*, July–August:4–9.

Ferguson, K. E. 1990. Women, feminism, and development. In *Women, International Development, and Politics: The Bureaucratic Mire*, ed. K. Staudt, pp. 291–303. Philadelphia: Temple University Press.

Guillet, D. 1992. *Covering Ground.* Ann Arbor: University of Michigan Press.

Jahan, R. 1995. *The Elusive Agenda: Mainstreaming Women in Development.* London: Zed Books.

Jiggins, J. 1994. *Changing the Boundaries: Women-Centered Perspectives on Population and the Environment.* Washington, DC: Island Press.

Lynch, B. 1993. *The bureaucratic transition and women's invisibility in irrigation.* Proceedings of the 27th Annual Chacmool Conference, Calgary, Alberta.

Manrique, N. 1996. Two faces of Fujimori's rural policy. *NACLA Report on the Americas,* 30:39–43.

Ore, Maria Teresa. 1989. *Reigo y Organizacion: Evaluacion Histórica y Experiencias Actuales en el Peru.* Lima: Tecnología Intermedia (ITDG).

Starn, Orin. 1994. Vehuraceay y el retorno a los Andes. *Que hacer* (Lima). Sept.–Oct.

Chapter 7

Rural Women and Irrigation: Patriarchy, Class, and the Modernizing State in South India

PRITI RAMAMURTHY

University of Washington
Seattle, WA 98195
USA

Abstract *Irrigation is the major strategy used by "modernizing" states in India and throughout the Third World to raise agricultural productivity and surpluses. This paper shows that irrigation is not gender-neutral, focusing on how canal irrigation affects women's work and lives in Andhra Pradesh, India. First, it delineates the particular consequences for women of state-sponsored irrigation. It then focuses on transformations in the sexual division of labor, workloads, and the labor process for women of different classes and castes and shows how the economic and physical burdens of agricultural intensification have fallen most heavily on women of agricultural labor and marginal cultivator households. It concludes by suggesting policy measures that can meet poor women's basic livelihood needs and points out that only working class women's organizations will be able to change the preoccupation of the state with modernization, the inequitable distribution of resources, and the stranglehold of patriarchy.*

Keywords: agricultural labor, class, irrigation, modernizing state, patriarchy, rural women, sexual division of labor.

This article analyzes the effects of irrigation on women's work and lives in rural south India. The article focuses on irrigation because it has been the most important component of the Green Revolution package (high-yielding varieties of seeds, chemical fertilizers, and pesticides being other components). Irrigation continues to be used as a strategy by modernizing states in India and throughout the developing world in attempts to replace traditional peasant agriculture with scientific technology so as to raise productivity and agricultural surpluses.[1] Despite the warning by Ester Boserup as far back as 1970 that there are constraints on *women* realizing the benefits of agricultural modernization, empirical studies on the gender-specific effects of irrigation are virtually nonexistent (Agarwal, 1981). Based on fieldwork in rural south India, this article is an attempt to redress the substantial gap in the data base.

Theoretical Considerations

Because *patriarchy, class,* and *state* are all disputed terms, I begin by defining them and theorizing the interplay between these terms.

Patriarchy "encapsulates the mechanisms, ideology and social structures which have enabled men throughout much of human history to gain and maintain their domination over women" (Ramazanoglu, 1989, p. 33). Patriarchy expresses not only the power of men over women but also the ideological legitimization of this power as natural, normal, and just (Eisenstein, 1979). According to Althusser (1971), *ideology*[2] is a relatively coherent ensemble of representations, values, and beliefs that reproduce existing relations of production by interpellating subjects in imaginary relations to the real. As processes, rather than systems of ideas, ideologies have a material existence and are intermeshed with both the way production systems are organized and the operation of male dominated gender relations.

In accord with Marx, *class* is used as an analytical category. To the extent that the means of production are owned by some historical groups in society and not others, and to the extent the owners control both the decisions about and the fruits of production, the production system is organized along class lines (Kohli, 1987). Women cannot be understood as constituting a separate sex class, because if they were, then differences between women of different classes would not be recognized.[3]

In accord with Alfred Stepan (1978) and Theda Skocpol (1979), the *state* is defined as a set of administrative and coercive institutions headed by an executive authority that structures relationships not only between civil society and public authority but also within civil society itself. Modernizing states or Third World states can be distinguished by the facts that they are committed to deliberate development or planned socioeconomic change and that this goal is widely accepted as legitimate by the politically relevant strata (Kohli, 1987).[4]

Structural theories of the state (Poulantzas, 1978; Skocpol, 1979) have not considered gender as central to their analyses. Socialist feminist theory, however, recognizes that, to the extent that the state is the site for the systematic concentration of man's power, it codifies, institutionalizes, and legitimizes patriarchy (MacKinnon, 1983). The repercussions of the male state for women in modernizing states have only just begun to be studied,[5] yet the findings are unequivocal: "Gender is at the heart of state origins, access to the state, and state resource-allocation. States are shaped by gender struggle; they carry distinctive gender ideologies through time which guide resource allocation decisions in ways that mould material realities" (Parpart and Staudt, 1988, p. 6).

The implicit assumption of modernization policies such as the policy of agricultural development through the introduction of large-scale irrigation schemes is that such state intervention is gender-neutral. In fact, the way the male point of view frames an experience is the way it is framed by policy (MacKinnon, 1983). Thus, patriarchal ideology that defines women primarily as dependent beings has provided the rationale for colonial and postcolonial male officials to assume that the benefits from irrigation accruing to men will automatically benefit women and children as well.[6]

This article shows that this is not true. First, it shows that the distribution of irrigation benefits within the household reflects the fact that the household is a hierarchical social unit and embodies relations of subordination based on gender.[7] Second, it shows the differential effects of irrigation on women of different classes. Third, it shows that changes in the sexual division of labor consequent to the introduction of irrigation cannot be seen only as technically determined transformations. Based on socialist feminist theory, both the social relations of production—the mode of appropriation of surplus labor and the social distribution of the means of production—and patriarchy are taken into account in understanding why women continue to be exploited and oppressed (some more than others) even after the introduction of irrigation.[8]

The Research Context

The southeastern region of Andhra Pradesh lies in the uplands of the Deccan plateau (in contrast to the fertile, long-irrigated deltas) and is characterized by sparse (29 in. annually) and uncertain rainfall, 77% of which falls during the 5 months of the southwest monsoon (June–December). Consequently, the area is dry and drought-prone. Without irrigation, only a rain-fed crop is cultivable.

The canal that is the focus of this study was built in 1956. It flows off the Tungabhadra river and provides water to cultivators in 80 villages in either the first (June–December) or the second (December–April) cropping season. The cropping pattern is diversified. In the first season of 1984–1985, for example, there were 15,000 acres of paddy, 8000 acres of cotton, and 1000 acres of vegetable crops. In the second season, other than cotton (which is a 7- to 9-month crop), there were 21,000 acres of groundnut and 2500 acres of paddy. In addition, in the first season, about 40,000 acres of sorghum and 30,000 acres of tobacco, cotton, groundnut, and other crops were grown with monsoon rainfall.[9]

Patriarchy is a powerful system in southeastern Andhra Pradesh as in the rest of contemporary India. It draws historical and cultural legitimacy from the norms of *streedharma* (proper behavior for women) codified by Manu in the first century A.D., which hold that a woman should obey her father when she is young, her husband when she is married, and her son when she is old (Bardhan, 1986). Families in the region are structured as patrilineal-patrilocal units, with males controlling land, capital, and the female labor process. However, as in other regions in south India and in contrast to north India, patriarchal domination does not include proscriptions on women's participation in fieldwork. The operation of patriarchy in structuring the form and extent of women's work participation varies with class and caste, as is shown subsequently.

Three classes can be distinguished in southeastern Andhra Pradesh, based on their relationship to land, the principal means of production. The distribution of land is highly skewed. Five percent of all landowners (Group 1) are rich cultivators holding 20–200 acres of land each; they own 25% of the land. Twenty-five percent of landowners (Group 2) are middle-income cultivators holding 5–20 acres each; they own 40% of the land. Seventy percent of all landowners (Group 3) are smallholders or marginal peasants holding less than 5 acres each; they own 35% of the land. Fifty-four percent of the population (Group 4) are agricultural laborers who own no land. Groups 1 and 2 form a class of owner-cultivators with landholdings from which they can obtain a surplus over their minimum subsistence requirements and employ wage labor. Group 3 consists of a class of owner-cultivators who are unable to meet livelihood requirements solely from their lands and must engage in wage labor to balance the family budget. Group 4 consists of a class of laborers who are separated from the means of production and must earn their livelihoods solely from wage labor.

According to the Census of India (1984), the number of owner-cultivators in the *taluk* (administrative subregion) through which the canal flows was 25,000 (Groups 1, 2, and 3), of whom only 24% were women, indicating that the men have a stranglehold over the most valuable means of production. In contrast, of the 40,000 agricultural laborers (Group 4), 55% were women.

The ranking of groups according to social hierarchy or caste broadly parallels the classifications. The rich cultivators are of the higher or upper-middle Reddy or Kamma castes. The Reddys have been the dominant caste in the region for many generations.

They are the old elite, few in number but owners of a disproportionate amount of land and wielders of enormous economic, political, and social power. The Kammas migrated into the region from the delta area of Andhra Pradesh in the mid-1960s. They number only in the thousands but are the new entrepreneurs. Soon after the introduction of the canal, they bought irrigated land cheaply from naive native cultivators, and they now cultivate high-value crops for profit. Middle and marginal cultivators are of the middle and service castes (Kurruva, Boya, Golla, Muslim, and so on), and agricultural laborers are from the lowest and the untouchable castes (Mala, Madiga, Teliga, and so forth). Brahmins, the highest caste, are numerically and politically insignificant in the region.

The research focused on three villages at the tail end of the canal system, all of which had *niti sangams* or water associations, and one village at the head end, without a *niti sangam*. In each village, rich, middle, and marginal cultivators and landless laborers were interviewed. Within each household, both women and men at various stages of the life cycle were questioned.[10]

Implications for Women of Irrigation as a State Intervention

Historically, canal irrigation in south India has been managed as a state enterprise. Since independence from the British colonial government in 1947, the modernizing state has continued this tradition. Irrigation was seen as a means "to get out of this rut of poverty" by Nehru, India's first prime minister, and numerous new canals have been built, one of them the system studied.

The consequences of state-sponsored irrigation development for women are manifold. First, thousands of women were employed as construction workers when the canal was built. In construction, as in agriculture, the sexual division of labor is explicit: women carry headloads of earth or concrete, sieve sand, and so forth, whereas men dig, mix the concrete, and perform other such tasks. Women are paid a lower wage than men, and this is justified by an overall ideology of gender that considers the jobs women do as lighter. The government Committee on Fair Wages upholds this view, arguing that "where women are employed on work exclusively done by them or where they are admittedly less efficient than men, there is every justification for calculating minimum and fair wages on the basis of the requirements of a smaller standard family in the case of a woman than in the case of a man" (quoted in Swaminathan, 1987, p. WS37). Thus, the state institutionalizes and legitimizes patriarchy.

Second, local labor was not used to build the canal. Instead, the Class I (big) private contractors (always men), to whom the Irrigation Department awarded the contracts, employed labor agents (also always men) who supplied the labor—women and men—from unirrigated villages where living conditions were even more bleak. By employing migrant laborers, contractors were (and still are) able to pay them less and make them work longer hours than local laborers, holding them captive in makeshift camps at the work site. Work conditions were, and still are, especially oppressive for women, who continue to be responsible for child care, finding fuel and water, and cooking, tasks that are more difficult in locations far from home.

Third, when plots of land were localized or given rights to water by the state, rosters were drawn up of those with title to the land, the majority of whom were and still are males. By placing another critical resource in the hands of men, the state only reinforced patriarchal domination.

Fourth, as a result of state policy, water at the tail end of the canal is scarce, but not all tail-end villages are affected equally.[11] Tail-end villages, where the *peddamanshulu* or Big Men—and they are always men—are able to organize a *niti sangam* or water association, are much better off than villages without them. This is because the *peddamanshulu* lobby, negotiate, and bribe Irrigation Department staff and officials to get water for their villages. Tail-end villages that do not organize *niti sangams* receive little water and have not been able to intensify agricultural production. They remain dry farming economies and are unable to support the increasing population base. The consequences of this for women of the lowest classes and castes is migration in search of livelihoods. These are bound to be in technologically deprived sectors, whether agriculture, construction, petty commodity production, or the informal sector where working conditions and wages for women are more insecure, contingent, and arduous than for men, who do not have to shoulder a double burden of work.

Fifth, even in the tail-end villages where there are *niti sangams*, the economically significant and socially valued work of "coping with the bureaucracy"[12] is performed exclusively by men. The Irrigation Department, established in colonial times, continues to be an all-male enclave—from the chief engineer to the ditch tender. The officers are separated from their field staff and marginal cultivators by their class; they are also distanced by the location of their offices far from most canal villages. Only the *peddamanshulu* or rich peasants have the status, class characteristics, transportation, and political savvy to deal with irrigation engineers and other government officers. At critical junctures in the crop season when water gets very scarce, they may be accompanied to a government office by a large contingent of marginal cultivators in a show of force. Women are conspicuously absent from this public domain.

Sixth, the ability of the *peddamanshulu* to spend time and resources organizing village activities, such as the collective provisioning of water, is predicated on the fact that their wives or mothers are not only performing the work of childrearing and all the work of managing the home but are also out there in the fields supervising day-to-day operations. This puts an extra and unacknowledged burden on these women.

Seventh, it is community men who are hired for the irrigation jobs that have been created since the introduction of the canal; these include the jobs of water guards, common irrigators, or laborers employed on contract for the task of irrigating plots. Female-headed households are most likely to employ men on contract to do the task of irrigating their plots, because night irrigation is perceived as being particularly dangerous for women, and even during the day, irrigating one's plot may involve fighting with upstream neighbors. Although women often irrigate their family plots, the ideology of gender is apparent in remarks such as "this plot is easy enough for a woman to irrigate" or as an explanation for breaking out of turn, "the woman of the family irrigated; what does she know about the *niti sangam* schedule of turns?"

In concluding this section, it is pertinent to point out that the gender implications of state-sponsored irrigation systems such as this canal have been characterized as unplanned fallouts (Agarwal, 1988). In contrast, in numerous other irrigation projects, such as Mwea, Kenya (Hanger and Moris, 1973) and Mahaweli, Sri Lanka (Schrijvers, 1988), gender biases are an explicit part of the plans and include attempts to "integrate women in development."[13] In either case, the consistent workings of patriarchy within the state serve to reinforce patriarchy in civil society, with adverse consequences for women.

Transformations in the Sexual Division of Labor, Workloads, and Control of the Labor Process

Women in Agricultural Labor Households: Technical Organization of Labor

"All our lives are a summer time." This statement by Maryamma, a woman of the lowest agricultural labor class and caste (Madiga), sums up the close dependency of finding employment on the availability of water: during the dry summer months there is very little work available. In the days before canal irrigation, the only source of moisture was rainfall, and this is still the case for the majority of plots in the area. A comparison of the technical organization of labor under rainfed and irrigated conditions demonstrates the process of transformation.

The rainfed cycle of agricultural production usually begins in June or July and ends in December or January. Preparation of the land for sowing can begin only after the rains, as the black cotton soils in the area cannot be worked unless first moistened. As is the practice in all other parts of the country, only male labor is used for ploughing, though women may assist in clearing the fields and in manuring. Working the bullock team and seed drill to sow groundnut and sorghum is also an exclusively male job; a woman may follow the team dispensing seed. Rainfed cotton, however, is sown by women, as each seed is planted individually. Women prepare tobacco nurseries and transplant the seedlings; these are almost exclusively women's jobs, as is the process of debudding to allow for greater leaf growth. Debudding tobacco is a particularly offensive task that leaves a blackish poison on the hands; women usually skip their midday meal when they are doing this work so as not to ingest the poison. Harvesting the leaves and stringing tobacco garlands are also considered female jobs. Then both men and women work on curing the leaves for 3–4 months.

After planting, there is little work for women on cotton, groundnut, and sorghum until the harvest; it is the men who are involved in hoeing and fertilizing. If the rains are good and weed growth is excessive, women may be employed once for weeding. Cotton is picked using women's labor, with the number of individuals hired varying with the quantity of cotton to be picked. *Gumpus* or groups of women and a few men usually take responsibility for harvesting groundnut and sorghum. For groundnut, harvesting entails uprooting the plant, plucking the pods off the stems, and binding the stems. Women are again employed to clean the nuts and bag them. For sorghum, harvesting entails reaping, cutting the cobs off the stems, and binding the sheaves. Both men and women are employed to thresh and winnow the grain.

The irrigated agricultural production cycle starts with the release of water into the canal by the authorities of the Irrigation Department. The main irrigated crop in the first season is transplanted high yield variety (HYV) rice paddy. Ploughing the paddy fields and building and repairing bunds are all male tasks; leveling work is done by women. Nurseries are tended for 25 days, after which women's labor is used exclusively for transplanting. Men procure and apply water, broadcast fertilizer, and spray pesticide. Weeding, which is done at least twice, is women's work. Harvesting operations—reaping and binding, followed by threshing and winnowing—are carried out by groups of women and men.

The next irrigated crop to be sown is hybrid cotton of both the long- and short-duration varieties. In addition to the agricultural operations necessary to grow rainfed cotton, women work on hand weeding (after the men hoe) and basal fertilizer application three or four times during the crop cycle. Harvesting irrigated cotton employs approximately three times the number of woman-days required to pick rainfed cotton. The most labor-intensive crop to be grown is hybrid cotton for seed. Although it is grown on less than 5% of the irrigated land, it employs mainly children almost continuously for 4–5 months to cross-fertilize each flower of one variety with the pollen of another.

The most significant irrigated vegetable crop is onion. Again men plough the fields, build the furrows, and irrigate the plots. Women transplant the seedlings into the furrows. Weeding is done two or three times during the crop cycle by the women and involves a lot of work, because hoeing is not possible. Harvesting—uprooting the bulbs, trimming the stems, and bagging—is done by groups consisting of both men and women.

Irrigated groundnut is the major crop in the second season. In addition to the agricultural operations necessary to grow the rainfed crop, women shell twice the quantity of seeds and weed after hoeing by the men once or twice. Harvesting, which is by groups, employs at least twice the number of woman-days as required by rainfed groundnut.

Thus, under both rainfed and irrigated conditions the technical organization of labor feeds off the sexual division of labor that assigns women of the lowest class and caste to labor-intensive, backbreaking, and sometimes hazardous tasks.

Women in Agricultural Labor Households: Women's Workloads

Table 1 summarizes the labor required to grow the different crops. The transformation in women's work as a result of irrigation is dramatic. In contrast to the 25 woman-days for sorghum, paddy requires 53 woman-days per acre. Compared with 44 days for rainfed cotton, irrigated cotton requires 112 woman-days per acre, and compared with 23 days for rainfed groundnut, irrigated groundnut requires 45 woman-days per acre.

More generally, changes in the technical organization of labor as a result of irrigation mean that even marginal cultivators must hire labor for transplanting and harvesting. The specific cropping pattern adopted in the area, historically and currently, has not lent itself to heavy mechanization and consequent unemployment for women.[14] Also,

Table 1
Women's Labor Demands for Major Crops

Crop	Labor Demand (in Woman-Days per Acre)
Rainfed sorghum	25
Rainfed tobacco	55
Rainfed cotton	44
Irrigated cotton	112
Irrigated cotton seed	1891[a]
Rainfed groundnut	23
Irrigated groundnut	45
Irrigated paddy rice	53
Irrigated onion	125

[a]Includes children.

unlike in monocrop rice and wheat areas, there is mixed cropping in the region of Andhra Pradesh studied. The various cropping cycles overlap, and seasonal peaks and troughs in employment opportunities are not as marked. These changes have led to an increase in the availability of work for women agricultural laborers. But to find employment over a longer period of time, women must be skilled at various agricultural operations for many crops.

Women agricultural laborers are employed for approximately 20 days a month from August to April. But as Mencher and Saradamoni (1982) point out, the number of days a woman works varies even within the same village depending on factors such as age, marital status, husband's contribution, and so forth. Additionally, during periods of peak labor demand, the length of the working day is increased, but this bars women with infants, who cannot leave their children alone from 8 A.M. to 7 P.M.

Women in Agricultural Labor Households: The Labor Process

In the hierarchy of laborers (permanent, regular casual, and seasonal casual), women are at the lower end (Sen, 1982). Unlike men, who may be employed as *ghasaghadlu* or attached laborers for a year, women can never be sure of being continuously employed. It is inconceivable that they could be employed as attached laborers, because that involves practically full-time work and would leave them with no time to meet the needs of their families. Thus, the ideology of gender, which assigns primary responsibility for the work within the household to women, has a direct effect on their participation in paid work outside the house. That is, although women work as hard as men, they do not have the security of regular, assured work.

The insecurity attending women's work is intensified by changing relations of production as traditional networks of responsibility fall apart. For generations, women of the lowest class and castes have worked as agricultural laborers or coolies for those who belong to privileged classes and castes. Although the form of coolie work, referred to by the women as *kashtam* (hardship), and overt production relations continue, the significance of these production relations has undergone dramatic change (Mies, 1986). For example, according to Ramlamma, an elderly woman whose husband has been a *ghasaghadlu* for more than 30 years, in the days before the canal "practically the whole village" used to be employed by the biggest landowning family in one of the villages.[15] All those employed had the right to do certain jobs for which they would receive remuneration, usually in kind. Today, only about 50 women are employed by the family during the agricultural season as regular casual workers or those who are "free to work for others if her permanent employer has no work" (Mencher and Saradamoni, 1982). Of these 50 women, only about 10 retain a right to work around the house for the *sahukar* or patron during the summer or monsoon when there is no work in the fields. These women occasionally pay for the relative security of their work by accepting wages that are below what others receive during times of peak labor demand. The remaining 40 women are in a position where traditional relations do not exist and they are not free wage laborers. The rich peasant can be fairly sure of his labor supply during the agricultural cycle, and he need not be burdened by a wage bill during the slack months. But the laborers are in no position to bargain for adequate wages to compensate for those off-seasons. The emergence of women as wage laborers has denied them patronage in slack months.

The chances of a woman being a regular casual laborer are higher if she is married or related to a male attached laborer, because recruitment of labor by large and middle-size cultivators is through their attached labor. Thus, within agricultural laborer households, some are more disadvantaged than others.

The most frequent form of employment is seasonal casual work for which individual women may receive a daily wage or a piece rate. For some operations, such as transplanting or harvesting, women may receive a share of the amount her *gumpu* or group has contracted. On days when a woman does not find work, she receives no income, because employers are not obligated by law or custom to support daily casual workers. Neither are they obligated to provide pensions, support for health care or child care, occupational safety devices (like gloves), or even drinking water and toilet facilities at the workplace.

Not only is the labor market for women agricultural laborers insecure and fluctuating and the working conditions abysmal, but also gender disparities in wages are extreme. The daily wage rates or equivalents paid per person per day for casual labor in 1984–1985 are presented in Table 2. It is apparent that the highest wages are received for the group activities of transplanting and harvesting, where in a narrow window of time a lot of work must be accomplished. *Gumpus* have varying numbers of people depending on the size of the plot to be worked. Usually a group consists of related households, and it rarely cuts caste lines. Both men and women form groups, and the women are paid the same rates as the men.

Women are paid a lower wage for all other tasks where the sexual division of labor is explicit. The justification men give for this is that women are weaker than men and cannot do as much work as men; therefore, their productivity is lower. Besides, they said, "from the time of our ancestors that is the way things have been." Wage discrimination upholds a vision of the male as the primary breadwinner and the male wage as the main source of family income. Thus, the ideology of patriarchy and the sexual division of labor both work to undervalue women's labor.

Piece rates are paid for jobs such as plucking groundnuts, shelling groundnuts for seeds, and stringing tobacco garlands. A woman's productivity is gauged by how much work she has been able to accomplish, regardless of the time it has taken. Such a system of measuring work means women spend long hours doing intensive and arduous tasks. Yet it is these very jobs that are defined as light because they are done in the landowner's home or the village threshing floor, not in the fields. The social stigma of doing light work has been effective in keeping the number of men doing these jobs low; this is in the interests of the higher classes, who would have to pay higher piece rates if the men were employed. The Equal Remuneration Act of 1976, which stipulates that men and women be paid equal piece rates when the work is "demonstrably identical" does not take into account the fact that often there are no male counterparts with whom women can claim equality. Thus, the state does not overcome the patriarchal prejudices or structural inequalities in defining certain work as exclusively women's work.

Since the introduction of the irrigation-HYV package, all wages except the harvesting of foodgrains—paddy and sorghum—are paid in cash and offer no hedge against inflation. When there is an increase in demand for labor during planting or harvesting, the money wage rate increases. Peak demands for labor are also met by using migrant laborers from unirrigated regions who stay until the end of the harvest, thus dampening potential increases in wages caused by temporary scarcity. However, this does not mean that wages are determined only by the market forces of demand and supply; there seems to be tacit agreement among the *peddamanshulu* or Big Men as to how much should be paid. The average daily wage paid to men and women hovers around the minimum necessary for family subsistence. Paternalistic relationships with cultivators increase the

Table 2
Wage Rates or Equivalents per Person per Day for 1984–1985

Month	Female Male/Children/ Group	Individual Wage or Equivalent in Rupees	Piece or Contract Rate in Rupees	Activity
May	Female	3.00	—	Cotton picking
	Female	4.00–5.00	1.25/dabba	Groundnut plucking[a]
	Male	4.50–5.00	—	Clearing fields
	Male group	4.00–5.00	1.00/yard	Desilting canals
June	Female	3.00	—	Leveling and manuring paddy fields
	Female	1.50–3.00	0.25/seer	Shelling groundnut for seeds[b]
	Male	5.00	—	Clearing fields
	Male	15.00	15.00/acre	Building canals and drains for paddy fields
July	Child	3.50–4.00	—	Cross-fertilizing cotton[c]
	Female	5.00	—	Transplanting onion, sowing groundnut and cotton
	Female group	10.00–14.00	100.00–140.00/acre	Transplanting paddy[d]
	Male	8.00–10.00	—	Ploughing paddy, sowing groundnut
	Male group	10.00–20.00	10.00–15.00/acre	Irrigating cotton
August	Female	4.00	—	Weeding paddy, onion, cotton; basal fertilizer for cotton
	Female	5.00	—	Tobacco transplanting
	Male	8.00–10.00	—	Fertilizer application, hoeing
September	Female	5.00	—	Weeding onion, transplanting tobacco; basal fertilizer for cotton, sowing sorghum
	Male	8.00–10.00	—	Sorghum sowing, hoeing; fertilizer application
October	Female	5.00	—	Cotton picking, weeding, fertilizing, sorghum sowing
	Male	8.00–10.00	—	Sorghum sowing, hoeing, canal cleaning
November	Female	5.00	—	Cotton picking, tobacco debudding
	Female	3.00–6.00	0.50/seer	Groundnut shelling
	Male and female group	18.00–22.00	150–180 seers paddy/acre	Harvesting paddy (transplanted)[e]
	Male and female group	9.60–12.00	48–60 seers + 30 seers paddy/acre	Harvesting Threshing paddy (broadcast)
December	Female	4.50–8.75	0.30–0.35/kg	Cotton picking[f]
	Female	6.00–12.00	0.30/dornam	Tobacco garlands[g]
	Male	8.00–10.00	—	Groundnut sowing, irrigating
	Male and female group	5.00–8.00		Onion harvesting
	Male and female group	6.00–7.00	60.00–70.00/acre	Groundnut harvesting
January	Female	5.00	—	Groundnut sowing, weeding, cotton picking
	Male	8.00	—	Groundnut sowing, irrigating
	Male and female group	6.75	54 seers sorghum/acre	Sorghum harvesting

Table 2
Wage Rates or Equivalents per Person per Day for 1984–1985 (*Continued*)

Month	Female Male/Children/ Group	Individual Wage or Equivalent in Rupees	Piece or Contract Rate in Rupees	Activity
February	Female	4.00–5.00	—	Groundnut weeding, cotton picking, tobacco curing
	Male group	10.00–30.00	10.00–14.00/acre/ wetting	Irrigating
	Male and female group	6.00–12.00	9.00/tractor trip	Desilting canal
March	Female	4.00–5.00	—	Tobacco curing, cotton picking
	Female	5.25–8.00	1.75–2.00/dabba	Groundnut plucking and cleaning
	Male and female group	7.00–9.00	140.00–180.00/acre	Groundnut uprooting and plucking[h]
April	Female	5.00	—	Tobacco curing, cotton picking
	Female	6.00–11.00	2.00–2.75/dabba	Groundnut plucking, cleaning
	Male	8.00–10.00	—	Transporting harvest
	Male and female group	10.00–13.00	200.00–260.00/acre	Groundnut uprooting and plucking

[a] A dabba is an empty oil tin used as a standard measure. Average number of dabbas a woman can pluck in a day is 3–4.

[b] Average number of seer shelled per day is 6–12.

[c] This activity employs children every day until January/March. However, less than 5% of the gross cropped area was used for seed cotton production.

[d] A group of 20 can transplant 2 acres per day.

[e] A group of 20 can finish all harvesting operations except winnowing in a day. Conversion factors: after milling, the rice yield from paddy is 66% by weight. 1 seer = 0.93 kg. Open-market price of rice in November 1984 was Rs. 4.00/kg.

[f] In most villages, cotton picking is paid for as a daily wage. In one village, a rate per kg was paid. On average a woman can pick 15–20 kg per day.

[g] One woman can sew 20–40 dornams (garlands) per day.

[h] A group of 20 can uproot and pluck 1 acre of groundnut per day.

security of livelihoods and the chances of being able to borrow for emergencies. The system of patriarchy exacerbates relationships of exploitation between the classes, introducing gender inequities in the already inadequate wages paid to both men and women; as a result, no family savings are available, which further increases the dependence of agricultural labor on the rich and middle-class peasants, to whom agricultural laborers must turn for loans.

Women in Cultivator Households

It could be hypothesized that with irrigation and the intensification of agricultural production, patriarchal proscriptions are reshaped, and women of landowning classes withdraw from fieldwork as the new culture makes women's work invisible by restricting it to the home, while simultaneously endowing men, whose wives are now defined as dependent housewives, with greater prestige (Mies, 1986). Contrary to these expectations, in the region studied neither women from marginal nor middle-class cultivator households have withdrawn from fieldwork. Except for the Kamma and Komati castes, this did not seem to be as important for a man's status and prestige as in other regions of India (Epstein, 1962).

In fact, as we have seen, with the introduction of irrigation and changes in the technical organization of labor, even the smallest cultivator needs to hire nonfamily labor at certain times in the crop cycle. To minimize the cost of hired labor, no longer paid in kind but in cash wages, family labor is used as much as possible. Moreover, the financial intensity of irrigated agriculture, the purchase of labor and HYV seeds, chemical fertilizers, and pesticides, instead of their organic substitutes, entails a higher investment outlay. Because these households do not have investable surpluses, they are forced to borrow. Often this is from the rich cultivators at usurious rates of interest (24–60% per annum). Even if the borrowings are from the nationalized banks, the financial intensity of the new mode of production forces the women of these households out into the wage labor market. While the men of these households busy themselves with grain processing and sale and land preparation during the dry season, the women migrate to irrigated villages to work as wage laborers. Thus, as in the case of agricultural laborers, patriarchy and structures of exploitation combine to push women from marginal cultivator households out to find work, swelling the ranks of the labor reserve and depressing wages. Yet without their contribution, it is unlikely that family subsistence needs would be met.

The contribution of women of middle-class cultivator families, unlike that of women agricultural laborers, is relatively invisible. In fact, these women play a critical role in the supervision of day-to-day agricultural operations. This entails managing the labor force and seeing that workers are provided with supplies such as seeds for sowing or fertilizer. In addition, women cultivators often join in field activities, pushing the pace of coolie women and correcting them as they work alongside. In the process these women deny one woman agricultural laborer a day's work and contribute to the exploitation of women of the lowest classes. Women of these households are not compensated for their work either as supervisors or as laborers, but clearly without their work it would be impossible for their families to maintain their class positions. When the men of these classes go to the fields to supervise coolie work, they do not actually soil their hands or do women's work, because that is culturally unacceptable. Only women in the very rich households are not involved in supervision these days (although they were just two generations ago): a male *ghumasta* or supervisor is hired in addition to many attached laborers.

The case of women from the Kamma caste is noteworthy because although they form less than 5% of the population, they are aspiring for the position of a new rural elite. The case also demonstrates the way patriarchal ideology works hand in hand with masculinism to assert caste and class superiority. Patriarchal domination by Kamma men limits women to work within the household, such as caring for the cattle. Kamma men are highly critical of local men for being lazy and letting their womenfolk work in the fields; they question the very manhood of these men. Currently the economic, political, and social control of the old elites and the material necessity for their women to be out working in the fields, while the men are organizing the *niti sangam* or building other political alliances at the village level, continue to support non-Kamma norms as the dominant ideology.

Thus, in a region where patriarchal domination did not include proscriptions on women's participation outside the home, the economic and physical burdens of agricultural intensification have fallen heavily on women in marginal and middle-class cultivator households; consequently, there has been little reshaping of preexisting patriarchal norms.

Conclusions and Implications for Development Policy

When asked whether the canal had benefited them, women agricultural laborers and marginal cultivators replied "no." Although they acknowledge that the demand for labor has increased, it still entails *kashtam* or hardship. That is, working and living conditions are still intolerably tough. The wages they receive in cash get spent immediately on daily consumption, usually at the discretion of the men.[16] The women said they could not question the right of the men to assert control over the cash income because "we cannot depend on our uncertain income to raise a family" and "it is the man of the house who gets loans during emergencies." In the days before canal irrigation, they said, whatever both men and women earned as daily laborers or attached laborers was in kind, and the grain could be stored for household consumption throughout the year.

Women of marginal cultivator households complained that not only has their work load on their own farms increased, but so has the investment necessary to engage in agricultural production. As a result, they are forced to work as agricultural laborers on the lands of others. The HYV rice they grow is sold, and an inferior variety of rice is bought for household consumption. At the end of each season, loans are repaid with high rates of interest, leaving no investable surplus, and the cycle of poverty and toil continues.

In contrast, the women of the larger landowning classes and castes said they had more work but had benefited from the canal. But it is the men who control the surpluses that are made, particularly because the women do not earn cash incomes. The men of these classes went further to say: "not only have we benefited, but agricultural labor has benefited too." First, these men pointed out that real wages of agricultural workers have increased. They calculate that before the canal the average male attached laborer got 5 quintals of sorghum and Rs. 500–1800 in today's prices, whereas today the average attached laborer gets Rs. 2800–4200. In fact, the issue of whether real wages have increased or decreased as a result of agricultural intensification is debatable.[17] Second, they argued that agricultural laborers and marginal cultivators have shifted their consumption habits from inferior grains like sorghum to superior and more costly grains like rice. In fact, although rice is socially considered superior to sorghum, it is much less nutritious.[18] Third, according to the men of the larger landowning classes, the disposable income of laborers and cultivators has increased. In fact, a substantial part of the increase in disposable income is spent by men of all classes and castes on personal consumption: tea, beedis (cigarettes), toddy (alcohol), movies, gambling, and womanizing. This spending is taken for granted as normal male prerogatives by most women, even though it means that household earnings are diminished (see also Mencher and Saradamoni, 1982). According to one male informant, the quantity of alcohol consumed in his village has increased fourfold in the last 15 years!

A third set of assessments about the benefits of canal irrigation in general come from urban, elite, male development planners within or in complicity with the state:

> Due to irrigation, India has been in a comfortable position with regard to the availability of foodgrains over the last ten years or so. *Ignoring for the moment the problem of hunger among the income deficient households of the economy,* the country's granaries are now more than full and there is enough grain stored up to prevent any famine which may occur in the foreseeable future. In this context, no less significant is the fact that the pace of grain output (growth) has been noticeably higher than the rate at which the population has been increasing (emphasis added) (Dhawan, 1988, p. 12).

Thus, according to this view, state investments in irrigation have paid off handsomely, and India has been successful in its efforts to modernize.

In contrast to the gender-blind, class-biased assessments of irrigation presented above by those who make macro policy, this article highlights the extent to which the official development strategy of modernization through irrigation has increased the drudgery in the lives of poor rural women in a specific micro setting. But can the state as a patriarchal institution possibly implement irrigation schemes that are more favorable to women? As Parpart and Staudt have said, "(i)t would be a mistake to overdetermine the state's relationship to women" (1988, p. 15). States are neither monolithic nor unchanging. Whether for reasons of populism or competitive party politics, the state in India "is not in a position to leave completely unmitigated the stresses of inequities generated by the capitalistic growth processes in a socially and economically stratified context (Bardhan, 1989, p. 24). More important, although women may appear powerless in comparison with the state, recent experience with various grassroots movements in India reveal women as actors who recognize, defend, and advance their own interests.

The women's movement in India consists of many diverse organizations that occupy a range of political and social spaces. In the context of irrigation, some of these organizations have an important role to play in reshaping irrigation policy and projects so that they are more favorable to women in the short run by (1) demanding that technical decisions, such as whether to invest in female labor-intensive earthen tanks, dams, and canals or in skilled male labor-intensive tubewells, be considered by irrigation policy planners; (2) exerting pressure on the state to enact and implement legislation that will provide women migrant construction workers and agricultural laborers with adequate and equal minimum wages, better working conditions, and social services; (3) insisting that the cropping patterns suggested by the state include crops that will increase women's employment and not attract mechanization; (4) calling for changes in the structure and staffing of the Irrigation Department so it is decentralized and employs women; and (5) pushing for the distribution of irrigation rights to women of landless and marginal cultivator households.[19]

But, while building women's labor-intensive schemes or increasing the participation of women within existing structures are essential beginnings, they will not be enough. What is needed is a radical shifting of priorities that politicize the effects of irrigation on poor rural women. Working-class women's organizations, such as the Chipko movement,[20] will be most effective in tackling the issues central to the daily life of poor rural women in irrigated regions—food, insecure employment, low and inequitable wages, abysmal working conditions, indebtedness, gender disparities and the double burden of work, and exclusion from decision-making bodies like *niti sangams*. In the process, they would exchange the preoccupation of the state with modernization for a commitment to change the unequal division of labor between the sexes, to meet basic subsistence needs, and to ensure an equitable distribution of resources, higher standards of living, and a better quality of life. Such a redefinition of development is imperative not only to change the conditions under which poor rural women work and live in irrigated regions of the world but for society itself.

Acknowledgments

Fieldwork was undertaken during 1983–1985 toward a doctorate from Syracuse University. I was funded by a doctoral research grant from The Ford Foundation, for which I am grateful. I thank two anonymous reviewers, Nancy Peluso, and especially Rajeswari Mohan for their comments on an earlier version of this article. The views expressed are my own.

Notes

1. Irrigation is described as "the lynchpin in the entire effort directed towards rural development in India" (Dhawan, 1988, p. 14). The government of India is planning to spend Rs. 400 billion to double irrigated acreage from its present level by the turn of the century.

2. I do not mean to obscure differences in ideology (see Boswell, Kaiser, and Baker, 1986), but for the purpose of this article I focus on the main contribution of Marxist thought.

3. For a detailed discussion of this issue see Ramazanoglu (1989, pp. 101–103).

4. In contrast, Western democratic states can be characterized as having a managerial commitment to development (Kohli, 1987).

5. See Deere and Leon (1987) on Latin American states, Parpart and Staudt (1988) on African states, and Agarwal (1988) on Asian states.

6. According to modernization theory, irrigation and adoption of the Green Revolution package of seeds, fertilizer, etc., is supposed to lead to the following general benefits: (1) an increase in agricultural output through increases in yields of existing crops, gross cropped area, cropping intensity, and shifts to higher yielding strains of crop, and (2) an increase in the demand for labor time through increases in gross or net cultivated area, in yields and therefore in harvesting operations, in the overall care and supervision necessitated by the new varieties, and in the number and kinds of agricultural operations as results of the changes in the cropping pattern (Agarwal, 1981). As a consequence of (1), the increase in output, the income of cultivating households is supposed to increase; as result of (2), the increased demand for labor time and real wages, the income of agricultural labor households is supposed to increase. Further, irrigation projects per se are expected to generate employment during project construction, for maintenance and water management, and through secondary employment effects (Meinzen-Dick, 1987).

7. This article does not discuss the effects on women of domestic water supply schemes or water-related diseases and other health issues. From field observations it is possible to say that the water table has risen in the area, and drinking water is more accessible than in the past.

8. Although I recognize the importance of the social relations of reproduction, unfortunately this was beyond the scope of my initial study. From field observations it was clear that women have primary responsibility for childrearing and domestic work. Moreover, even though the physical work for women outside the house has increased since the introduction of irrigation, there have been no changes in the division of labor between the sexes within the home.

9. Only about one-third of the total crop area in each village is actually zoned (localized) for irrigation by the state. The remaining area is cultivated with a rainfed crop in the first season and left fallow in the second season. Head-end villages tend to be monocropped with paddy in the first season. In tail-end villages, mixed cropping is the norm; as landholdings are fragmented, owner-cultivators may grow onion on one of their plots and chilies or cotton on another, for example.

10. The research on women presented here is part of a larger study on irrigation and equity that deals with the formulation and implementation of irrigation policy, the relationship between

the state bureaucracy and farmer associations, and the distribution of irrigation benefits within village communities.

11. The state built, in south India, canal systems extensive in length to protect as many villages as possible from drought if the monsoons failed. Water was not scarce until the late 1960s. Since then, with the process of incorporation into the market economy and the introduction of Green Revolution technologies, the demand for water has increased dramatically and is the cause of much conflict.

12. "Coping with the bureaucracy" is borrowed from Lees (1986).

13. In the Mahaweli, for example, contrary to traditional Sinhalese inheritance rules, which prescribe that both sons and daughters should inherit paddy lands, only one heir can be nominated, and it is usually the son. Furthermore, married women are not entitled to plots. And although women are working harder on both paddy plots and home gardens, they have no independent access to money. The ideology of women as dependent housewives is reinforced through scheme-sponsored training programs to teach women skills such as needlework and poultry raising. Little information on extension or cooperative credit, which may enhance women's agricultural productivity, is provided to them (Schrijvers, 1988).

14. There are a few rice mills, but there has been little tractorization, introduction of irrigation pumps, and so forth; in the limited areas where these and fertilizer/pesticide sprayers have been introduced, they have displaced women's labor.

15. The population of this village 30 years ago was approximately 200 families and today is approximately 500 families (average family has five members).

16. Mencher and Saradamoni (1982, p. A165) estimated that "the ratio of female to male contributions is at least 1:1 and in most cases it is higher" in agricultural labor households and "a little over half to well over half household income" in marginal landowning families. These conclusions held in this case study as well.

17. Some researchers have contended that real incomes have decreased (Agarwal, 1988). Others (Chambers and Harriss, 1977; Lal, 1976) have shown that real incomes have increased. According to Government of India (1979, 1981) reports, in Andhra Pradesh annual real earnings over the period 1964-1965 to 1974-1975 decreased from Rs. 246.80 to Rs. 198.20 for men but increased for women from Rs. 88.40 to Rs. 104.80. It is noteworthy that these government reports are for the whole state of Andhra Pradesh, not the specific research area, and that very often they are biased against women (Agarwal, 1985).

18. According to Gopalan (1981), milled rice contains 6.8 g of protein, 0.6/100 g minerals, 10 mg calcium, and 3.1/100 g iron in comparison with 10.4 g protein, 1.6/100 g minerals, 25 mg calcium, and 5.8/100 g iron in jowar (sorghum).

19. The divorcing of water rights from land rights and their allocation to the rural poor has been effective in lift-irrigation schemes in Pune and Chandigarh districts organized by nongovernmental agencies. For details see Chambers, Saxena, and Shah (1989).

20. For details on Chipko and other mass-based women's movements in India, refer to Bardhan (1989) and Everett (1986).

References

Agarwal, B. 1988. Neither sustenance nor sustainability: Agricultural strategies, ecological degradation and Indian women in poverty. In *Structures of patriarchy: State, community, and household in modernising Asia*, ed. B. Agarwal, pp. 83-120. New Delhi: Kali for Women.

Agarwal, B. 1981. Water resource development and rural women. New Delhi: Ford Foundation (mimeograph).

Agarwal, B. 1985. Work participation of women in the Third World: Some data and conceptual biases (Special Issue, Review of Agriculture). *Economic and Political Weekly* 20(51 & 52):21-28.

Althusser, L. 1971. Ideology and ideological state apparatuses (notes towards an investigation). In

Lenin and philosophy and other essays, ed. L. Althusser, pp. 127-186. New York: Monthly Review.

Bardhan, K. 1989. Agricultural growth and rural wage-labor in India. *South Asia Bulletin* 9(1):12-25.

Bardhan, K. 1986. Women: Work, welfare and status. Forces of tradition and change in India. *South Asia Bulletin* 6(1):3-16.

Boserup, E. 1970. *Women's role in economic development.* London: George Allen and Unwin.

Boswell, T., E. Kaiser, and K. Baker. 1986. Recent theories in Marxist theories of ideology. *Insurgent Sociologist* 13(4):5-22.

Census of India. 1984. *Andhra Pradesh: Revised figures of population.* New Delhi: Government of India.

Chambers, R., and J. Harriss. 1977. Comparing twelve south Indian villages: In search of practical theory. In *Green Revolution? Technology and change in rice growing areas of Tamil Nadu and Sri Lanka,* ed. B. H. Farmer, pp. 301-322. London: Macmillan.

Chambers, R., N. C. Saxena, and T. Shah. 1989. *To the hands of the poor: Water and trees.* New Delhi: Oxford and IBH.

Dhawan, B. D. 1988. *Irrigation in India's agricultural development: Productivity, stability, equity.* New Delhi: Sage.

Deere, C. D., and M. Leon. 1987. *Rural women and state policy: Feminist perspectives on Latin American development.* Boulder, CO: Westview.

Eisenstein, Z. 1979. *Capitalist patriarchy and the case for socialist feminism.* New York: Monthly Review.

Epstein, T. S. 1962. *Economic development and social change in south India.* Manchester: Manchester University Press.

Everett, J. 1986. We were in the forefront of the fight: Feminist theory and practice in Indian grass-roots movements. *South Asia Bulletin* 6(1):17-23.

Gopalan, C. 1981. *Nutritive values of Indian foods.* Hyderabad: National Institute of Nutrition.

Government of India. 1981. *Rural labour enquiry 1974-75, final report on employment and unemployment.* Chandigarh: Ministry of Labour.

Government of India. 1979. *Rural labour enquiry 1974-75, final report on wages and earnings.* Chandigarh: Ministry of Labour.

Hanger, J., and J. Moris, 1973. Women and the household economy. In *Mwea: An irrigated rice settlement in Kenya,* ed. R. Chambers and J. Moris, pp. 209-244. Munich: Weltforum Verlag.

Kohli, A. 1987. *The state and poverty in India.* Cambridge: Cambridge University Press.

Lal, D. 1976. Agricultural growth, real wages, and the rural poor in India. *Economic and Political Weekly,* Review of Agriculture, June 1976, pp. A 47-61.

Lees, S. 1986. Coping with bureaucracy: Survival strategies in irrigated agriculture. *American Anthropologist* 88:610-622.

MacKinnon, C. 1983. Feminism, Marxism, method, and the state: Toward feminist jurisprudence. *Signs* 8(4):635-658.

Meinzen-Dick, R. 1987. Labor demand and employment generation in irrigation systems. Ithaca, NY: Cornell University. (Irrigation Studies Group Working Paper).

Mencher, J., and K. Saradamoni. 1982. Muddy feet, dirty hands: Rice production and female agricultural labour. *Economic and Political Weekly,* Review of Agriculture 17(52):A 149-167.

Mies, M. 1986. *Patriarchy and accumulation on a world scale: Women in the international division of labor.* London: Zed.

Parpart, J., and K. Staudt, eds. 1988. *Women and the state in Africa.* Boulder, CO: Lynne Rienner.

Poulantzas, N. 1978. *State, power, and socialism.* London: New Left Books.

Ramazanoglu, C. 1989. *Feminism and the contradictions of oppression.* London: Routledge.

Schrijvers, J. 1988. Blueprint for undernourishment: The Mahawweli development scheme in Sri

Lanka. In *Structures of patriarchy: State, community, and household in modernising Asia*, ed. B. Agarwal, pp. 29-51. New Delhi: Kali for Women.

Sen, G. 1982. Women workers and the Green Revolution. In *Women and development*, ed. L. Beneria, pp. 29-64. Geneva: International Labor Office.

Skocpol, T. 1979. *States and social revolutions: A comparative analysis of France, Russia, and China*. Cambridge: Cambridge University Press.

Stepan, A. 1978. *The state and society: Peru in comparative perspective*. Princeton: Princeton University Press.

Swaminathan, P. 1987. State and subordination of women. *Economic and Political Weekly* 22(44):WS34-39.

Epilogue

This epilogue has two purposes: (1) to update the empirical analysis with results from brief field restudies in 1992 and 1994 and (2) to delineate the contours of the markedly changed policy environment regarding "women in water," critique it, and highlight the continued need for gender-centered irrigation research.

Gendered Transformations in Irrigation: Southeastern Andhra Pradesh, India

The Changes. The starkest change I found on my return to the area in the summer of 1992 was the near absence of canal water in the tail-end villages where I had done fieldwork. They were now dry, dependent on rainfall and, where possible, lift irrigation. Canal-irrigated acreage in the tail-ends (beyond the midpoint of the Main Canal) had decreased by 97% (or about 17,000 acres) in 1991–92 compared to 1984–85. The situation had not improved as of December 1994. According to official statistics, water inflows had decreased by 38% since 1986–87.[1] Simultaneously, Main Canal–irrigated acreage plummeted from a high of 66,362 acres in 1986–87 to an all-time low of 42,381 acres in 1992–93. Disaggregating the figures reveals three trends: (1) the first season was less severely affected than the second season, (2) on each distributary, even those with offtakes at the head-end of the Main Canal, less water was reaching the tail-ends, and (3) on distributaries with offtakes beyond the midpoint of the Main Canal little or no water was available. The Main Canal has thus become both much shorter and fatter.

A second change was the difference in the cropping pattern. In comparison to 1984–85, when first-season canal-irrigated acreage consisted of 65% paddy, 34% cotton, and 1% groundnut and other crops, in 1991–92 the breakdown was 56% paddy, 25% cotton, 8% sunflower, 1% groundnut, and 10% other crops. In the second season, the canal-irrigated acreage was 8% paddy, 25% cotton, 66% groundnut, and 1% other crops in 1984–85, but 7% paddy, 18% cotton, 25% groundnut, and 50% sunflower and other crops in 1991–92. In general, at the head-ends, paddy predominates. Villages that still get some canal irrigation have moved to less water-consumptive crops, and in the tail-end villages mainly dry crops are being raised.

A third change has been the spontaneous development of farmer-run lift irrigation groups in villages located on the banks of the river.[2] Many villages now have three or four pumping stations, each irrigating 50–100 acres every season. In the villages I restudied, the main lift-irrigated crop is paddy (55% of total area); cotton (17%), groundnut (16%), sunflower (4%), and vegetables (8%) are also grown. The lift-irrigation groups are organized mainly by the larger farmers and are run by them. Costs are apportioned per acre and a water manager is appointed to keep the pump running and distribute water. These changes have important implications for women as users of irrigation and as laborers and cultivators.

Implications for Women of Changes in the Organization of Irrigation. Irrigation water, which was already a scarce resource in 1984, has all but dried up in tail-end villages. The *niti sangams*, or water associations, that used to be organized by village Big Men are no longer formed, because the potential for conflict is too high. Cultivators who used to benefit from their existence must now do without water or battle for it with head-enders. Women, especially de facto or de jure household heads, are particularly disadvantaged. Often, they have no choice but to employ roving labor "gangs" of men who charge Rs 200 per acre each time they irrigate a plot. Only men continue to be employed to undertake irrigation at night, because that is still considered too dangerous for women. A few women are now visible negotiating for water at the Irrigation Department offices and some are active in the lift-irrigation groups, but such "public" roles continue to be very much the exception, not the norm. In general, women are involved in irrigating fields whereas the management of irrigation is still done by men, whether the Irrigation Department or cultivators.

Transformations in the Sexual Division of Labor, Workloads, and Control of the Labor Process. Given the continuing patterns of the sexual division of labor in Southeastern Andhra Pradesh, female laborers have been the most adversely impacted by the decrease in canal-irrigated acreage. These women, who are either landless or belong to marginal cultivator households, must work on land belonging to others to ensure family survival. In general, irrigated agriculture continues to require more work than rainfed agriculture, and much of this work—sowing/transplanting, weeding, harvesting—is still defined as women's work. In particular, the decrease in paddy and groundnut cultivation, both generally women's labor-intensive crops, has decreased the demand for female labor. The increase in sunflower cultivation has not compensated for this decrease because sunflower uses relatively little women's labor. Like irrigated hybrid *jowar*, sunflower sowing is done by a male-driven bullock team with a seed drill; men also hoe and fertilize the crop. Occasionally women may weed sunflower, but otherwise their main role is at harvest. To about 15 woman-days per acre for rainfed sunflower, irrigation may add 8 days' work. In comparison to 23 woman-days per acre for irrigated sunflower, I calculated that irrigated paddy rice demanded 53 woman-days and irrigated groundnut 45 woman-days per acre.

Rainfed agriculture has also undergone a change in the region as cultivators of all classes have become more dependent on it. The main crop substitutions on rainfed lands are the replacement of tobacco, *jowar*, and groundnut by cotton and sunflower.[3] The results for female laborers are mixed. On the one hand, female labor demand has gone down. Tobacco, for example, is relatively female labor intensive (55 woman-days per acre) compared to rainfed sunflower (15 woman-days per acre) and demands labor throughout the crop cycle rather than in a narrow window of time at harvest. On the other hand, some rainfed crops are now being cultivated as intensively as irrigated crops. Rainfed cotton, for example, has undergone a transformation in the technical organization of labor. Women's labor is still used for sowing cotton seeds, weeding, basal fertilizer application, and, of course, most of all for picking. The most dramatic change is the increase in the use of pesticides on both rainfed and irrigated cotton, three, four, or even more times during the crop cycle. Although the new hand pump used to spray pesticides is operated by men and spraying is defined as a "skilled job," with the increases in fertilization, pest control, and supervision, rainfed production is up to 60–75% of irrigated cotton production. Women are employed on a daily basis as casual labor for cotton picking. They also find employment in fertilizing and weeding cotton. My 1984 estimate of 44 woman-days per acre of rainfed cotton compared to 112 woman-days per acre of irrigated cotton will have to be revised up by about 40 days to account for this change.

In terms of the labor process, women are still employed as daily casual laborers recruited mainly through networks of male relatives who work as permanent laborers for large landowners. Female wage rates were Rs 12–15 per day in 1993–94 compared to Rs 20–25 per day for males. When compared to Rs 3–6 in 1984–85 (Table 2), this represents a marginal increase in the real daily wage rate. Children are now paid Rs 10–11 per day compared to Rs 3–4 in 1984–85.[4] Male permanent laborers now get Rs 6,000–10,200 per year compared to Rs 2,800–4,200 in 1984–85. For group (*gumpu*) work in paddy harvesting, threshing, and winnowing, daily wage equivalents continue to be higher than average wages. Paddy harvesting, however, is still paid in both kind and cash. Groundnut harvesting, which used to be compensated at the rate of Rs 200–260 per acre, now pays Rs 450–500 per acre, or a daily wage equivalent of Rs 22.50.

Conversations with laborers and marginal cultivators suggest transformations in the social organization of labor that, although tentative at this point, are important enough to map.[5] In villages that used to get canal irrigation but are now dry there is a lot more labor out-migration, especially in the second season, from December onward. The large stream of labor in-migration from noncanal areas that used to be an annual occurrence during the groundnut harvesting season in April has completely stopped. Nowadays, male out-migration is for casual waged work in stone quarrying, river sand collection, or market-yard or construction jobs in the nearest urban or semi-urban center. Female out-migration is within a relatively smaller radius and mainly for agricultural labor work or fuel gathering. In villages that still receive irrigation, the pattern of men working on their own lands while women work both on their own lands and that of others continues. However, in both cases there seems to be a movement by women to organize into groups mainly to demand higher wages, but also, in the case of paddy rice, to ensure continued payment in kind. Several large cultivators have complained about the shortage of labor during peaks in the crop cycle and how "demanding" and "contentious" labor groups and negotiations had become.[6]

Four other recent trends with consequences for poor women's lives are noteworthy. The substitution of *jowar* with sunflower and cotton has meant wages in foodgrains are no longer paid (except for paddy) and the consumption of coarse grains by the lower classes and castes has dramatically dropped. Milled rice is now consumed by everyone, but it is less nutritious than either *jowar* or hand-pounded rice (rich in vitamin B). Women did report, however, that rice was a lot easier and less time consuming to cook. Second, because neither cotton nor sunflower can be used as fodder, whereas the groundnut they replaced could, women are spending longer hours foraging to feed their cattle. In extreme cases, they have had to buy fodder. Third, the potential health hazards from the indiscriminate use of chemical pesticides (and fertilizers) are frightening, although currently complaints are voiced only in terms of the immunity pests are developing. Fourth, in all the villages where less or no canal water is now available women are burdened with finding and fetching water for domestic use—drinking, washing, cooking, and cattle. In some villages the problem has become so acute that the government is instituting well- or river-sourced drinking water schemes.

Class Consequences. In 1994, as in 1984, women in all but the largest cultivator households continued to both labor and supervise laborers working on their fields. Again, there seems to have been no reshaping of patriarchal norms to proscribe women from working outside their homes. Whereas initially only the larger landholders had the financial capability and risk-bearing potential to make the high investments required to cultivate irrigated cotton for example, by 1994 every landholder was cultivating a commercial crop. The process of commodification now ensures that all size classes purchase inputs and sell outputs in the market, rather than self-provision. Certainly the cost of capital is higher for smaller farmers, so that they are left with smaller surpluses at the end of the season. As a result the relative inequality in the distribution of land and other assets has not improved. Yet, the number of small farmers leasing-in land seems to have increased; potentially, their surplus production should increase as well but, of course, not without the labor of the women in these households. For the most part, men continue to control the allocation of household income except in de facto or de jure women-headed households. Thus, much of the burden of intensification and capital accumulation continues to be borne by women's labor.

Several recent scholarly writings bear out similar general trends but historically specific and complex relationships between women and irrigation in the context of The Gambia (Carney, 1993), Senegal (Nation, 1994), Tunisia (Andes, 1994), Sudan (Osman, 1995), Lesotho (Riley and Krogman, 1993), Pakistan (Hewitt, 1991), Nepal (Neupane, 1991), Malaysia (Hart, 1992) and Indonesia (MacPhail and Bowles, 1989). Women's resistance strategies, the relationship between irrigated agriculture and fertility, and the correlation between female labor intensification and children's morbidity and mortality are some of the issues illuminated by these writings, which beg further cross-cultural and historical research.

Gender and Water Resource Policy[7]

In contrast to the policy environment of the late 1980s, in the mid-1990s the importance of "women in water" is widely acknowledged by the international donor community.[8] To varying degrees, depending on the specific donor, the "women in water" initiative is couched within a broader thrust toward integrated water resource management (IWRM). As elaborated by the World Bank (1993), IWRM encompasses "the treatment of water as an economic good, combined with decentralized management and delivery structures, greater reliance on pricing, and fuller participation by stakeholders" (p. 10).

Potentially, IWRM could increase the pragmatic interests of women by valuing their work in water management. Yet, in its possibly narrow recognition of women's activities as users of water for domestic consumption, IWRM not only disregards women's activities as agriculturalists but reinforces patriarchal ideologies that naturalize women's domestic labor. Further, as Green and Baden (1995) have argued, IWRM categorizes "women" as a separate, already constituted, vulnerable group whose interests must be added on to those of men. This ignores the very significant differences within "women" due to class, caste, age, and ethnic or religious affiliation as well as the fact that women are not separate from men but very much embedded in unequal relationships with them.

IWRM could also potentially increase the benefits to women by involving them in the decentralized planning and delivery of irrigation services. Women's participation in water user associations can be empowering to them. However, this disregards the structural, ideological, and pragmatic constraints to women's "public" participation. More invidious is the attempt to entrust community systems to women because they are "less likely to migrate, more accustomed to voluntary work, and better trusted to administer funds" (World Bank, 1992, p. 113). Such attempts to impose participation may only add to women's already excessive workloads and, far from empowering them or changing their strategic interests, further reinforce gender inequalities.

Another major thrust of the new IWRM policy is the pricing of water to reflect its value as a scarce resource. The move to formalizing property rights in water as a first step to the development of water markets is fraught with difficulty for women, given historical patterns of exclusion and their indirect, contingent, and negotiated patterns of resource usage (Green and Baden, 1995). It is a real possibility that men will decide to sell high-priced water or divert its use from food to commercial crops or nondomestic uses. Ultimately, as noted by Green and Baden (1995), the pricing of water resources must take into account intra-household differentials in power and access to disposable income. Theoretically, there is a contradiction that the "efficient" working of market principles, stressed by IWRM, should rely on nonmarket institutions (water user groups) and women's voluntary labor.

Finally, IWRM seeks to sponsor environmental protection and conservation measures. Too often, the burden of these measures may also increase women's unpaid family labor. However, theoretically there is no reason why women should be better conservers of natural resources just because they are active users of them (Agarwal, 1992).

The move toward IWRM is worldwide—even on the Main Canal, which was the focus of this article. Amazingly, research reports on the canal with a women in development (WID) focus were produced in 1994. The reports, funded by the Dutch government, aimed at "determining the role of women in irrigated agriculture and water management," "identifying the characteristics of women-farmer leaders," and "finding out the training conditions preferred," so as to increase the participation of women in water user associations. The water user associations (among other initiatives, such as changes in the cropping pattern) have been mandated by the World Bank on its adoption of the canal for a rehabilitation loan. The main findings of the reports are as follows: the women interviewed were fairly inarticulate, they assumed leadership is synonymous with maleness, and, for them, there was an "absolute unacceptability of training in leadership as an excuse to be relieved from their families." The findings illustrate some of the problems involved in doing participatory research in a social context embedded in relationships of gender inequality, and they could quite easily be used to ignore women in water issues. In fact, there is some indication that the reports are already gathering dust. Gender analysts in water resource management will have to theorize and develop better empirical understandings not only of women's roles in water but of the politics of women's knowledge production, how local knowledge can be transformed to problem solving, and the opportunities (and limitations) for action. Intervention in the larger debate about water resource management from a gender perspective is crucial at this point in time.

Additional Notes

1. The Irrigation Department blamed the head-end state of Karnataka for reducing water inflows. Tail-end farmers were more inclined to believe that illegal head-end paddy cultivation was the cause.

2. The Main Canal is diverted 89 miles upstream off the left bank of the eastward flowing river and then runs roughly parallel to it. It flows on a ridge and feeds 40 distributaries mainly aligned southward, that is, sloped down toward the river. Villages at the tail-ends of distributaries, therefore, tend to be on the banks of the river but high above it.

3. Interestingly, in other parts of Andhra Pradesh *jowar* is referred to as a "woman's crop" (*aadola panta* in Telugu), and cotton and other commercial crops are considered "men's crops."

4. Children are still employed mainly for cotton cross-fertilization in the production of hybrid seeds.

5. I will be returning to the field for nine months in 1997 to study these transformations in depth.

6. In the Malaysian (Hart, 1992) and Gambian (Carney, 1993) contexts, these processes have been interpreted as embodying patterns of women's resistance to processes of commodification. Clearly, more work is necessary in the south Indian case before such generalizations are possible.

7. For a detailed discussion on integrated water management as it relates to gender, see Green and Baden (1995). The first part of this section draws on their excellent analysis.

8. Institutions such as the World Bank, the International Irrigation Management Institute, the International Food Policy Research Institute, the Overseas Development Administration, and the Dutch, Swedish, and Norwegian development agencies have all expressed the need for a new approach to irrigation. The United Nations Conference on Environment and Development also includes a chapter on the subject in its policy statement.

Additional References

Agarwal, B. 1992. The gender and environment debate: Lessons from India. *Feminist Studies*, 18(1):119–158.

Andes, K. L. 1994. *Fertility and farming: A comparative political economy of childbearing in two agricultural communities in Tunisia.* Ph.D. diss., Northwestern University.

Carney, J. 1993. Converting the wetlands, engendering the environment: The intersection of gender with agrarian change in the Gambia. *Economic Geography*, 69:329–348.

Green, C., and S. Baden. 1995. Integrated water management: A gender perspective. *IDS Bulletin*, 26(1):92–100.

Hart, G. 1992. Household production reconsidered: Gender, labor conflict, and technological change in Malaysia's Muda region. *World Development*, 20(6):809–823.

Hewitt, F. 1991. *Women in the landscape: A Kakoram village before "development."* Ph.D. diss., University of Waterloo.

MacPhail, F., and P. Bowles. 1989. Technical change and intra-household welfare: A case study of irrigated rice production in South Sulawesi, Indonesia. *Journal of Development Studies*, 26:58–80.

Nation, M. 1994. *Gender, irrigation and development in the Upper Valley of the Senegal river.* Ph.D. diss., University of Toronto.

Neupane, S. 1991. *The nutritional status of children in irrigated and nonirrigated villages in Nepal.* Ph.D. diss., Cornell University.

Osman, E. 1995. *The fertility impact of rural development projects: The case of the Rahad irrigation project, Sudan.* Ph.D. diss., Brown University.

Riley, P., and N. Krogman. 1993. Gender related factors influencing the viability of irrigation projects in Lesotho. *Journal of Asian and African Studies*, 28:162–179.

World Bank. 1992. *World Development Report.* New York: Oxford University Press.

World Bank. 1993. *Water Resources Management: A World Bank Policy Paper.* Washington, DC: The World Bank.

Chapter 8

Women as Rice Sharecroppers in Madagascar

LUCY JAROSZ

Department of Geography
University of Washington
Seattle, WA 98195
USA

Abstract *This case study, drawn from a rice-growing region in Madagascar, demonstrates how gender and class differences shape individual access to and control of productive resources. Production strategies differ among the women and men who crop rice on shares and are primarily distinguished by class position and gender. Single women invariably share out the land they own to male croppers, whereas men of all classes may sharecrop land from or to other men. Only wealthy male farmers implement sharecropping as an accumulation strategy. Wealthy female farmers are concerned with mobilizing male labor power in their sharecropping strategies. Poor, landless, female heads of households are the only persons in this study who cannot and do not crop rice on shares and are the most disadvantaged and poorest.*

Keywords: gender, class differences, sharecroppers, rice, Madagascar, Green Revolution impacts.

The image of a starving African child is a widely accepted Western media representation of a continent in crisis. The dimensions of this crisis are environmental, demographic, economic, and political. Second in land area only to Asia, Africa has the potential resources to produce more food than its current population requires, but food production has been steadily declining since 1970 (Okigbo, 1988). Issues of food security are tied to the use and management of natural resources. How African men and women use and manage agricultural resources—land, water, tools, seed, capital, and labor—is a critical issue in understanding the dimensions of the current crisis as it relates to the broader issue of the social dynamics of natural resource use in agricultural production in Africa.

Sharecropping and Resource Access

Sharecropping is one of the most ancient and widespread relationships that shapes resource use and management in the developing world. Sharecropping can refer to sharing not only the crop but also agricultural tasks. It ensures access to land, equipment, labor, and capital in exchange for a negotiated share of the crop (Swindell, 1985). In Ecuador, Mexico, India, Thailand, Malaysia, the Philippines, Ghana, the Sudan, and Madagascar, sharecropping is a fundamental form of rural labor organization and agricultural production. At one time or another, sharecropping has existed in virtually all parts of the world (Byres, 1983).

From his study of sharecropping in Ghana, the Senegambia, the Sudan, and Lesotho, Robertson (1987) argued that sharecropping uniformly and equitably facilitates economic development and agrarian capitalism in Africa. He concludes that because of their dynamism and flexibility, share contracts can ensure that young people have access to land and capital while providing elderly peasants with access to labor power. Thus, individually negotiated share contracts can efficiently mediate intergenerational access to productive resources and eventually culminate in their transfer to former sharecroppers.[1]

The impacts of gender and class difference on sharecropping relations are not addressed by Robertson's theory. In my study of the evolution of rice sharecropping in Madagascar, I did not find that sharecropping uniformly facilitated economic development according to an intergenerational, linear trajectory. Sharecropping as a resource access strategy in Madagascar does not uniformly impede or facilitate a transition to agrarian capitalism; rather, its effects are uneven and ambiguous, increasing productivity for some and leading to economic stagnation and dispossession for others.

In this case study drawn from north-central Madagascar, I examine the roles of gender and class in regional rice production in order to demonstrate that sharecropping is one strategy male and female household heads use to gain access to resources, and in so doing, reinforce relations structured by kinship, gender, and class.[2] I found that class and gender shape, and are shaped by, sharecropping strategies and formulas. This interactive process is one response to worsening economic conditions and changes in the agrarian economy.

Rice Sharecropping in Alaotra, Madagascar

The Democratic Republic of Madagascar is an island nation of 226,558 square miles. The island is made up of five major ecological zones: the cool, rolling hills of the centrally located High Plateau, which runs almost the length of the island; the dense humid tropical forest descending sharply to the east coast; the huge area of savanna grassland to the west of the High Plateau; a large arid region in the south; and the humid northwest, where export crops such as coffee and vanilla are grown (Heseltine, 1971).

Over three-quarters of its 11.2 million citizens live in rural areas. Per capita income is approximately $270 per year; it is one of the 12 poorest countries in the world. Foreign exchange is based on the agricultural exports coffee, vanilla, and cloves. The national economy has been disastrously affected by plummeting world coffee prices and the widespread use of synthetic vanilla. Failed and badly planned industrialization projects added to the growing debt problem as the socialist policies of President Didier Ratsiraka increasingly concentrated resources in industry to the neglect of agriculture. Madagascar teetered on the edge of bankruptcy in 1981. Rice imports tripled between 1980 and 1982, and oil price hikes further depleted meager export earnings. Madagascar's economy has been under the management of the World Bank and the International Monetary Fund since 1981. Its deficit has been cut in half, although debt servicing consumes 55% of export earnings. In 1986, food took up 70% of the average household budget. Rice, the basis of the Malagasy diet, is both a subsistence and a cash crop. The price has increased 700% since 1974.

The Alaotra region is a fertile lacustrine basin at an elevation of approximately 800 m and situated 160 km northeast of the national capital, Antananarivo, at the north end of the High Plateau. It is the site of Lake Alaotra, the island's largest lake. Presently, Malagasy farmers cultivate 90,000 hectares of irrigated rice here; 30,000 hectares are part of an intensive irrigation development project begun in 1960. The project was designed to modernize and increase rice production through the construction of a network of dams and canals and the introduction of high-yield rice varieties and of transplanting techniques. The area is watered by 30 rivers, and the region has long been known as Madagascar's granary, supplying 10–15% of the island's rice (see Figure 1). The large-scale development project was to aid in the increase of national rice production.

Men and women farm dryland rice using fire and spades (*angady*) to clear the land. Maize, manioc, peanuts, and vegetables are also grown on dryland fields. Swamp rice is grown in the marshy areas surrounding the lake, and irrigated rice within and outside the boundaries of the development project demands the use of plow agriculture for fields larger than 1 ha. Wealthy farmers cultivate their fields with tractors.

Sharecropping forms an important part of Alaotra's agricultural history. At the turn of the century, the French colonial state created large concessions for French companies and settlers on the most fertile lands and pushed indigenous farmers onto reserves. Wage workers, forced laborers, and sharecroppers cultivated these concessions. In 1960, the

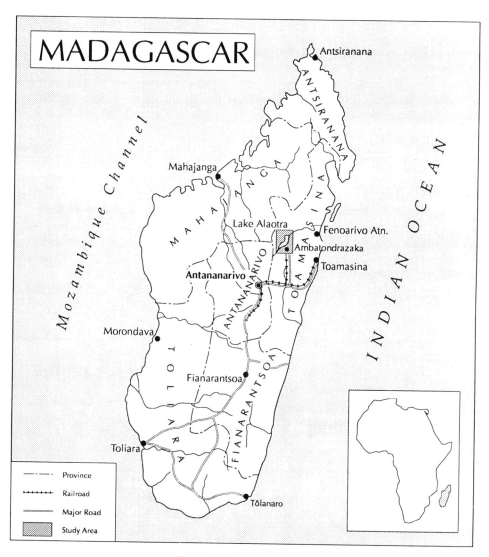

Figure 1. Map of Madagascar.

year Madagascar declared its independence from France, surveys revealed that approximately half the region's population was composed of landless sharecroppers (LaPierre, 1964; SCET, 1963).

There are myriad different sharecropping contracts that may exist between individuals, but the most common is that landlords provide the land and seed and sharecroppers provide the labor. At Alaotra, extra labor is hired for transplanting in November and December and for harvest in May and June, and these costs are divided. Seasonal hired labor costs may be split in half, or the landowner may pay for transplanting and have the sharecropper pay for harvest help. The rice crop is divided in half (*misasabokatra*) at harvest time when both parties are present. Contracts usually are oral agreements.

In principle, the state has been opposed to the practice of sharecropping, viewing it as an exploitation of the landless disadvantaged at the hands of absentee landlords (*Charte de la Révolution Socialiste Malagasy*, 1975). Despite a national ban on the practice of sharecropping in 1973, between 40 and 50% of the agricultural lands at Alaotra are cultivated by sharecroppers (Phillips et al., 1986). My interviews with 181 farmers reveal that nearly 44% are involved in some form of sharecropping arrangement (Jarosz, 1990). More than half of these cases involve kin-based sharecropping.

One of the most striking things about sharecropping at Alaotra is that sharecroppers are not solely landless or nearly landless peasants; members of the wealthy and middle classes also sharecrop the land of others. These landowners are generally smallholders, poor or middle-class farmers who are indebted, elderly, female, or the descendants of former slaves.

Women Who Sharecrop

Marital status and class[3] are key determinants of women's sharecropping practices at Alaotra. Sharecropping is virtually imperative for female-headed households farming one or more hectares of irrigated rice, but no mention is made of women sharecroppers in any of the official reports or in contemporary research on rice agriculture. Indeed, most sharecroppers are generally thought to be male, and women are considered as their wives and not as sharecroppers in their own right.

Married women who have inherited lands near their ancestral villages and then have moved away to live near their husbands' lands generally sharecrop their own lands with brothers, uncles, or nephews who live near their fields.[4] If a wife and husband remain near her ancestral village, she will work her land with her husband and children. Upon divorce, the husband claims two-thirds and the wife one-third of the property accumulated since their marriage.

Widows or divorced women with children over 15 years old who have adequate equipment are able to cultivate their lands with the help of their sons and daughters. Divorced women or widows who have no children at home, or who have small children, invariably sharecrop their irrigated and swamp rice fields, generally with male kin living nearby. For example, a 38-year-old married woman was abandoned by her husband 2 years ago. She says he has a drinking problem. She has four children, all under 15 years old. She owns 10 ha of marshland but does not have adequate male household labor power or capital to cultivate it independently. She sharecrops with her brother, a local cattle merchant living in a nearby town. She provides the land and seed and pays for transplanting labor. Her brother pays for plowing and harvest labor. She hires five women for a week's transplanting and has three male workers who are paid by sharecropping 2 ha of her dryland fields. She cultivates the dryland crops and decides the cultivation methods and calendar for her land. By sharecropping with her brother and the farm workers she hires, she obtains access to male labor power and the capital necessary for production.

I interviewed 29 women who were heads of their households; nearly one-third sharecropped their lands with men. These women shared out as little as 1 ha and as much as 20 ha. Half of these women (one-sixth of those interviewed) sharecropped with male kin. Women provide the seed and land; men provide the labor. Of the women who cultivated their lands directly, most possessed small dryland fields of less than 1 ha.

Typically, middle-class women sharecrop their land in order to get the male labor power and capital necessary for production. The field preparation tasks of clearing, plowing, and harrowing are socially defined as male tasks. The sharecropping rationale is only slightly different for middle-class men.[5] Generally, they sharecrop their lands with other men in order to obtain the means of production—the use of plows, cattle, or tractors for field preparation—or to split the seasonal labor costs and thereby minimize capital outlay. However, even if female-headed households possess the necessary cattle and equipment for field preparation, they still must have access to male labor power, a problem that does not generally plague male-headed households.

Plowing takes the physical strength and presence of at least two men. One controls the plow, and the other drives a team of two or four zebu cattle. Plowing can be dangerous work, because many zebu are temperamental and unpredictable. Men have been badly gored in the fields. I never saw women plowing the fields; other agricultural tasks such as weeding, transplanting, and harvesting were performed by both men and women, although transplanting and weeding are generally considered women's work. Peasants who crop on shares are generally responsible for clearing and preparing the fields, a task defined as men's work, and therefore taking in land to sharecrop is an avenue closed to female-headed households that cannot lay claim to adequate household male labor power or that do not possess the capital to hire workers.

Wealthy widows and divorced women are generally the most powerful and most economically autonomous. They can exercise total control over farm decisions. T., for example, married at age 27. Her husband drove a bush taxi (*taxi brousse*) in this area, and she first met him when she made the long trip north from Antsirabe to visit her cousin living in Alaotra. Her husband inherited 50 ha of irrigated rice fields upon his parents' death. He was one of two children, and his sister had moved to the neighboring Comores Islands. In the beginning, the couple sharecropped their lands, because they had no agricultural equipment. They saved money and bought harrows, plows, cattle, and eventually a tractor. After the death of her husband a few years ago, the 42-year-old widow invited family members from the south-central highlands, which are overcrowded and very poor, to immigrate to Alaotra to sharecrop her lands. There are now 11 adults working as sharecroppers on her lands in a small community of more than 50 people, including children. She says she is happy with this arrangement. Sharecropping provides land and rice to her extended family, provides her with labor power, and makes her much less vulnerable as a woman alone. She oversees the farm operation rather than being directly involved in the day-to-day tasks.

Unlike wealthy and middle-class male farmers, female farmers do not take in additional land as sharecroppers in order to accumulate rice, lay claim to their sons' household labor power, or extend their farming areas as men do. Women of all classes do not seek out lands to sharecrop as a means of accumulation or to provide land for landless sons; rather, they invariably share out their land to male sharecroppers.

One divorced woman with two small children rents land. She has a small business buying and selling rice sacks, geese, and poultry. She provides the seed and works the rented lands with her brother; they divide production costs and the harvest in half. Sharecropping is a means for her to obtain the male labor power necessary to put her rented lands into production.

Sixty-six percent of the women I interviewed were either landless or nearly landless, owning 1.5 ha or less of rainfed fields. This is not enough land to feed their households adequately. Regional landholdings average between 4 and 6 ha. Average household size is about four people, three people less than the national average of seven. Landless female-headed households emerge as the most vulnerable and marginalized households of Alaotra's agriculturally active population. Women heading these households work irregularly as seasonal laborers or laundresses or husk rice for a day's meal. Many are descendants of slaves. Their incomes are inadequate, and they and their families live hand to mouth. The poorest are homeless. One young women and her two tiny children live in a neighbor's tool shed. She has no direct access to a cooking fire, and she and her children are ill and malnourished. Single, landless men can earn $66 per year as permanent farm workers who room and board on the farms of their employers. Landless women without children can earn $40 per year as domestic workers who room and board with their employers. Landless men can sharecrop dryland or marshland, options that are unavailable to landless female-headed households. These households do not have adequate household labor power or capital to fulfill standard sharecropping contract obligations unless the household heads are economically autonomous business or professional women. Their economic independence, one result of education beyond the primary level, enables middle-class landless women to sharecrop and/or hire male labor, an option closed to poor women.

Married women in middle-class households are more involved in farm work than are rich women. Depending on the size of the family's irrigated and swamp rice holdings and the availability of family labor, middle-class women are largely responsible for planting, weeding, and harvesting in the rainfed fields if most of the male labor is devoted to livestock production and working in irrigated and swamp rice fields larger than 3 ha. Women also supervise or work alongside transplanting and rice harvest crews. With the advent of plow agriculture, transplanting, and other technological innovations brought about by the development project, many older women say that these changes have meant a decrease in their participation in irrigated rice agriculture. Formerly, women directly participated in all aspects of rice cultivation, including handling the cattle when the cleared swampland was prepared for broadcast seeding by trampling. Now their role is confined to cooking for husbands, male relatives, and crew members during labor bottlenecks as well as sometimes providing some field labor during this period.

Rich women employ one or more domestic workers. These workers are young, single girls from struggling peasant families. Domestic wage workers perform household and seasonal farm labor tasks. Wealthy women oversee household workers, ensure food preparation and delivery to family members or crews in the field, and supervise seasonal female transplanting crews. Their activities in rice agriculture have been progressively marginalized. This marginalization helps explain why rich and middle-class female-headed households invariably share out irrigated and swamp rice lands. Technological innovation has encouraged, indeed necessitated, the increase in rice sharecropping in this region for female-headed households. Lack of access to plows, harrows, and cattle for field preparation has resulted in the resurgence of sharecropping practices among middle-class and poor male-headed households.

Women's workloads have increased over the past 20- to 30-year period across all classes. In order to help provide for their households during the preharvest season, many sell horticultural products, baskets, mats, geese, chicken, and fish. Although their incomes from these activities are not as substantial as those coming from irrigated rice, they are essential for family production needs as well as providing cash to pay for rice production costs, most notably for labor for transplanting on fields within the development project's boundaries.

The farm women of Madagascar are working more but are increasingly marginalized within irrigated rice agriculture (Byres, Crow, and Ho, 1983; Sharma, 1985). Similar findings have been made concerning women and Green Revolution technologies in India. Wealthy and middle-class women participate less directly in rice agriculture. Cooking takes up the majority of their time, as does work in rainfed fields that produce manioc, maize, and rice. Poor women are locked into seasonal farm labor and work as laundresses and domestics. They do not earn enough to feed and shelter themselves and their children adequately. Poor women work shoulder to shoulder with male family members in household fields and for wages in rice production. Of all rural women, they are the most directly involved in irrigated rice cultivation but have the least control over productive resources.

Poor female-headed households with access to small dryland fields cultivate them with household labor. Neither plows nor cattle are required to cultivate these small fields. These households grow a combination of manioc, maize, and dryland rice and have small vegetable gardens. They have insufficient land to ensure food security for more than a few months' time and must buy food beginning in November or December until the dryland rice is harvested in March and April. Aside from a few weeks of transplanting work in November and December, there are few income-earning opportunities for women between January and March. They and their children go hungry. This period coincides with the height of malaria season, and there is little or no money available for medicine.

Middle-income and poor women who have access to land spend more time on dryland fields than on rice production. Rice is a man's crop at Alaotra. Said one Sihanaka man: "Women know only pots and pans. They know nothing of the fields." Green Revolution technologies have moved women out of all but seasonal labor in irrigated and swamp rice production. Women increasingly devote themselves to market gardening and subsistence production on rainfed fields, signaling a trend in the spatial segregation of the gender division of labor. The gardens and rainfed fields ensure a small cash flow during the preharvest season as well as an essential food supply for the family until the irrigated and swamp rice harvests in May and June.

Gender Differences

Landless poor men sharecrop swamp rice or dryland fields where rice yields are lowest. Lands are available from single women or from rich and middle-income male farmers who are short of labor power for timely field preparation. This option is closed to landless poor women. The poor who have land may sharecrop it with middle-income or rich farmers or city dwellers, because they have neither the capital nor the equipment to cultivate it themselves.

Landless professionals, wage workers, and government officials rent lands to sharecrop with other peasants. In a sort of reverse proletarianization, wage workers either rent or sharecrop land in order to have access to rice. Low-level male government workers form part of this group. Their salaries are often too low to enable them to feed all of their children. Many earn $40 per month, as chauffeurs, for example. This is not enough to feed a family of seven or eight. Workers use sharecropping as a strategy to supplement their incomes by sharecropping with landlords owning 4 or 5 ha who have fallen upon hard times and do not have enough money or equipment to work all their lands during a cultivation cycle.

Middle-class male landlords sharecrop to increase their land area, help provide for landless, married sons, and strengthen their claims on their sons' labor. Middle-class female-headed households sharecrop to obtain labor power, equipment, or capital. Peasants in financial difficulties—the result of medical expenses, funerals, or bad harvest years—offer their lands to sharecroppers. The competition for these lands is fierce because of high rates of regional immigration and landholding monopolies by the rich and powerful. It is common for indebted peasants to lose their lands to rich and politically influential sharecroppers.

The rich loan money to poorer farmers at usurious rates of 100% and take payment in paddy at harvest time. Urban professionals and government officials also participate in this practice, called *vary maitso* (green rice). Rich landlords look for lands to sharecrop with sons, but single women in this group generally look for sharecroppers to provide labor power. It is an accumulation strategy and a way to lay claim to sons' labor power for rich men; it is a means to mobilize labor power for rich women.

Individual strategies and forms of sharecropping differ according to class-based relations of production and according to gender differences, particularly among the rich and poor. Kin-based sharecropping relations are found in all classes. These relationships are not necessarily transitional, uniform, or just. The form depends upon the family's class position. Gender is a key factor in explaining sharecropping among brothers and sisters. Brothers may sharecrop together, but I did not find cases of sisters doing so.

Family and class relations determine various forms of sharecropping in contemporary Alaotra and imply differing power relations. There are important differences in the relations of production and power between, say, interfamily, small-scale commodity production, capitalist enterprises seeking more land, and destitute female-headed households. Gender is one of the most important independent variables, cross-cutting kin and class relations, found in the variety of sharecropping relationships. A singular focus upon the phenomenon of sharecropping can actually mask important differences in the relations of production that are constituted through kin, class, and gender at Alaotra.

A general typology of sharecropping relations at Alaotra must include considerations of class, land ownership, gender, kinship, and individual motivations and strategies common to each group (see Table 1).

Gender makes a difference in every kind of sharecropping relationship except among the landless middle class, for whom the same strategy drives both men and women workers, professionals, and small-scale merchants. Rich female landlords sharecrop their lands, but they do not look for additional lands to sharecrop with sons to extend land area or as a means of paying for tractors, as males do.[6] Poor and middle-class women cannot look for additional lands to sharecrop and are barred from this option for resource access that is open to men in these groups. Women's primary concerns are the mobilization of male labor power for field preparation and access to equipment and capital. Sharecropping can involve kin relations for members of all classes who have land. I found no examples of kin-based sharecropping among the landless, although this type is certainly possible.

Table 1
A General Typology of Sharecroppers at Alaotra

Class	Landed/ Landless	Gender Difference	Kin-Based	Individual Strategies
Rich	Landed	Yes	Yes	Men: To loan money and receive paddy at harvest at 100% interest; fathers look for additional lands to sharecrop with sons; to extend land surface; to pay for tractors Women: To mobilize male labor power
Middle	Landless	No	No	Men and Women: To obtain rice; to supplement wages
Middle	Landed	Yes	Yes	Men: Access to capital, equipment, labor, or more land Women: Access to labor, equipment, capital
Poor	Landless	Yes	Yes	Men: To obtain rice; to supplement seasonal wage Women: No cases recorded
Poor	Landed	Yes	Yes	Men: Access to equipment, capital, land Women: Access to labor

Conclusions

Not everyone benefits from sharecropping, and it does not necessarily confer upward mobility upon younger sharecroppers in cases of family-based sharecropping. My findings reveal that the wealthy and middle classes benefit both socially and economically from cropping on shares. Dentists and lawyers loan money to farming acquaintances for a share of the crop. Wealthy males generally take in land for sharecropping as an accumulation strategy and as a way to solidify local patron-client relations. Tractor owners crop in from poorer landowners experiencing financial difficulties. Sharecropping is most egalitarian between members of the same class or the same family within the rich and middle-income groups. The young and landless, poor, female-headed households, and descendants of former slaves are at the greatest disadvantage and subject to the most severe exploitation. Rich and middle-class divorced women and widows who have access to land and some equipment inevitably crop on shares in order to have access to male labor power or the capital necessary to hire labor for plowing, transplanting, and harvesting. These relationships may be advantageous to the women or highly exploitative, depending on the tenor of the social relations between the women and the sharecroppers. Outcomes are very different, for example, for single women who share out their small parcels to wealthy male patrons than they are for financially strapped middle-income men who share out to the same patrons, because women universally have less authority in all societies, and single women are particularly vulnerable. For this reason, women prefer cropping on shares with kin, but there is no guarantee that even these relationships will be equitable.

This case study demonstrates that power relations springing from kinship, class, and gender categories are pivotal in determining who gains and who loses in cropping rice on shares in Madagascar. Sharecropping as a resource access strategy undercuts the food security of middle-class and poor landowners who largely depend on their harvest to provide for their families. Only half of the harvest is available to meet their food and subsistence needs. This invites near-famine conditions for the poorest smallholders. The situation is bleakest, however, for the poor and landless. Men in this category are generally exploited by wealthy and powerful landlords, who often hold them in debt bondage as sharecroppers. Women in this category are even more vulnerable in that they must depend solely on seasonal wage labor to provide food for themselves and their families. Poor, landless, female-headed households live under the worst conditions, and their plight demands special policy consideration.

Notes

1. Lehmann (1986) reached a similar conclusion about sharecropping and economic development in Ecuador.
2. This case study is based upon dissertation research conducted in the Lake Alaotra region of Madagascar from 1988 to 1989. I surveyed the farmers in an ancient town of 277 inhabitants and in a town of 381 inhabitants that was established in the wake of the intensive irrigated rice project in the 1960s. I obtained a general outline of class structure for this area by surveying household access to cattle, plows, land, and capital and also learned about the various sharecropping contracts within each village. I then selectively interviewed farmers in seven other villages to gain a comparative understanding of the development of sharecropping among recent immigrants, indigenous inhabitants, and areas considered wealthy or impoverished by extension agents and researchers working in the area. I spoke with approximately 180 Alaotra residents about farming practices, land tenure, family organization and structure, land-use patterns, and farm work.
3. I have categorized male and female sharecroppers as poor, middle-class, and wealthy. These categories are based on access to agricultural resources, employment patterns, and educational levels. Poor households are landless or nearly so and have no cattle. Household members have little or no formal education and work as seasonal laborers, domestics, or sharecroppers. Middle-class households have access to combinations of irrigated, swamp, and dryland fields larger than 10 ha, cattle, and some equipment. Household members grow various crops for cash and subsistence. Children from these households have secondary or postsecondary educations; many do not return to the farm. Landless middle-class households have one or two members working for wages in jobs requiring education beyond the primary level or who are small-business owners such as shopkeepers or traders. Wealthy households' landholdings exceed 50 ha. They possess large numbers of cattle, some in herds of 500 head, tractors, and automobiles. Members can be merchants, cattle traders, or government officials. Their children attend universities.
4. The Sihanaka people make up the majority of the region's population and approximately 2% of the national population (Covell, 1987). Sihanaka sons and daughters inherit land bilaterally.
5. Fortmann (1984) made a similar point regarding class similarities between men and women in her study of rural households utilizing plow agriculture in Botswana.
6. One rich woman used her tractor to haul wood in order to make extra money to make the payments on the tractor.

References

Byres, T. J. 1983. Historical perspectives on sharecropping. *Journal of Peasant Studies* 10:7–40.
Byres, T. J., and B. Crow, with M. W. Ho. 1983. *The Green Revolution in India*. Case Study 5. Milton Keynes: Open University Press.

Charte de la Révolution Socialiste Malagasy. 1975. Atananarivo, Madagascar: Imprimerie d'Ouvrages Educatifs.

Covell, M. 1987. *Madagascar: Politics, economics and society.* London & New York: Frances Pinter.

Fortmann, L. 1984. Economic status and women's participation in agriculture: A Botswana case study. *Rural Sociology* 49(3):452-464.

Heseltine, N. 1971. *Madagascar.* London: Pall Mall Press.

Jarosz, L. 1990. Rice on shares: Agrarian change and the development of sharecropping in Alaotra, Madagascar. Ph.D. diss., Department of Geography, University of California, Berkeley.

Lehmann, D. 1986. Sharecropping and the capitalist transition in agriculture: Some evidence from the highlands of Ecuador. *Journal of Development Economics* 23:333-354.

LaPierre, W. 1964. Les transformations de la société rurale dans la région du Lac Alaotra. *Civilisation Malgache* 1:203-224.

Okigbo, B. N. 1988, September-October. Finding solutions to the food crisis. *Africa Report,* pp. 15-18.

Phillips, L. C., J. C. Riddell, J. L. Stanning, and T. K. Park. 1986. *Land tenure issues in river basin development in sub-Saharan Africa.* LTC Research Paper 90. Madison: Land Tenure Center, University of Wisconsin-Madison.

Robertson, A. F. 1987. *The dynamics of productive relationships: African share contracts in comparative perspective.* Cambridge: Cambridge University Press.

SCET (Société Central pour l'Equipement du Territoire). 1963. *Etude des périmètres de l'Anony et Sahamalaoto.* Antananarivo, Madagascar: SCET.

Sharma, M. 1985. Caste, class and gender: Production and reproduction in North India. *Journal of Peasant Studies* 12(4):57-88.

Swindell, K. 1985. *Farm labor.* Cambridge, London, and New York: Cambridge University Press.

Epilogue

This case study demonstrates how gender and class differences shape sharecropping strategies and outcomes in Madagascar. The influence of feminist theory on postmodernism development studies has resulted in challenges to universal, essential categories: "women" and "Third World women" (Mohanty, 1991; Ong, 1994). Local, specific, and historically informed analyses anchored in specific cultural and geographic contexts have been emphasized (Marchand and Parpart, 1995; Thomas-Slayter and Rocheleau, 1995). An awareness of the simultaneity and multiplicity of roles, identities, and practices that make up social relations and individual experience has enriched feminist analysis and critical theory in the study of gender and development (Kabeer, 1994). This case study is situated within this interdisciplinary literature, which encompasses development studies and feminist theory. It reveals how gender differences are cross-cut and shaped by the multiple social categories that comprise individual identity. It also reveals that neither women nor men are inexorably winners or losers in sharecropping strategies, but that outcomes depend on social positioning and context. This perspective indicates the need for tightly focused, regionally specific development policies that focus on impoverished families headed by women.

This case study emphasizes gendered relations as power relations and demonstrates how social relations shape, enable, or constrain resource access and control. Research detailing struggles and conflicts over resource control and access as they concern gender relations is a necessary and needed extension of this study, as is the role of gendered forms of knowledge about agricultural practices. There has been little research on gender relations and resource access in the Malagasy context, and this, as recent research (Escobar, 1995) indicates, is central to understanding local forms of organization, participation, and empowerment and the ways in which gender shapes availability and access to natural resources.

Additional References

Escobar, A. 1995. Encountering development: The making and unmaking of the Third World. In *Princeton Studies in Culture/Power/History*, eds. S. B. Ortner, N.B. Dirks, and G. Eley. Princeton: Princeton University Press.

Kabeer, N. 1994. *Reversed Realities: Gender Hierarchies in Development Thought*. London: Verso.

Marchand, M. H., and J. L. Parpart, eds. 1995. *Feminism/Postmodernism/Development, International Studies of Women and Place*. London: Routledge.

Mohanty, C. T. 1991. Under Western eyes: Feminist scholarship and colonial discourses. In *Third World Women and the Politics of Feminism*, eds. C. T. Mohanty, A. Russo, and L. Torres, pp. xx–xx. Bloomington: Indiana University Press.

Ong, A. 1994. Colonialism and modernity: Feminist re-presentations of women in non-Western societies. In *Theorizing Feminism: Parallel Trends in the Humanities and Social Sciences*, eds. A. C. Hermann and A. J. Steward, pp. 372–381. Boulder, CO: Westview Press.

Thomas-Slayter, B., and D. Rocheleau. 1995. *Gender, Environment, and Development in Kenya: A Grassroots Perspective*. Boulder, CO: Lynne Rienner.

Chapter 9

Subsistence and the Single Woman Among the Amuesha of the Upper Amazon, Peru

JAN SALICK

Department of Botany
Ohio University
Athens, OH 45701

Abstract *Among the Amuesha (Yanesha) indigenous people, single women practice agriculture differently and opportunistically construct subsistence strategies differently than most tribal members. Single women cross gender-defined boundaries that traditionally determined participation in subsistence. Single women cultivate small lowland fields cut from young secondary regrowth. Crop density is high while diversity is lower because crops needing heavy labor are avoided. Crop rotation is delayed, resulting in fields that are older than normal. Home gardens are important. In a broader context, creative subsistence strategies are opportunistically developed using child labor, barter, and cash. Off-farm employment may be important. These trends may represent acculturation or some creative development within the larger trend of feminization of agriculture in Latin America.*

Keywords Agricultural ecology, feminization of agriculture, subsistence agriculture, women in agriculture, women in development.

Introduction

Women's roles in subsistence vary greatly among indigenous Amazonian peoples.[1] The Amuesha of Peru's Upper Amazon region maintain fairly traditional gender roles: men hunt, fish, raise cattle, and do the heavy agricultural labor and cultivate corn crops; women's participation is generally planting, weeding, and harvesting the subsistence crops (Barclay, 1985). Single Amuesha women, however, cross the gender-defined boundaries that traditionally have determined participation in subsistence.

Single women became known to me because they do agriculture differently. They consistently differ from the coherent ecological patterns of Amuesha agricultural practices (Salick, 1989; Salick and Lundberg, 1990). Initially, I asked the questions, "Who are these people, and why are they different?" My interest led me to wonder, "How *do* they make ends meet?" and, in a larger context, "What roles do single women play in subsistence, in development, and in social and economic evolution?" These questions, and my initial data on single women's agriculture, are the basis for this paper. I will concentrate on Amuesha women, with or without children, who are largely economically independent.

Female heads of households are a group mostly absent from the literature on indigenous peoples. This is not surprising, since there has been limited analysis of single women in Latin America as a whole. Buvinic, Youssef, and Von Elm (1978) explore "Women headed households: The ignored factor in development planning." Bourque and Warren (1981) discuss a female underclass in the Peruvian Andes, made up of impoverished widows, single mothers who have no access to land, or illegitimate female children with no land rights. Ashby (1985) suggests that abandonment of women may be the major cause for the "feminization of farming" in Latin America.

One problem in analyzing single women is that it is an unwieldy social category. For example, in Ashby's categorization of "types of rural women in Latin America," she differentiates women by social class and type of farm enterprise. Single women cross all social classes and significantly modify the types of farm enterprises she describes. Additionally, women are alone for many reasons: widowhood, divorce, abandonment, "grass widowhood" (left at home by itinerant husbands), or merely not marrying. Single women may have a diversity of dependency relationships—children (dependents), extended family (dependees), or they can be relatively independent. There are also diverse cultural contexts in which single women are found. In the case of the Amuesha, these range from relatively traditional subsistence agriculture to modern (albeit usually impoverished) urban living. The unique circumstances of, and profound consequences resulting from, single women in agriculture make it a meaningful subject of investigation in Latin America.

Background

The Amuesha are an indigenous group of about 5000, inhabiting the tropical rainforests of east-central Peru (Figure 1). Linguistically, the Amuesha are classified as pre-Andean, and are distinct from their nearest neighbors, the Campa (Wise, 1976). Preliminary archaeological excavations in the Pichis (Allen, 1968) and Palcazu (Jimenez, 1987) valleys date extensive Amuesha habitation in these areas to perhaps 4000 B.P. At the time of European entry into Peru, the Amuesha also inhabited the valley regions of Pozuzo, Oxapampa, and Chanchamayo. Over the last 50 years, however, they have mostly retreated to the Palcazu Valley (Smith, 1977), where my study is located. The basis of the Amuesha economy is subsistence agriculture, supplemented by fishing, gathering, and (to a lesser extent) hunting (Barclay, 1985; Smith, 1977). Women participate in all these activities to varying degrees. Nonetheless, the role of single Amuesha women is unique.

Methods

To uncover and analyze the varied roles of single women in Amuesha society, I followed a sequential line of enquiry. I first separated a small data set of single women from my studies of Amuesha agriculture (Salick, 1989; Salick and Lundberg, 1990) and reviewed the reasons why these single women added variance to my statistical and quantitative analyses. I then interviewed a larger set of single women about their general strategies of subsistence and their perceived prospects.

Tactic 1

The first tactic came from Salick (1989) and uses identical methods. I made a broadscale ecological survey (e.g., Mueller-Dombois and Ellenberg, 1974) of Amuesha agricultural fields in the lower watershed of the Palcazu Valley (Figure 1). Thirty-one families were interviewed, their home gardens censused, and their 65 slash-and-burn fields

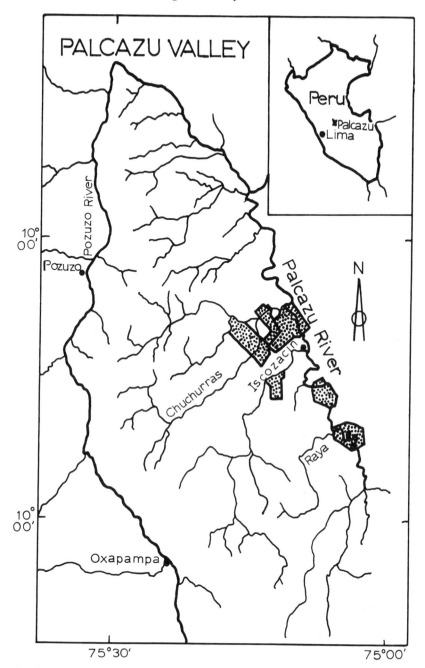

Figure 1. The Palcazu Valley, located in east-central Peru at the headwaters of the Amazon. Amuesha native communities in which agricultural fields were measured are indicated by stippled areas on the map.

sampled. Food crops (including flavorings, colorings, and stimulants) and plants used in indigenous technologies (e.g., weaving, fishing, and poisoning) were analyzed; medicinal or magic plants were not. The informal interviews of the 31 families covered agricultural practices, field and homestead histories, household membership, extended family membership, ages, birthplaces, subsequent movement, labor, commercial enterprises, and community roles (Salick and Lundberg, 1990).

Amuesha swiddens were sampled and described using ecological vegetation sampling methods and analyses (Mueller-Dombois and Ellenberg, 1974). Within each field, I laid stratified random quadrats of 2 m × 5 m, covering 5–10% of the total field area. All crops and edible volunteer plants in each quadrat were inventoried. Measurement of height, estimated percentage cover (including plants rooted outside the quadrat), distance to another plant of the same species, and the nearest neighbor species of each plant were also recorded. From these data I calculated density, α, height, and β diversity,[2] and cover. Plant communities are described using polar ordination, with presence weighted by density; this is a commonly used vegetation analysis method for ordering plant communities based on similarity of composition.

This quantitative database allowed me to identify samples one standard deviation from the mean in the statistical analyses, plus those samples generally termed "outliers" in vegetation ordination. Single women, although limited in this first study to five widows, consistently were found at the extremes of the normal distributions. The subsequent interviews confirmed and elaborated on these initial quantitative leads.

Tactic 2

The second tactic involved interviewing 25 single Amuesha women. Those interviewed included subsistence farmers, an herbalist, an *ayahuasquera*,[3] a craftswoman, a research assistant, a cook, local domestics, Lima domestics, and a secretary. The interviews were informal, of ½–6 h duration, and extensive. They included subjects covered in the interviews of farmers in my first study: agricultural practices, field and homestead histories, household membership, extended family membership ages, birthplaces, subsequent movement, labor, commercial enterprises, and community roles. Except in a few cases, interviews were conducted with the assistance of one or two Amuesha single women who understood my motives and interests; they were invaluable in elucidating replies and in maintaining a natural, informal dialogue. Interviews were conducted in various situations, from the informants' agricultural fields while working to my kitchen table over coffee. The interviews brought personal insights and perspectives to the study of single women and subsistence.

Results

Uniqueness of Single Women's Subsistence Agriculture

There are several statistical and quantitative variables that show the agriculture of single Amuesha women to be aberrant from that of the rest of the tribe. Statistics for crops and field architecture in all Amuesha agricultural systems (Table 1; Salick, 1989) and in Amuesha single women's systems are compared (Figure 2 and Table 2) to demonstrate where the statistics for single Amuesha women fall outside the standard deviations. Single women cultivate few extensive upland fields, which are typically extensive rice/

Table 1
Comparative Statistics for Crops and Field Architecture in Amuesha Agricultural Systems across the Population in General (Means ± Standard Deviations from Salick, 1989)

		Lowland			Upland	
Parameter	Beach: Bean, Nonsuccessional	Maize, Early	Cassava, Transition, Middle	Plantain, Late	Rice, Early	Cassava, Late
Number of samples (n)	6	15	9	21	5	9
Field size × 100 m^2	6 ± 4	15 ± 14	11 ± 5	13 ± 8	38 ± 29	42 ± 23
Planting distance (m)	0.56 ± 0.23	1.13 ± 0.46	1.12 ± 0.37	4.39 ± 0.65	0.56 ± 0.19	1.17 ± 0.47
Crop diversity (H')	1.88 ± 1.31	2.26 ± 0.93	3.19 ± 0.36	2.83 ± 0.55	0.55 ± 0.95	1.70 ± 1.22
Height diversity (H')	1.13 ± 0.95	1.52 ± 0.93	2.21 ± 0.48	2.39 ± 0.47	0.85 ± 0.76	1.26 ± 0.84
Percentage cover	92 ± 40	61 ± 25	42 ± 19	52 ± 23	60 ± 34	31 ± 23
Density (stems/m^2)	3.37 ± 1.54	0.95 ± 0.46	0.63 ± 0.18	0.49 ± 0.23	3.04 ± 1.71	0.55 ± 0.23

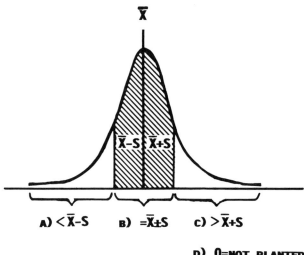

Figure 2. Categories assigned to single women's plots based on the relative values of their statistical variables (see Tables 1 and 2). Single Amuesha women's plots are recorded as either (a) $< x - s$, less than the mean minus a statistical standard deviation, (b) $= x \pm s$, within a standard deviation of the mean, (c) $> x + s$, more than the mean plus the standard deviation, or (d) 0, when the field type was not planted.

cassava plantings after the felling of tall forest. Their lowland fields are smaller than average and are cut from young secondary regrowth. Fields of single women show lower crop diversity within and among fields (α and β diversity) compared with other Amuesha fields. Crop density and crop cover tend to be higher in single women's fields.

Succession. The ecological ordination of Amuesha fields by vegetational species composition (Figure 3; Salick, 1989) separates upland rice and cassava fields and defines a cropping sequence for alluvial lowland fields. Several fields of single women differ ("X"s on Figure 2), less because of their positioning than the age of the field at the given stages. Single women tend to crop fields longer, repeating successional stages, so that the fields of single women are older than the crop successional stages and ages on the figure, which are for fields of the Amuesha as a whole. Some stages, especially the final plantain stage, may be skipped altogether. Overall, succession is much slower and field abandonment less frequent. A bean patch cleared from a portion of an old cassava field represents a case of retrogressive succession. The low number of samples or the absence of several field types among single women (Table 2, $n = 0$ or 1) substantiates these differences.

The retarded succession and delayed field abandonment are due to the ability of single women to provide abundant light labor. Single women are very capable of weeding or cultivating their fields, and this may allow them to defer succession 1–3 years. Once a field is opened, it is continuously cropped much longer than normal: from 3 to 7 years or more, compared to 2–4 years for most Amuesha lowland agriculture (Salick, 1989). This avoids opening new fields, but results in more time spent weeding increasingly infested fields. Usually a lowland field goes through a continuous series of stages: from maize to an intercropped, cassava transitional stage, to plantains, and then to a

Table 2
Agricultural Plots of Single Amuesha Women[a]

Parameter	Beach: Bean, No Succession	Alluvial Lowland, Delayed Succession			Upland, Delayed Succession	
		Maize, Early and Repeated	Cassava, Middle and Late	Plantain, Seldom Planted	Rice, Seldom Planted	Cassava, Early and Late
Number of samples (n)	1	3	5	1	0	1
Field size × 100 m²	$< x - s$	$= x \pm s$	$< x - s$	$< x - s$	0	$< x - s$
Planting distance (m)	$< x - s$	$= x \pm s$	$< x - s$	$< x - s$	0	$< x - s$
Crop diversity (H')	$> x + s$	$= x \pm s$	$< x - s$	$< x - s$	0	$= x \pm s$
Height diversity (H')	$> x + s$	$= x \pm s$	$< x - s$	$< x - s$	0	$= x \pm s$
Percentage cover	$> x + s$	$> x + s$	$> x + s$	$> x + s$	0	$> x + s$
Density (stems/m²)	$> x + s$	$> x + s$	$> x + s$	$> x + s$	0	$> x + s$

[a] Agricultural plots of single Amuesha women often fell outside the standard deviations indicated in Table 1. See also Figure 2. Single Amuesha women's plots were: $< x - s$, less than the standard deviation; $> x + s$, greater than the standard deviation; $= x \pm s$, within the standard deviation; or 0, when fields of that type were not planted by single Amuesha women. The fact that some field types are underrepresented ($n = 0$ or 1) is significant in that single women are not practicing those types of agriculture.

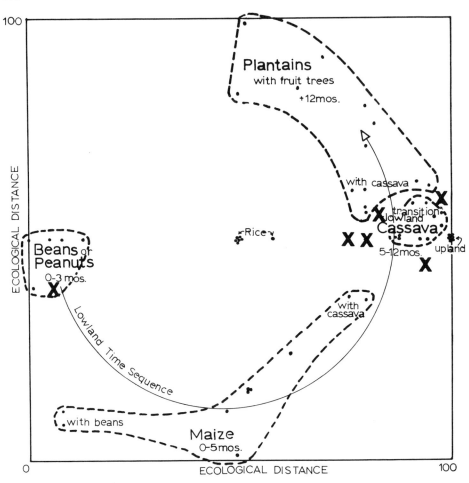

Figure 3. Single women's fields are aberrant compared with the ordination of Amuesha fields by similarity of species occurrence from Salick (1989). Standard Amuesha agriculture displayed six field clusters: (1) bean or peanut fields, (2) maize fields potentially intercropped with beans or cassava, (3) lowland, transitional cassava fields, (4) plantain fields, often intercropped with cassava or fruit trees, (5) rice fields, and (6) upland cassava fields. Single women's agricultural fields not following the general trends are marked with heavy Xs; the placements of the fields are less aberrant than their ages. Fields of single women often showed retarded succession and no plantains, as indicated by the five Xs on the right, where fields are much older than the usual 5–12 months noted on the figure. The X on the left represents a bean field cleared from a cassava field already cropped for several years, demonstrating retrogressive succession. Lack of fields is also evidenced: there are no single women's fields in the upper plantain quadrat without cassava; nor are there among the centered rice fields, because such fields are not a part of single women's agriculture.

managed, in-planted fallow. The single woman wishing to prolong each stage and to avoid plantains will keep the cassava field well intercropped for as many years as it produces and for as long as weed invasion can be kept back. On good alluvial soils, this may be an extended period of up to 7 years or longer. I do not know the effects of such continuous cropping on the soils or on subsequent regeneration, but the impact on single women's time expenditure can be severe.

Planting Strategies. The informal interviews in this first analysis helped explain the special attributes of single women's subsistence agriculture. Without men to cut and burn primary forest and old secondary forest, women tend to take on the time-consuming (but less arduous) tasks of clearing brush and continually cropping and weeding older fields. Recuperation of soils and elimination of weeds, insects, pests, and pathogens (the presumed reasons for fallowing fields) are thought to be quicker on better alluvial soils, allowing farmers to productively clear bush fallows rather than tall forest. As a result, single women most often concentrate their efforts on small lowland fields. If male labor is temporarily available to them through labor barter, the visit of a relative, or the return of an itinerant husband, the first major task requested is aid in clearing a new field. Some exceptional single women clear their own fields, felling primary forest trees of 3 m girth or more.

Cropping in a lowland field may start with a very dense planting of maize to feed animals, especially important because chickens and eggs are sources of income for the single woman. However, the women are seldom able to sustain production, because maize requires new fields every year. To overcome this problem, single women sometimes take extreme measures: one woman planted maize inside one of my experimental plots, and another planted on the airstrip. These two women had to rely on sympathy—both mine and the community's—to successfully harvest their maize.

A single woman tends to plant very densely, using all available microsites for varied appropriate crops. In spite of this, crop diversity within fields (α diversity) may be low due to failure to plant some standard crops that need heavy labor. For example, banana, plantain, and palm suckers require that large planting holes be dug, and that large suckers be dug up and cut from parent plants and transported to a field. Rice is an upland, extensive crop as cultivated by the Amuesha, needing much mature forest cleared, and so is seldom planted by single women. Cassava is their staple. Other traditionally male crops, such as barbasco, coca, cacao, and coffee, are also often absent from the fields of single women. Animal production by single Amuesha women is not as common as it is in households with male labor. Single women may raise chickens or ducks; less frequently, they tend pigs or sheep.

Home Gardens. The other major form of agriculture for single women is the home garden. In general, the Amuesha have fairly simple home garden cultivation, concentrated on fruit production and dietary diversity (Salick, 1989). When other forms of agriculture are restricted, the home garden can take on special importance and can contain much greater diversity (Salick and Lundberg, 1990). Such is the case for single women.

Single women tend not to change their homesteads. Among the Amuesha, household movement is culturally dictated by death or illness in the family, by controversy with neighbors, or by a need for access to new fields (Smith, 1977). However, women seldom undertake building a new house alone. Many widows break with tradition and stay in the same house that they shared with their husbands. Also, within the community single women avoid controversies and display subservient behavior, appealing to neighbors' sympathies to resolve conflict; thus, they avoid moving their house for this reason, as well. One of the agricultural results of this stationary behavior is a diverse and densely planted home garden, with great variety of fruits, herbs, flowers, medicines, minor crops, and—unique to single women—major crops. Single women depend on home gardens for much more than the average family, both supplementing and supplying a significant proportion of their family's diet with this relatively intensive method of cultivation. Despite this above-average dependence, total field area cultivated per household is lower for single women (these and α diversity data from Salick and Lundberg, 1990).

Subsistence with Opportunism, Child Labor, Barter, and Cash

The Amuesha single women described subsistence strategies that range far beyond agriculture. They stressed subsistence traits that are found generally among the Amuesha (Salick and Lundberg, 1990), but that are especially important to these single women: highly developed skills of opportunism, heavy reliance on their children, and food acquisition through barter and purchase.

Single women are particularly attuned to new opportunities, in whatever form they may come. Single women experiment with new crops, new cultivation techniques, and new varieties. They are willing participants in development projects, if the projects are sincerely designed to incorporate women. Any male labor is always appreciated: the visit of a cousin, the illegal logging of a piece of forest, the cleaning of the airstrip, and the clearing of a plot for my own experiments were all opportunistically taken advantage of to start new agricultural fields. Novel goods and services are often supplied by single Amuesha women, to satisfy many needs: a development project's need for a cook, foreign consultants' appetites for indigenous handicrafts, the Amuesha community's new attraction to hair permanents, and an externally sponsored cultural revival of ayahuasca healing groups.

Among the Amuesha (and most subsistence farmers), child labor and education in traditional farming skills go hand-in-hand. Single women, who face continual labor and time shortages, may develop a heavy dependence on child labor for help in cultivating fields, or for bartered or wage labor. In extreme cases, schooling for children may be neglected by single women because of a lack of resources and a need for labor, or an indentured status for a child may be negotiated.

Even in the native communities, few single women are able to subsist on agricultural production alone. Bartered labor and goods are the most traditional method of obtaining goods or services within the native communities. Barter is seldom formal: a bucket of cassava beer, for example, is not considered worth a fixed amount of labor or corn in return, but rather adds to an accumulated debt of the receiver or is subtracted from a debt accrued by the donor. In this way, women use various products to repay debts, including cassava beer, prepared food, eggs, chickens, baskets and other handicrafts, or medical attention. Women usually grow and collect a number of common medicinal plants that they use to cure their families. Such home remedies are occasionally bartered, and one single Amuesha woman is known to be so accomplished that she practices her herbalist art as a profession. This is a traditional avenue for single women. Labor can also be bartered, so that single women and their children may cultivate other people's fields in return for food, land clearing, and so forth.

There are less traditional ways of getting by in the modern world, such as selling goods and labor, for those Amuesha who rely heavily on a cash economy. Single women in particular, because of agricultural and time constraints, purchase foods such as rice, noodles, tunafish, oil, and sugar. In order to generate the income to buy these goods, they may sell the same products that they barter. They also may sell very different goods and services. One widow learned to give hair permanents, for which she was paid. Another single woman works in a popsicle factory outside the valley. Many single Amuesha women work as domestics in the Palcazu Valley, in nearby towns, or in Lima, where at least 30 Amuesha women are now working, mostly as domestics. These women rely almost entirely on a cash economy, although they are very disadvantaged as wage laborers. One single woman was taken as a teenager to the United States to provide childcare. Now that she has returned, she works as a trilingual (Amuesha-Spanish-English) secretary for a donor agency—a very unusual success story.

Discussion

As a basis for discussing the role of single women, it is worth briefly looking at the broader context of women's role in general. The woman's role in subsistence among the Amuesha is well covered by Barclay (1985), and I would differ from her analysis on only a few points. I have found less division and more shared labor between men and women in planting, cultivating, and harvesting crops than she reports. For example, she stated that the cultivation of cassava is clearly the woman's domain. I have found many men planting, weeding, harvesting, and managing germplasm in cassava fields. I suspect that Barclay has a more traditional view of the Amuesha than I, since she has worked with traditional Amuesha communities for many years, antedating recent changes. Also, her interests focus more on the traditional cultural context, whereas mine are focused on present-day agricultural practices. One female informant told me that in the past women and men had separate tasks, but that now there is more work to be done, and all must pitch in and share. In general, I have found more shared labor among the Amuesha than in most other tribes described in the literature. As a general trend for Latin America, Ashby (1985) points out there is great heterogeneity in the work roles performed by women in agriculture: "The sex-specificity of tasks appears to be diverse and flexible or responsive to changing labor market relations." It is little contested that Amuesha society is undergoing such change (Barclay and Santos, 1980; Salick and Lundberg, 1990; Smith, 1983). Still, most domestic work falls within the Amuesha female province; hunting, cattle grazing, forestry, cutting and burning, and heavy cultivation are within the male province.

Notably, the roles assumed by single Amuesha women fall outside even such broadly defined women's roles. Single women take on all agricultural tasks, participating in clearing and burning new fields and in planting and cultivating all crops, regardless of traditional gender associations. However, single women modify agricultural practices to avoid heavy tasks. Similar trends in women's agriculture have been noted elsewhere. In the Caribbean, women farmers neglect heavy male tasks so that land goes out of cultivation, terraces and irrigation systems deteriorate, and production falls back to levels of subsistence manageable by women as the principal source of family labor (Chayney, 1983). In the sierra of Peru, Bourque and Warren (1981) also observed that women do not participate in heavy tasks such as breaking ground, plowing, and opening irrigation channels.

Time seems to be a universal constraint among single women. Agriculturally, time constraints have led Amuesha single women to plant fewer, small fields that are more intensively cropped. Within the family, time constraints precipitate a dependence on child labor. Time constraints (along with entrance into cash economies) result in dependence on purchased food, which may be nutritionally deficient (I shared several meals of noodles with catsup). This trend has been found in other Latin American societies (Carloni, 1984; Stavrakis and Marshall, 1978). Ashby (1985) points out that time constraints will also affect women's acceptance of technical change.

Female Amuesha wage laborers might be in an analogous situation to those studied in the Peruvian Andes by Bourque and Warren (1981). Like their highland counterparts, single Amuesha women who are removed from their cultural context and work within the national economy represent a female underclass. To avoid this disadvantaged position, most single Amuesha women try to maintain their cultural context. Young single women without children work in urban environments for a period in their maidenhood, in order to learn about the larger world. If they maintain their family connections and return to their culture to marry and raise children, they do not permanently join the underclass. The economic marginalization of Amuesha society has less direct effects on individual women if they maintain their cultural identity.

The urban migration of young Amuesha women to work as domestics is apparently part of a larger trend. "Rural women in Latin America are more likely to find urban employment, albeit at very low wages, than men; in certain groups they have higher rates of rural–urban migration than men" (Ashby, 1985, citing Singh, 1980, and Youssef et al., 1979). The effect of this migration on young Amuesha society is marked, though the results are not yet clear.

Trends toward acculturation among the Amuesha are especially noticeable among single women. More than merely lacking male labor, single women may also lack both a community role and a cultural context. Traditionally, single women were probably either remarried or absorbed into the tightly knit family–community support systems of Amuesha clans (Smith, 1977). Women's work was abundant, and extra hands were probably more valuable than the cost of the extra mouth to feed, since the society produced the staples of life in abundance (Barclay, 1985). I know of traditional Amuesha households that incorporate up to three single women with children into an integrated work and social frame. Now, however, few single Amuesha women can depend on the weakened support structures and family ties. The modern disruption of traditional Amuesha society is perceptively discussed by Barclay and Santos (1980) and Smith (1983). With dissolution of communal support structures and shared wealth, hallmarks of traditional Amazonian cultures, single women are strikingly affected. They could be permanently relegated to the female underclass, like their highland counterparts (Bourque and Warren, 1981).

Single Amuesha women farmers may be part of another large trend in Latin America: the feminization of farming. If a major factor in the increasing importance of women in farming is the abandonment of women by men (Ashby, 1985), this would put single women at the core of this general phenomenon. Although I have found farming and subsistence practices of single women to be different from the Amuesha norm, it is important not to totally remove single women from the larger context of women in agriculture. Single women might be viewed within a continuum, based on their degree of independence from men (this includes grass widows left at home by itinerant husbands). Among the Amuesha, peon labor takes men out of the domestic scene for extended periods. In the 1980s, a large development project was siphoning off a noticeable proportion of the male work force and was fairly unresponsive to Amuesha agricultural cycles. Buvinic et al. (1978) and Deere (1982) see the *de facto* female-headed farm, where men are seasonal migrants or primarily engaged in off-farm labor, as an extreme case of the tendency for women to be more heavily involved in agricultural production. I would say that single women in agriculture represent the extreme case, followed by female-headed farms, and in turn followed by the range of women in agriculture—all are taking active part in the feminization of farming in Latin America.

The magnitude of these trends for women—acculturation, economic marginalization, feminization of agriculture—is impossible to estimate from the available information (e.g., Ashby, 1985). There is a similar lack of information on demographic trends among single women, especially among the Amuesha. Censuses often include single females under some household classification, or they totally miss them. Among the Amuesha, single women are often withdrawn and subservient, making them very difficult to interview. Male census takers find no one home or find women unresponsive to their many insensitive questions. My original study (Salick, 1989) was field (rather than household) based, and thus also undersampled those single women who cultivated few fields.

Sampling problems aside, the single women I interviewed felt that their numbers were increasing. They gave numerous causes, including lower mortality during childbirth, higher labor-related male mortality, increased male peregrination through modern transportation, and increased divorce and abandonment. It is undetermined whether these factors in fact have increased the number of single women.

Additionally, my informants repeatedly complained of weakened cultural traditions and support structures. There may have been a reciprocal support between women and the society as a whole, where women could depend on the society for support, and the society depended heavily on women. Barclay (1985) states, "Amuesha women have always been the conservative nuclei of culture and of equilibrium against violent change."[4] The loss of this conservative influence on Amuesha society is probably most noticeable among single women. Those in the older generation (30+ years) of single Amuesha women tend to be very conservative, retaining traditional dress and reclusive ways to a greater degree than their married sisters. In contrast, younger single women work in Lima and dress in modern fashion to the extent permitted by their economic status. They also tend to marry outside the Amuesha culture, further lessening their ties to traditional methods of subsistence.

Although these trends suggest acculturation, a modified view presents a more positive role for single women. Ashby (1985) points out that the feminization of farming implies an increase in women's decision making over production inputs, including choice of technology. Single women might represent a source of Amuesha cultural evolution. With their ready acceptance of new opportunities, single women might test the breadth and limits of environmental choice open to women in their subsistence culture; they are a great source of industry and creativity. This industry and creativity might be constructively tapped by both indigenous people and development experts alike.

Notes

1. The roles of women in Amazonian cassava cultures and the division of labor between men and women have been addressed in various ethnographic contexts by many authors: Basso (1973), Bergman (1969, 1980), Carneiro (1961), Clay (1984), Denevan (1971), Goldman (1963a, 1963b), Harner (1972), Hurault (1965), Johnson and Johnson (1975), Montgomery and Johnson (1977), Murphy (1960), Murphy and Murphy (1974), Nietschmann (1973), Siskind (1973), Smole (1976), Wagley (1977), Wagley and Galvao (1949), and Whitten (1976).

2. Shannon–Weiner diversity index: $H' = -\Sigma(P_i)(\log_2 P_i)$.

3. *Ayahuasquera*: a woman who imbibes the hallucinogen ayahuasca (*Banisteriopsis caapi*, Malpighiaceae) for the purposes of divining illness and curing patients.

4. "Las mujeres Amuesha siempre han sido nucleos conservadores de cultura y de equilibrio ante el cambio violento" (Barclay, 1985).

References

Allen, W. L. 1968. A ceramic sequence from the Alto Pachitea, Peru: Some implications for the development of tropical forest culture in South America. Ph.D. thesis, University of Illinois, Urbana, IL.

Ashby, J. A. 1985. Women and agricultural technology in Latin America and the Caribbean. In *Women, agriculture, and rural development in Latin America,* ed. J. A. Ashby and S. Gomez. Cali, Colombia: CIAT.

Barclay, F. 1985. Analisis de la division del trabajo y de la economia domestica entre los Amuesha de la Selva Central. Lima: PEPP/AID.

Barclay, F., and F. Santos. 1980. La conformacion de las comunidades nativas Amuesha en Amazonia Peruana. No. 5, Lima.

Basso, E. B. 1973. *The Kalapaso Indians of central Brazil.* New York: Holt, Rinehart and Winston.

Bergman, R. W. 1969. Shifting cultivation in the high rain forest: The Chirripo Indians, Costa Rica. M.S. Thesis, Department of Geography, University of Wisconsin, Madison.

Bergman, R. W. 1980. *Amazon economics: The simplicity of Shipibo Indian wealth.* Syracuse, NY: Syracuse University.

Bourque, S. C., and K. B. Warren. 1981. *Women of the Andes. Patriarchy and social change in two Peruvian towns.* Ann Arbor, MI: University of Michigan Press.

Buvinic, M., N. H. Youssef, and B. Von Elm. 1978. *Women-headed households: The ignored factor in development planning.* Washington, DC: International Center for Research on Women.

Carloni, A. 1984. *The impact of maternal employment and income on the nutritional status of children in rural areas of developing countries.* Rome, Italy: United Nations Subcommittee on Nutrition.

Carneiro, R. L. 1961. Slash-and-burn cultivation among the Kuikunu and its implications for cultural development in the Amazon Basin. In *The evolution of horticultural systems in native South America: Causes and consequences,* ed. J. Wilbert, *Antropologica,* supplemento no. 2, pp. 47–67. Caracas: Editorial Sucre.

Chayney, E. M. 1983. Scenarios of hunger in the Caribbean: Migration, decline of smallholder agriculture, and the feminization of farming. Women in International Development working paper no. 18, Michigan State University, East Lansing.

Clay, J. W., ed. 1984. *Women in a changing world.* Cambridge, MA: Cultural Survival.

Deere, C. D. 1982. The division of labor by sex in agriculture: A Peruvian case study. *Economic Development and Cultural Change* 30:795–812.

Denevan, W. M. 1971. Campa subsistence in the Gran Pajonal, Eastern Peru. *Geographical Review* 61:496–518.

Goldman, I. 1963a. *The Cubeo Indians of the Northwest Amazon.* Urbana: University of Illinois Press.

Goldman, I. 1963b. Tribes of the Vaupes-Caqueta region. In *Handbook of South American Indians,* Vol. 3, *The tropical forest tribes,* ed. J. H. Steward. New York: Cooper Square Publishers.

Harner, M. J. 1972. *The Jivaro: People of the sacred waterfalls.* Garden City, NY: Natural History Press.

Hurault, J. 1965. *La vie materielle des noirs refugies boni et des indiens Wayana du Haut-Maroni (Guyana Française): Agriculture, economie et habitat.* Paris: Orstrom.

Jimenez, J. 1987. An archeological reconnaissance of the Rio Palcazu. Willay No. 25, Peabody Museum, Cambridge.

Johnson, O. R., and A. Johnson. 1975. Male/female relations and the organization of work in a Machiguenga community. *American Ethnologist* 2(4):634–648.

Montgomery, E., and A. Johnson. 1977. Machiguenga energy expenditure. *Ecology of Food and Nutrition* 6:97–105.

Mueller-Dombois, D., and Ellenberg, H. 1974. *Aims and methods of vegetation ecology.* New York: Wiley and Sons.

Murphy, R. F. 1960. *Headhunters heritage: Social and economic change among the Mundurucu Indians.* Berkeley: University of California Press.

Murphy, Y., and R. F. Murphy. 1974. *Women of the forest.* New York: Columbia University Press.

Nietschmann, B. 1973. *Between land and water: The subsistence ecology of the Miskito Indians, Eastern Nicaragua.* New York: Seminar Press.

Salick, J. 1989. Ecological basis of Amuesha agricultural systems, Peruvian Upper Amazon. *Advances in Economic Botany* 7:189–212.

Salick, J., and M. Lundberg. 1990. Variation and change in Amuesha agriculture, Peruvian Upper Amazon. *Advances in Economic Botany* 8:199–223.

Singh, A. M. 1980. The impact of migration on women and the family: Research, policy, and programme issues in developing countries. *Social Action* 30:181–200.

Siskind, J. 1973. *To hunt in the morning.* New York: Oxford University Press.

Smith, R. C. 1977. Deliverance from chaos for a song: Preliminary discussion of Amuesha music. Ph.D. thesis, Cornell University, Ithaca, NY.

Smith, R. 1983. *Las comunidades nativas y el mito del Gran Vacio Amazonico,* Documento 1. Lima: Asociación Interétnica de Desarrollo de la Selva Peruana (AIDESEP).

Smole, W. J. 1976. *The Yanoama Indians: A cultural geography.* Austin: University of Texas Press.

Stavrakis, O., and M. L. Marshall. 1978. Women, agriculture, and development in the Maya lowlands: Profit or progress? *International conference on women and food.* Arizona: University of Arizona and Consortium for International Development.

Wagley, C. 1977. *Welcome to tears. The Tapirape Indians of central Brazil.* New York: Oxford University Press.

Wagley, C., and E. Galvao. 1949. *The Tenetehara Indians of Brazil, a culture in transition.* New York: Columbia University Press.

Whitten, N. E. 1976. *Sacha Runa. Ethnicity and adoption of Ecuadorian Jungle Quichua.* Champaign: University of Illinois Press.

Wise, M. R. 1976. Apuntes sobre la influencia Inca entre los Amuesha, factor que oscurece la clasificación de su idioma. *Revista de Museo Nacional (Lima)* 42:355–366.

Youssef, N., M. Buvinic, J. Sabstand, and B. Von Elm. 1979. *Women in migration: A third world focus.* Washington, DC: International Center for Research on Women.

PART THREE

WOMEN'S KNOWLEDGE, WORK, AND STRATEGIES FOR SUSTAINABILITY

Photograph courtesy of Carolyn Sachs.

Chapter 10

Women and Livestock, Fodder, and Uncultivated Land in Pakistan

CAROL CARPENTER

Environment and Policy Institute
East-West Center
1777 East-West Rd.
Honolulu, HI 96848
USA

Abstract *This article presents a model of the relationships in rural Pakistan between women and livestock, fodder, and uncultivated land, and their importance to farming households in rainfed and irrigated farming areas. In rainfed areas, fodder gathered by women from public, uncultivated land is a key natural resource, making possible the manure production that is essential to agriculture. This resource is reduced in irrigated areas, where agriculture has encroached upon uncultivated land. Women in landed households in irrigated areas can harvest fodder from their own fields, but women in land-poor households must contend with fodder and fuel shortages that compel them to burn dung. This key natural resource, the fodder gathered by women from wastelands, has not received the attention it deserves. As a result, it and the land whose fertility it helps replenish are vulnerable to development efforts that overlook its existence.*

Keywords: women, gender division of labor, livestock husbandry, fodder, common-property resources, manure, agricultural sustainability, agricultural intensification, Pakistan.

Introduction

This paper presents a model of the relationships that exist in rural Pakistan between women and livestock, and thus between women and an important and often overlooked source of fodder, uncultivated land. These relationships are described in the first section. According to the gender division of labor on farms in Pakistan, women are largely responsible for the livestock portion of the household economy, as men are for agriculture. Women's and men's parts of the household economy are not separate, however; in fact, agriculture and livestock husbandry are linked so closely that any limitations on women's ability to find fodder for livestock affects agricultural productivity and sustainability and thus the viability of the farm household as a whole.

The second part of this article contrasts the ways in which the relationships between women and livestock, fodder, and uncultivated land have developed in rainfed and in irrigated farming areas. In rainfed areas, public uncultivated land (which may be village commons or state land) is a key source of fodder. Fodder, transformed by livestock into manure, is essential to agriculture.

In irrigated areas, where agriculture has progressively encroached upon uncultivated public land, most fodder must of necessity come from land that is privately owned and cultivated. In this way the intensification and expansion of agriculture create severe fodder shortages for landless and marginally landed households. As a result, women in these households must labor much harder to provide fodder than women in rainfed areas. More important, shrinking common-property resources also result in fuel shortages for these households. This means that the dung produced by the livestock these women feed and some of the dung produced by their landlords' animals ends up being burned in their cookstoves.

The contrast between rainfed and agricultural areas is important because it reveals the impact of agricultural intensification on an overlooked natural resource: the fodder gathered by women from uncultivated land. Perhaps because uncultivated fodder is exploited primarily by women, in a culture that strongly values women's seclusion, it has not received the research and development attention it merits. More important, this natural resource is particularly vulnerable to development programs that overlook its existence, and this may have serious consequences for cultivated as well as uncultivated land.

The ideas for this article were generated by a review of ethnographic literature on women's patterns of livestock production in the rainfed and irrigated areas of Pakistan, much of which is unpublished and not widely available. Ethnographic literature on North India, Nepal, and Bangladesh was also reviewed.[1] Analysis was limited to literature on agricultural as opposed to pastoral groups, although this distinction is often difficult to make in Pakistan, as is discussed subsequently. The model presented in this article is supported by 3½ years of experience in Pakistan. During this time I observed land-use patterns, farming systems, and fodder-gathering activities and interviewed rural women and men about livestock, fodder shortages, and uncultivated land.

The purpose of presenting the model in this article is to provide a framework that will stimulate thinking and research and more appropriate development planning concerning the links forged by women between uncultivated land and the rural household economy.

Women and Livestock Husbandry

In the farming households of Pakistan it is typical for women to be primarily responsible for livestock husbandry and men for agriculture, with some exceptions. The cultural association between men and plowing is so strong that the general care of draft animals often falls to men. Similarly, the association between women and fodder is so strong that women typically harvest fodder crops (World Bank, 1989). Women are most closely linked to milk-producing livestock, and men to production of cash and staple food crops.

This division of labor is in marked contrast to that of pastoral societies, in which men are culturally linked to and primarily responsible for livestock. As Hewitt (1989) argued for northern Pakistan, in the gender division of labor in pastoral societies, men specialize in livestock husbandry and women in agriculture.[2] But there is a great deal of variation in the gender division of labor between pastoral groups, even in Pakistan, and pastoral women cannot generally be said to specialize in agriculture.

Societies that are culturally pastoral—that is, that view livestock as a man's affair—may exhibit a pastoral division of labor even if agriculture is economically predominant. For example, among the previously pastoral groups in the irrigated Punjab (the *janglees*, the original inhabitants of the area before the introduction of irrigation), men collect dung and milk livestock, tasks that are characterized as female for all other farming households in this area (Khan and Bilquees, 1976). Women in this group are not allowed to touch a cow's teats, the pastoral version of the farmer's stricture against women touching a plow (Khan and Bilquees, 1976).

In culturally agricultural societies in Pakistan, then, raising livestock is what women do, in the same way that raising crops is what men do. This is well documented in a study of villages by Freedman and Wai (1988) in the Northwest Frontier and Punjab provinces. They concluded that "livestock production is largely a woman's job," based on findings indicating that more than 20% of the average woman's day[3] is spent on livestock operations, especially (in order of importance) preparing ghee (clarified butter), selling animal products to villagers, collecting manure, cleaning animal sheds, milking animals, gathering fodder, cleaning animals, caring for animals giving birth, and watering animals (Freedman and Wai, 1988, pp. 27, 31). They also noted that women play a major role in operations that bring farming products to animals and animal products to crops, including collecting manure and applying it to fields; collecting green fodder from maize, rapeseed, and pulses; and cutting weeds with a sickle to feed livestock.

Freedman and Wai's conclusion confirmed previous studies. Naveed-i-Rahat's (1981) research in a village in the Punjab hills shows that women cut grass for fodder, take cattle out to graze, make clarified butter and sell it, and breed buffalo, cows, and goats for sale. According to Khan and Bilquees's (1976) study of a village in the irrigated Punjab, a typical wife of a tenant farmer spends 25% of her workday (3.75 hours) collecting, carrying, and preparing fodder; 12% (1.75 hours) tending animals (including watering and feeding them and moving them between stall and courtyard); 7% (1 hour) milking and churning; and 5% (0.75 hour) collecting dung and making dung cakes. They concluded that "tending the animals is mainly a woman's job" (p. 257). Qadri and Jahan's (n.d.) survey of the irrigated Sind showed that women spend 29% of their workday performing activities related to livestock, including cutting and bringing fodder (14%), collecting dung (4%), churning milk (4%), feeding animals (3%), and milking them (3%). According to Ishrat (1981), in the rainfed Hazara hills of the Northwest Frontier Province women spend 28% of their time (6.75 hours) in animal care, including 3 hours collecting fodder, 1.5 hours milking and churning, 1 hour making dung cakes, 45 minutes preparing fodder, and 0.5 hour cleaning cattle sheds.[4]

Tending milk animals means much more than duties completed or hours invested to women in Pakistan, more than activities completed or hours spent. Women profess to like raising milk animals, taking the same pride in them that men take in ripening fields.[5] In Pakistan the birth of a calf and the death of a cow or buffalo are celebrated, in rituals that involve networks of fellow villagers (Naveed-i-Rahat, 1981). Owning a milk buffalo is a major source of *izzat* prestige, and being able to serve to guests milk and ghee that has not been borrowed or purchased is an important sign of this prestige (Merrey, 1983). Similar views are found in India. According to U. Sharma (1980), livestock-related work in India is classified as housework and female, not fieldwork and male, like agriculture. Women prefer housework to fieldwork, and they particularly like to raise livestock, feeling that it is both proper and emotionally interesting and rewarding, in contrast to agriculture, which is slightly improper and not, for women, valued. For a household to have its women working in the fields is a sign that its means are limited (even though very few households can afford to hire the labor that would keep their women entirely out of the fields); for a household to have milk animals, on the other hand, is a sign that it is prosperous.

Milk and milk products are the most visible and valued contributions that women make to household subsistence in rural Pakistan,[6] and are the primary traditional source of income for them. Women typically sell milk or ghee to other women in their village; 94% of the women in Freedman and Wai's (1988) study sold milk to their fellow villagers.[7] In some areas dung cakes are also sold. Women generally control the profit they earn from selling animal products (Freedman and Wai, 1988; Merrey, 1983). Men may or may not be aware of the income women earn selling milk or milk products (Naveed-i-Rahat, 1981; Stanbury, 1984). In any case, women tend to use the income they earn from the sale of animal products for the household (Naveed-i-Rahat, 1981). In some areas milk products (usually the buttermilk left over from processing clarified butter) are regularly given to laborers working for the household, as an informal part of the payment for agricultural labor.

Women as individuals rarely own livestock (with the probable exception of poultry) in Pakistan.[8] Large livestock, like land, are often said to be owned by men (Naveed-i-Rahat, 1981). Women, however, clearly join in decisions about what animals to buy and sell and for how much and use their family connections in other villages to locate potential buyers and sellers. Men may control the cash earned from the sale of large livestock, but decisions about how it will be used are usually made by men and women together (Naveed-i-Rahat, 1981; Supple et al., 1985). Women in some areas of Pakistan breed livestock for sale. In the Murree hills, women set the price and locate suitable customers through their kin ties to other villages. Because markets are male domains in Pakistan, the final transaction is made by a man, but men and women together decide how to spend the money (Naveed-i-Rahat, 1981). Cultural values aside, it would probably be accurate to say that large livestock are owned by households.

Women and Fodder

Women are primarily responsible for providing the main input for livestock production, fodder, and it is often the most time-consuming livestock task they perform. In Pakistan women may spend 14–25% of their working day, or close to 4 hours, collecting and processing fodder (Ishrat, 1981; Khan and Bilquees, 1976; Qadri and Jahan, n.d.).

Fodder is collected from both cultivated and uncultivated land, and women are the primary collectors. In rainfed areas, collecting fodder may involve cutting grasses or lopping leaves off trees that grow on farms and in the uncultivated areas near and between villages. In irrigated areas, harvesting fodder crops is typical. Fodder crops grown in Pakistan include mustard and rapeseed (usually intercropped with wheat), barley, sorghum, millet, maize (which has largely replaced sorghum and millet), sugarcane, and mung pulses (Supple et al., 1985; World Bank, 1989). Other sources of fodder from cultivated land, in both rainfed and irrigated areas, include crop residues and weeds. Residues typically used for fodder in Pakistan include wheat straw (*bhusa*), sugarcane leaves, and corn stalks. Weeding is actually fodder collection, and, given the importance of fodder in South Asian agriculture, it should be viewed as such. Allowing weeds to mature before removal is one strategy by which farmers in rainfed Punjab increase their supply of fodder (Supple et al., 1985). The fodder component of food crops can be enhanced in various ways, depending on need: by thinning crops deliberately sown thickly (especially corn), clipping (especially wheat during fodder shortages), or stripping (especially sugarcane, also during fodder shortages).

Livestock are fed various fodders. Supple et al. (1985) give a fodder schedule for rainfed areas, including wheat straw (used all year), dry sorghum or millet (used 6 months), green sorghum or millet (used 4 months), mustard (used 4 months), maize thinnings (used 2 months), and weeds (used 8 months) from agricultural land, plus green grass (used all year) and dry grass (used 6 months) from uncultivated lands.

The amount of labor involved in providing fodder for livestock, and the gender and age to which this labor is typically assigned, varies depending on whether animals are stall-fed or grazed. Stall-feeding requires more intensive labor than grazing, and although grazing is usually done by boys in agricultural groups or by men in pastoral groups, women provide most of the labor involved in stall-feeding. It should be noted that women often supervise the boys who tend grazing animals, and certain groups of women, especially girls, older women, and poor women, graze animals themselves (Naveed-i-Rahat, 1981). Stall-feeding tends to replace grazing as agriculture intensifies (shortening fallows and allowing fodder crops to be cultivated), common-property grazing areas are privatized or degraded, or patterns of male labor migration are established (Jodha, 1985; Lefebvre, 1990). The first two factors are important in irrigated areas, where it is typical for women in landed households to stall-feed their livestock. But stall-feeding is common in rainfed areas also, because of the third factor: in rainfed Pakistan, men have been migrating out of the area to work in the army and police since before World War I (Lefebvre, 1990). This became a tradition in these areas; in a rainfed village studied by Lefebvre 89% of the households have a man irregularly employed outside the village (and 43% of these men have been or are enrolled in the army or the police).

Stall-feeding, and the labor that farm households and especially women contribute to it, integrates livestock closely with agriculture. As Robinson (1987, p. 114) wrote,

> The area of noncultivated land required to sustain the productivity of cropland through conversion of fodder to manure will be considerably smaller where livestock are stall-fed than in areas where livestock spend most of the day or perhaps several months of the year grazing and browsing on noncultivated land.

Stall-feeding, that is, produces more usable manure from less uncultivated land.

Women and Uncultivated Land

Uncultivated, fodder-producing land is often perceived as women's territory; women may spend hours in these areas each day, and women's expertise encompasses these lands and their products. Women are primarily responsible for fodder collection from uncultivated areas in spite of the fact that these areas are considered dangerous by their societies, especially to women. This dangerous territory includes forests, conceptually equivalent uncultivated lands between villages, and the "narrow strips of scrubby, rough grazing land" that have replaced the forest in the Punjab (U. Sharma, 1980, p. 43). U. Sharma wrote (of northwest India) that:

> Jungle land, i.e., waste land lying between the cultivated fields of one village and the next, is like the *bazar* in that it is also a category of space which women should avoid, but for somewhat different reasons. It is avoided not because it is public, but because it is lonely. . . . Many women have to go to the jungle in the course of their work, to cut grass, to graze cattle, to cut wood. But they do not go there unnecessarily (1980, p. 42).

Women should avoid the forest, but in many areas[9] it is precisely women who must go into the forest. In some areas the forest is explicitly a female domain; in the hills of North India "fields, forests, and the areas between these and the villages are 'female,' whilst commercial areas and roads are considered 'male' domains" (Mehta, 1990, p. 27). In Pakistan, pastures and forests may also be female domains (Hewitt, 1989; Naveed-i-Rahat, 1981), and they are clearly also uncultivated, lonely places—places outside human society. In Pakistan's culture, women are viewed as more wild than men, with more uncontrollable appetites; the separation of the sexes called *purdah* functions to contain their potentially dangerous but fertile power. This view is congruent with the fact that women make uncultivated land, through livestock, into the manure that makes cultivated land fertile.

Uncultivated land from which fodder is harvested tends to be circumscribed and separated from cultivated areas. In the hilly, rainfed districts of Pakistan, several patterns occur: grass belts of commons land may ring hill villages, separating them from the forest; small tracts of grassland may occur inside the forest, often in ravines or on old landslides and especially on hilltops and mountaintops; or the bottom slopes of a hill may be cultivated and the upper slopes may be managed for fodder grasses. Irrigated villages in Pakistan are often surrounded by denuded hills used for grazing and fodder production. On the flat plains the unused lands lying between the cultivated fields of neighboring villages are fodder-producing lands. In irrigated areas the fodder shortage of landless households can be so severe that even roadsides become an important source of fodder.

Women typically respond to the danger associated with uncultivated places by collecting fodder in groups. In some areas the forest is one of the places (along with the village water source) where women most frequently interact with each other (Mehta, 1990; Naveed-i-Rahat, 1981). These groups are organized differently in different areas, from the friendship groups of rainfed Punjab (Naveed-i-Rahat, 1981) to the women from one compound in irrigated Punjab (Khan and Bilquees, 1976). Such groups provide for the cooperative management of fodder-producing lands and probably for training in land management skills.

It is important to realize that insofar as uncultivated land can be said to be managed for fodder production, it is being managed primarily by women. That is, most of the daily decisions about which grasses or tree branches to cut, and how, are made by women.[10] How women manage uncultivated land for fodder production is a very promising research topic. Freedman and Wai (1988) have established the fact that women who cut fodder have learned this skill from their mothers or other female relatives. Women in Nepal protect village forests by challenging encroachers from other villages and taking their fodder and fuelwood (Pandey, 1990). The most well-known example is the Chipko movement in India, in which women are protecting their forests from commercial exploitation.

Interdependence of Livestock Husbandry and Agriculture

Women's responsibilities for livestock husbandry must be considered in light of the strong linkage between livestock and farming, a linkage with roots in the "integrated economy of agriculture and livestock raising" described by Sopher (1980, p. 192) for the ancient Indian subcontinent. In Pakistan farmers need animals to plough their fields, carry their produce, provide dung to cook their food, provide nutritious milk products, bank their agricultural profits against their children's marriages, and insure against crop failure. Perhaps most important of all, farmers need livestock to fertilize their fields. As Robinson (1987, p. 104) wrote, most crop production systems "continue to rely upon the import of nutrients" from uncultivated land.

This integration of livestock husbandry and agriculture means that women's and men's parts in the household economy are not separate; they are linked so closely together that any limitations on women's ability to find fodder for livestock affects agricultural productivity and sustainability, and thus the viability of the household as a whole.

Pakistani farmers are graphically clear about the importance of their livestock. As a Punjabi woman, pregnant and carrying a load of fodder on her back, said: "I have to feed the animals, for our livelihood depends on them. We have to take more care of our animals than our children" (Khan and Bilquees, 1976, p. 259). Women in the irrigated Punjab say their daughters are kept from school because they cannot be spared from fodder collecting; as one woman said, when girls start reading books they neglect the animals (Khan and Bilquees, 1976).[11] In the Indian Punjab, similarly, livestock are said to be as important as land and sons (Stanbury, 1984).

There is ample evidence that these folk perceptions are well founded. Livestock perform a number of essential functions in farming systems apart from food production, especially dung production (Robinson, 1987). Combining the results of a number of studies in Nepal, Robinson concluded that it takes 7 ha of forest land to provide the fodder necessary to feed livestock so that they will provide sufficient dung to fertilize 1 ha of agricultural land. This is a striking illustration of the linkage between livestock and agriculture and of the magnitude of the resource exploitation that lies behind the use of livestock manure on cultivated fields.

The interdependence of livestock and agriculture has also been documented for Pakistan. Freedman and Wai noted that "crop and animal production cannot be considered as separate systems; . . . the interface between the two systems, is fundamental to the viability of the entire farming system" (Freedman and Wai, 1988, p. 16). These authors point out that the need to provide fodder for animals "strongly influences cropping strategies," using as examples the thinning of maize for fodder and the intercropping of wheat with rapeseed. A 1985 study of the rainfed farming systems of the Punjab (Supple et al., 1985) made the additional point that livestock serve as security against the agricultural risk associated with erratic rainfall. As noted above, women are primarily responsible for maintaining the interface between livestock production and agriculture; it is women who harvest weeds and fodder crops for the animals and women who collect manure for the crops (Freedman and Wai, 1988).

The collection and application of manure from livestock (and the collection and processing of dung into cakes for fuel) is primarily women's responsibility, though one in which men are involved. A general report on women in Pakistan says that women are primarily responsible for collecting dung and share responsibility with men for applying it to fields (World Bank, 1989). The application of chemical fertilizer, in contrast, is exclusively a male responsibility. As chemical fertilizers gradually replace organic ones, then, women have less and less to do with fertilizer. However, some manure is probably used on most farms in Pakistan, because manure is valued for qualities that chemical fertilizers are believed to lack. As Kurin (1983) wrote of the Punjab, farmers believe that chemical fertilizers increase production but "intoxicate" and "burn out" the soil. To prevent this, farmers temper chemical fertilizers by restricting their use, applying extra water, or mixing them with manure.

Collecting dung is an almost exclusively female task in Pakistan; in rainfed Punjab and Northwest Frontier Province, 91% of women said they collect dung (Freedman and Wai, 1988). In the irrigated Punjab, even high-caste women often collect dung themselves. In one study, for example, high-caste Rajput women showed a 52% participation rate for dung collection (Saeed, 1966).

To sum up the model described previously, women are primarily responsible for a flow of nutrients from uncultivated land—through fodder, livestock, and manure—to cultivated land, a flow that sustains the productivity of that land.

Women and Livestock in Rainfed and Irrigated Farming Areas

The relationships between women and livestock, women and fodder, and women and uncultivated land described above have developed differently in rainfed and irrigated farming systems.

Rainfed Farming Areas

The rainfed farming areas of Pakistan are in western Punjab province and Northwest Frontier province. In the farming systems in these areas the availability of uncultivated land, as Jodha (1986, p. 1172) argued for rainfed India, allows farming households to "devote all their land to food or cash crops rather than sparing part of it for fodder/fuel supplies." Farming households get all the fodder they can out of cultivated land, short of growing exclusively fodder crops. Separate fodder crops—sole cropping of mustard, oats, or barley—are grown only in high-rainfall, highly productive areas (Lefebvre, 1990; Supple et al., 1985).

Fodder production "is a major determinant of cropping systems in the area" (Supple et al., 1985, p. 34). A number of strategies maximize fodder production: food and cash crops are selected that can also be used as fodder crops; long-stemmed varieties of wheat are selected to produce wheat straw; wheat varieties that will tolerate clipping are selected; weeds are allowed to grow; weeds are cut rather than uprooted; maize is planted densely for thinning. With all this, if drought causes a shortage of fodder, farmers will cut their own food and cash crops for their animals (Merrey, 1983; Supple et al., 1985). In addition, agricultural land is often fallowed in order to provide grazing for livestock and simultaneously to provide manure for the next crop (Supple et al., 1985), because manure is clearly the most important product of animal husbandry in rainfed areas. The demand for fodder is high because agricultural production in these areas is almost entirely dependent on livestock manure as a source of nutrients.

In spite of all these strategies for maximizing the fodder output of cultivated land, and in spite of the availability of uncultivated land, farming households in rainfed areas usually do not have enough manure to fertilize all their fields every year. Most farms have both *lepara* fields, which regularly receive farm-yard manure and are thus more fertile and retain moisture longer; and *maira* fields, which are usually far from the villages and not manured by weeding and collecting fodder from uncultivated land (Freedman and Wai, 1988; Supple et al., 1985).[12] Fallowed *maira* fields are important sources of fodder grasses.

The high demand for fodder in rainfed areas and its importance for agriculture can be seen in another fact—men in some rainfed parts of Pakistan may contribute nearly 50% of the labor to two livestock-related tasks that are almost exclusively female in irrigated farming areas: collecting fodder from cultivated land by weeding and collecting fodder from uncultivated land (Freedman and Wai, 1988). Even with this help from men, women in this study put in an annual average of almost 2 hours each day collecting fodder from uncultivated areas, in addition to labor devoted to collecting fodder from cultivated fields, weeding, and performing various other agricultural tasks (Freedman and Wai, 1988).[13] In other rainfed areas, however, women get very little help with livestock-related work from men and spend as much as 3 hours daily just collecting fodder from uncultivated areas (Ishrat, 1981).

Irrigated Farming Areas

The main irrigated agricultural areas in Pakistan are the canal-irrigated districts of Punjab and Sind provinces. Most fodder in the farming systems in these areas comes from privately owned, cultivated land, and much of it comes from sole cropping of fodder crops. In a village in the Punjab studied by Merrey (1983, p. 163), "only a very small amount of fodder had been grown before canal irrigation, but since then its acreage has increased from about 12% of the winter acreage . . . to over 30%." As Merrey wrote, irrigation makes the cultivation of fodder crops both possible, because it greatly increases the productivity of cultivated land, and necessary, because it decreases the amount of uncultivated land.

Irrigation thus draws a line between households with sufficient land to grow fodder crops and those without. It allows those with enough land to intensify both agricultural and livestock production, by stall-feeding their animals with fodder crops.[14] But as irrigation makes previously marginal land productive, it increases the percentage of land under cultivation.[15] This reduces the amount of uncultivated land available for fodder production, creating a fodder shortage for those with insufficient land to grow fodder: marginal farmers, tenants, and landless laborers.

The present composition of the land-poor population in Pakistan's canal-irrigated districts reflects the history of the area. The oldest group of land-poor residents are the original inhabitants of the area (Khan and Bilquees, 1976). When the Punjab canals were constructed at the turn of the century, farmers, often from the old irrigated districts of what is now India, were settled to farm the newly irrigated land (Alavi, 1976). They were accompanied or followed by traditional service castes, including the tenants of landlords who were resettled (Khan and Bilquees, 1976), who are now the second group of land-poor residents. As their traditional caste specializations eroded, some of these groups came to rely on agricultural labor. This traditional caste structure was shaken up at the partition of India and Pakistan in 1947, when the Hindu castes evacuated their farms in Pakistan and Moslem refugees settled in their places (Alavi, 1976). In 1954 evacuee land was reallotted to refugees. Proof of land ownership in India was required to keep land, and, as Alavi wrote, one consequence of this was "a large increase in the proportion of the agricultural population that had to look to sharecropping or laboring as a principal means of livelihood" (1976, p. 323). Finally, in the old irrigated districts, land fragmentation through inheritance has reduced many farmers to marginal status (Alavi, 1976), and many of them both rent and own land. Rainfed areas in Pakistan also have some landless service-caste households, but to a significantly smaller extent. In one study in Pakistan, for example, the service-caste population of an irrigated village was 45%, whereas that of a rainfed village was only 15% (Lefebvre, 1990).

The subsistence of these land-poor groups often depends on a combination of agricultural labor and livestock production. Their livestock production depends on an insecure and dwindling fodder supply, what they can get out of large landowners, and shrinking common-property resources. Fodder provision under these difficult circumstances usually falls to women. Because women in Pakistan are underemployed as agricultural laborers, they specialize in livestock husbandry. The chronic fodder shortage affecting the land-poor does not necessarily make them give up their livestock. In Pakistan, land-poor groups may own more livestock than large landholders, and they may own cattle and buffalo as well as goats and sheep (Dove, 1990). Even landless households, as Merrey noted, "try to keep a cow or buffalo for milk, and beg or buy fodder for it" (1983, p. 385).

Women in households with large landholdings intensify their animal husbandry as men intensify agriculture. Burton and White (1984) suggested that in intensive agricultural systems that depend on livestock, an increasing percentage of livestock care (and a decreasing percentage of agricultural work) will be done by women. This seems to be true in some areas of Pakistan. In one survey in irrigated Sind, women from farming households with more than 7.5 acres of land spent 45 woman-days more per annum on livestock-related activities than women from households with 7.5 acres or less of land (Qadri and Jahan, n.d.).

By making it possible to grow fodder crops, irrigation makes *purdah* possible for those households with enough land to grow fodder crops.[16] There are two different kinds of women in irrigated societies, and they have different strategies for collecting fodder. The activities of women who can afford to be in *purdah* are limited to the house compound (where livestock are typically stalled), the area adjacent to it (where women often grow vegetables), and the family's own fields (where fodder is grown); they cannot go to neighbors' fields or uncultivated land. These women collect fodder from their own fields by weeding crops and harvesting fodder crops; they cannot collect it from uncultivated land (Khan and Bilquees, 1976; Stanbury, 1984).

Women who cannot afford to be in *purdah* in irrigated areas typically do some agricultural wage labor in other people's fields when they can find it, and try also to bargain or beg for the right to collect some fodder from these fields. They supplement this supply by collecting fodder from uncultivated areas, especially community-owned commons.

The interdependence of agricultural and animal production is complete on irrigated farms that are sufficiently large. Agricultural fields produce sufficient fodder for livestock, in addition to food crops, to produce sufficient manure to keep the fields producing food and fodder. In fact, the fields of large farms typically are producing a surplus of fodder. That is, they are producing some fodder for the animals of their landless laborers and tenants also, although not enough for their needs. Women in households with landholdings large enough to produce food and fodder are collecting fodder for milk, but more important, for dung to keep their land producing food and fodder.

This productive interdependence breaks down, however, for land-poor villagers. In Pakistan, when tenants bargain for a share of the food crop, they also bargain for the right to cut grass and weeds for fodder in their landlords' cultivated fields, and to purchase their landlords' fodder crops and crop residues at concessional rates (Khan and Bilquees, 1976). Laborers' wives in some areas are permitted to cut grass and weeds from landlords' fields for their own animals (Khan and Bilquees, 1976). These strategies do not, however, provide sufficient fodder, and women must also collect fodder from uncultivated areas (Khan and Bilquees, 1976). Resource-poor households depend on common-property resources much more than large farmers. Between 99 and 100% of the poor households in Jodha's (1986) survey collected fuel and fodder from commons, whereas only 11 to 28% of the nonpoor households did.

Shrinking common-property resources also mean shortages of fuelwood for the land-poor, and fuelwood shortages mean that livestock dung is burned in cookfires rather than being used to fertilize agricultural fields. Land-poor households burn the dung their animals produce and perhaps much of the dung from their landlords' livestock as well. In one village in the irrigated Punjab, laborers were supposed to turn over dung to their landlords to use for manure, but in practice their wives made dung cakes for fuel for their own use (Khan and Bilquees, 1976). This practice is probably widespread today. It decreases the amount of dung available for use as manure on the land of large landowners as well as small ones. Women in land-poor households are collecting fodder to produce milk and fuel, not to keep fields producing food.

Conclusion and Recommendations

This article has presented a model of the connection between women and livestock, and thus between women and an essential and often overlooked source of fodder, uncultivated land. Fodder is an important resource because it allows for the production of manure, which is necessary to sustain the productivity of cultivated land. Uncultivated land is a key source of fodder in rainfed areas, because cultivated land is not productive enough for both food and forage crops; it is a key source in irrigated areas because land-poor households do not have sufficient land to grow forage crops. And it is primarily women in both rainfed and irrigated areas who harvest fodder from uncultivated land.

The contrast that has been drawn in this article between rainfed and irrigated farming systems reveals the impact of agricultural intensification on this overlooked natural resource. The intensification of agriculture by irrigation is usually followed by the expansion of cultivation into uncultivated land, creating fodder shortages for land-poor households. This shrinking of "wasteland" resources also leads to fuelwood shortages for the land-poor, and these two shortages together decrease the availability of livestock manure, which is essential to the sustainability of agriculture on cultivated land. As Robinson wrote, "the expansion of cropland is at the expense of resources (forest, grazing land) upon which its own production is dependent" (1987, pp. 107–108).

Irrigation is not the only force responsible for the expansion of cultivated land at the cost of uncultivated, fodder-producing land. The amount of uncultivated land available for fodder production is decreasing even in many rainfed areas. Jodha (1985) argued that in rainfed India, population growth and increased commercialization have decreased the amount and quality of fodder and other common-property resources. Writing about the history of the Indian subcontinent, Sopher held overpopulation responsible for imperiling the traditional farming system of this part of the world. He argued that

> [a]s population has increased, more and more marginal land has been taken under cultivation and the amount of grazing land available per capita has declined together with a decline in the supply of firewood, thus drastically reducing the amount of dung available for use as fertilizer because of the priority given to using it for fuel. The working of the traditional agricultural system is therefore imperiled (1980, p. 193).

Mehta (1990) blamed the adoption of cash crops in the hills of North India, which resemble the Hazara and Murree hills of Pakistan, for the expansion of agriculture into forests and pastures, decreasing women's access to fodder and increasing their workloads, decreasing the livestock population, and jeopardizing the fertility of agricultural fields. These writers suggested that overpopulation and commercialization must be added to agricultural intensification as threats to the vital supply of fodder from uncultivated land.

In Pakistan, reforestation projects also threaten the supply of fodder from uncultivated land. These projects often target precisely the type of land that produces fodder, whether it is the privately owned patches of uncultivated land targeted by farm forestry projects, the degraded commons targeted by community forestry projects, or the degraded state forest preserves targeted by reforestation projects.

In rainfed areas, especially in the hill districts of Pakistan, fodder is often collected from land that is now national forest reserve land (but was, in some areas as recently as 1936, private or village land [Rauf, 1981]). These districts are characterized by a complex tradition of villagers' rights to fodder in protected and reserved state forests, which are legally recognized but often disputed in the courts. The villagers' forest use in these areas emphasizes fodder rather than tree production. Reforestation policies that exclude fodder from the forest—in Pakistan, monocropping *chir* pine (Ishrat, 1981)—have, not surprisingly, been less than successful. They have also created animosity between foresters and rural people. Respondents in a 1981 study in the Hazara hills said that the fact that the forest was no longer providing sufficient amounts of fodder was proof that the Forest Department was mismanaging it (Rauf, 1981). Village women in the Murree hills call foresters thieves. By reforesting the land with pines, foresters have in effect stolen a resource that is vital to these women and their households.

Development programs that make marginal land productive for agriculture by introducing irrigation or new cash crops, or reforestation projects that target fodder-producing land, are likely to increase demands on women's labor, decrease supplies of fodder and fuelwood, decrease supplies of manure, and in the long run endanger the productivity and sustainability of agricultural land and threaten the viability of rural households, especially land-poor households.

Once the agricultural and economic value of fodder gathered by women from uncultivated land is recognized, programs can be designed that will not only safeguard but enhance it. First of all, more research is needed on the linkages raised in this article, particularly on the role of livestock in agriculture and the role of women in animal husbandry. Second, research and development must be directed toward women. Such research could, for example, explore technologies to help women more efficiently process and store fodder and manure. Third, reforestation efforts should target small niches for fodder trees on private as well as public land. Fodder trees could be planted along roadsides and irrigation ditches, for example, or could be distributed to women from land-poor households for courtyard plantings. Fourth, to slow the decline in this and other natural resources, programs to slow the rate of population growth will ultimately be needed.

Finally, the grass, shrub, and tree fodders growing in uncultivated public land, including common-property resources and areas owned or protected by the state, must be protected and developed. This may include extension programs, which should be directed toward women. It may also include some means for preventing privatization and guaranteeing access, especially for women and land-poor households. Maintaining, if not increasing, the fodder productivity of uncultivated lands and protecting the access of women and land-poor households to them is essential to protecting (1) the economic position of rural women, (2) the viability of land-poor households, and (3) the productivity of agricultural land.

Acknowledgments

This article was written at the Environment and Policy Institute of the East-West Center, under the auspices of the Land, Air, and Water Management, Habitat and Society, and Social Forestry Writing Workshop programs. I acknowledge the support of Michael R. Dove, Lawrence Hamilton, A. Terry Rambo, and Jefferson Fox.

Notes

1. The farming systems of Pakistan and North India are very similar, especially the hill districts and the canal-irrigated Punjab in both countries. There are, however, important differences. The dairy industry is not as developed in Pakistan as in India, and, at the time I left Pakistan in 1989, no large-scale development projects existed in Pakistan. Furthermore, in much of India, castes take over the agricultural-pastoral relationship that is more commonly expressed in the gender division of labor in Pakistan. The farming systems of Nepal resemble the rainfed hill districts of Pakistan, but Nepal (like Pakistan's northern areas, not discussed here) has a strong mountain pastoral tradition. In Bangladesh, the farming systems resemble the irrigated parts of Pakistan, except that they seem to be more marginal; many families own neither land nor livestock.

2. It is interesting that even in this pastoral group women tend milk cows, which are as closely associated with them as yaks are with men (Hewitt, 1989).

3. It must be noted that most time allocation data available on rural women is from surveys of a mixture of pastoral, rainfed agricultural, and irrigated agricultural groups, and thus could not be used without ignoring the distinctions being made in this article.

4. Women spend a great deal of time on livestock-related activities in other parts of South Asia. In North India, according to M. Sharma (1989), women spend 3.5 to 4.5 hours per day in milk production–related work, with fodder collection alone taking 2 hours per day for two women. According to Kumar and Hotchkiss's (1988) study in Nepal, women spend an average of 1.3 hours just collecting fodder. In Bangladesh, according to Feldman, Banu, and McCarthy, "women take primary responsibility for cleaning and feeding small livestock, and poor women clean the stalls which house larger livestock" (1987, p. 14).

5. A friend of the author's, who grew up in a village in the Northwest Frontier Province but left to earn an American degree and now works in a city in Pakistan, keeps a milk buffalo in her yard.

6. Stanbury said of North India that "of all food items, milk and milk products are the most highly valued by the villagers and make up an important part of the diet, both nutritionally and for social and ceremonial purposes" (1984, p. 14).

7. In India it is more common to sell milk to agents who come to women's homes (M. Sharma, 1989).

8. In Bangladesh, in contrast, according to one study, women own 18% of the livestock (not including poultry), whereas men own only 7%; the remaining 75% are family-owned (Feldman et al., 1987).

9. In areas of Pakistan where *purdah* is conservatively defined, especially in Northwest Frontier Province, men seem to predominate in the collection of fodder from uncultivated land. In Freedman and Wai's (1988) study, 92% of the women from the Punjab villages said they collect fodder for livestock, whereas only 18% of the women from the Northwest Frontier Province villages said they did. It should be noted that this is survey data, not based on long-term observations of women's activities. The values of *purdah* may influence women's responses to questions from an outsider more than their behavior.

10. Moench called this *use-defined* management, defined as "management that occurs due to patterns of regular use" (1987, p. 25).

11. Note the implication that fodder shortages may have an adverse effect on the education of girls.

12. This distinction breaks down in the driest areas, where farmers apply manure to a different portion of their land each year in a cycle (Supple et al., 1985).

13. In Nepal, men contribute labor to fodder provision when the demands of agriculture allow, but women remain the main fodder collectors, even when their labor is also needed for agriculture. This means that women work more hours each day than men in busy agricultural seasons. During the July-to-September season, when labor needs are critical to both agriculture and grass fodder collection, the time men spend collecting fodder drops from an annual average of 0.45 to 0.1 hours daily as they invest more time (more than 4 hours) in their fields. Women compensate by increasing the time they spend each day collecting fodder, from an annual average of 1.33 hours to 2.4 hours, even though they are also putting 3.4 hours into agricultural fieldwork (Kumar and Hotchkiss, 1988).

14. See, for example, Stanbury (1984) on North India, where the animals are stall-fed in upper-caste homes.

15. When something happens to make communally owned marginal land become more productive, that is, worth cultivating, this generally sets off the process of privatization; furthermore, even when privatization is promoted to provide land for the landless, they typically lose access to it (Jodha, 1986).

16. Large landholding households in irrigated areas often own more land than they can work, even if their womenfolk were not in *purdah;* this fact, together with *purdah,* creates the market for agricultural labor by which small landholders and the landless survive.

References

Alavi, H. 1976. The rural elite and agricultural development in Pakistan. In *Rural development in Bangladesh and Pakistan,* ed. R. D. Stevens, H. Alavi, and P. J. Bertocci, pp. 317–353. Honolulu: University Press of Hawaii (for the East-West Center).

Burton, M. L., and D. R. White. 1984. Sexual division of labor in agriculture. *American Anthropologist* 86(3):568–583.

Dove, M. R. 1990. Approaches to the coevolutionary study of population and environment: Perception of and response to resource scarcity. East-West Center Population Institute Seminar Paper, Honolulu, HI.

Feldman, S., F. Banu, and F. E. McCarthy. 1987. The role of rural Bangladeshi women in livestock production. Working Paper 149. Office of Women in International Development, Michigan State University.

Freedman, J., and L. Wai. 1988. Gender and development in Barani areas of Pakistan. Unpublished report prepared for Agriculture Canada.

Hewitt, F. 1989. Woman's work, woman's place: The gendered life-world of a high mountain community in northern Pakistan. *Mountain Research and Development* 9(4):335–352.

Ishrat, F. 1981. Economic role of women in subsistence pattern of forest zone in Hazara. M.S. thesis, Quaid-i-Azam University, Islamabad, Pakistan.

Jodha, N. S. 1986. Common property resources and rural poor in dry region of India. *Economic and Political Weekly* 21(27):1169–1181.

Jodha, N. S. 1985. Population growth and the decline of common property resources in Rajasthan, India. *Population and Development Review* 11(2):247–264.

Khan, S. A., and F. Bilquees, 1976. The environment, attitudes and activities of rural women: A case study of a village in Punjab. *Pakistan Development Review* 15(3):237–271.

Kumar, S. H., and D. Hotchkiss. 1988. Consequences of deforestation for women's time allocation, agricultural production, and nutrition in hill areas of Nepal. Research Report No. 69. Washington, DC: International Food Policy Research Institute.

Kurin, R. 1983. Indigenous agronomics and agricultural development in the Indus Basin. *Human Organization* 42(4):283–294.

Lefebvre, A. 1990. International labour migration from two Pakistani villages with different forms of agriculture. *Pakistan Development Review* 29(1):59–90.

Mehta, M. 1990. *Cash crops and the changing context of women's work and status: A case study from Tehri Garhwal, India.* MPE Series No. 2. Kathmandu, Nepal: International Centre for Integrated Mountain Development.

Merrey, D. J. 1983. *Irrigation, poverty and social change in a village of Pakistani Punjab: An historical and cultural ecological analysis.* Ph.D. diss., University of Pennsylvania, Philadelphia, PA.

Moench, M. 1987. Forest degradation and biomass utilization in a Himalayan foothills village. Working Paper. Honolulu, HI: Environment and Policy Institute, East-West Center.

Naveed-i-Rahat. 1981. The role of women in reciprocal relationships in a Punjab village. In *The endless day,* ed. T. S. Epstein and R. A. Watts, pp. 47–81. Oxford: Pergamon.

Pandey, S. 1990. *Women in Hattisunde forest management in Dhading district, Nepal.* MPE Series No. 9. Kathmandu, Nepal: International Centre for Integrated Mountain Development.

Qadri, S. M. A., and A. Jahan. n.d. *Women in agriculture: Sind.* Islamabad: Women's Division, Government of Pakistan.

Rauf, M. A. 1981. *Forestry development in Pakistan: A study of human perspectives.* Peshawar, Pakistan: Pakistan Forest Institute.

Robinson, P. J. 1987. The dependence of crop production on trees and forest land. In *Amelioration of soils by trees,* ed. R. T. Prinsley and M. J. Swift, pp. 104–120. London: Commonwealth Science Council.

Saeed, K. 1966. *Rural women's participation in farm operations.* Lyallpur: West Pakistan Agricultural University.

Sharma, M. 1989, April 29. Women's work is never done: Dairy "development" and health in the lives of rural women in Rajasthan. *Economic and Political Weekly,* pp. 38–44.

Sharma, U. 1980. *Women, work, and property in north-west India.* London: Tavistock.

Sopher, D. E. 1980. Indian civilization and the tropical savanna environment. In *Human Ecology in Savanna Environments,* ed. D. R. Harris, pp. 185–207. London: Academic.

Stanbury, P. 1984. Women and water: Effects of irrigation development in a north Indian village. Working Paper 50. Office of Women in International Development, Michigan State University.

Supple, K. R., A. Razzaq, I. Saeed, and A. D. Sheikh. 1985. *Barani farming systems of the Punjab: Constraints and opportunities for increasing productivity.* Islamabad, Pakistan: Agricultural Economics Research Unit, National Agricultural Research Centre.

World Bank. 1989. *Women in Pakistan: An economic and social strategy.* Washington, DC: Author.

Epilogue

Since this article was first written, the visibility of women's economic activities in South Asia has greatly increased. The importance of these activities for efforts to conserve the environment is also now generally accepted.[1] Women in South Asia are now more likely to be incorporated into development projects in the forestry and agricultural sectors. With this progress, however, has come evidence that development of women's economic sphere brings with it unforeseen negative impacts on women.

The section first considers Pakistan, looking at recent research on women's decision-making authority (see Table E1) and then at the status of women in development efforts. Next, dairy projects in India and Nepal are examined to assess their negative effects on women. In the late 1980s the Parkistan Agricultural Research Council (PARC) sponsored two studies of rural women in the rainfed Punjab to gauge their decision-making authority concerning crop and livestock production (Bahar, 1987; Haque, 1986). The data from these studies fill a gap in the information available when this article was originally written[2] and support my assumption that women "manage" the natural resources they use.[3] The studies concluded that women have significant decision-making

Table E1
Women's Decision-Making Authority

Decisions	Decision-Making Authority		
	Female	Male	Shared
Livestock			
Number of breeds of animals kept	24%	29%	47%
Which animals to buy and sell	8%	58%	34%
Health care[a]	25%	33%	41%
Livestock feed			
Number of times and amount fed	84%	4%	11%
Purchase of feed	8%	58%	34%
Milk and milk products			
Milking	63%	26%	11%
Butter and ghee making	80%	6%	12%
Selling	74%	5%	18%
Manure collection	100%	0%	0%
Fodder			
Cutting, chopping, transporting	47%	32%	21%
Preserving	74%	5%	0%
Grazing	37%	16%	5%
Weeding (what and how much)	79%	6%	8%
Land allocation for fodder crops	5%	68%	16%

Sources: Compiled from Haque (1986) and Bahar (1987).

Note: Percentages in each row do not always total 100% because two other choices were used: "no one special" and "not pursued."

[a]This category includes decisions involving medical checkups, choice of curer, controlling disease, investing money in health care, and giving medicine, which were similar enough to be averaged (Haque, 1986).

authority in the areas of farm production that they dominate in terms of time invested. The research is summarized in Table E1. In spite of the fact that women do not usually own livestock in Pakistan, Haque found they have nearly as much decision-making authority as men over the size and composition of a farm's herd, and most such decisions are shared. Decisions about livestock health care follow a similar pattern. When it comes to marketing animals, however, men predominate, as they do in the purchase of feed. Women, however, make decisions about selling milk and milk products; this is possible because such commodities are sold to neighboring villagers, thus bypassing the male markets. Women make most of the decisions involved in their daily work, including feeding animals, milking and processing milk, and collecting manure. Decisions concerning the provision of fodder are more complex. Women weed the family fields for fodder and make decisions about this activity. Decisions about allocating crop land for growing fodder crops, however, are dominated by men. Women are in large part responsible for grazing decisions, but 29% (not included in the table) reported the decisions are made by "no one special"; probably the children, who do much of the fodder grazing in Pakistan. Decisions about preserving fodder are dominated by women; but decisions about the collection of fodder, although usually made by women, are often made by men or shared by both. This reflects the fact, noted in my original article, that the high demand for fodder in rainfed areas leads men to contribute significant amounts of labor to its collection.

The PARC studies, along with more recent research sponsored by the Pakistan Institute of Development Economics (PIDE; see, for example, Ibraz, 1993), continue a tradition of research begun in the early 1980s on rural women's economic activities. Such interest was sparked in 1979 when the Women's Division was established to include women in the Pakistan government's development planning.

In spite of this body of research and this commitment, however, a PIDE researcher in 1995 reported that "there has not been any significant improvement in the situation of women, particularly poor rural women" (Kazi, 1995, p. 79). The primary reason for this, she noted, is that development efforts have not effectively included the agriculture or forestry sectors. Only 1.9% of the projects funded by the Women's Division between 1979 and 1989 were in agriculture and only 1.3% in forestry (Kazi and Raza, 1992).[4] Rural women are thus, in 1995 as in 1989, "almost entirely excluded from training and extension programmes" in these sectors (Kazi, 1995, p. 85).

The story of nongovernmental initiatives in Pakistan is only slightly better. The Agha Khan Rural Support Programme (AKRSP) and similar organizations[5] recognize women's economic activities and have worked very hard to involve them in development. They have had some successes, especially the AKRSP, which has created women's groups focusing on poultry raising, nurseries, and vegetable growing. They do not, however, develop women's existing economic pursuits; they simply add to them. Thus they are criticized for mainly reaching women in relatively prosperous households, who have free time to invest in new activities (Kazi, 1995).[6]

In India, in contrast, development aimed directly at women's livestock activities has a long history. "Operation Flood," the government's huge program for developing the dairy industry, was launched in the late 1960s and is still expanding today. It is described as a "white revolution" to develop milk production, as the green revolution developed crop production. But recent criticisms suggest that this program depends on rural women's cheap labor to provide milk for male-controlled processing plants, which market it in urban areas (Sharma and Vanjani, 1993). Even when women themselves are given subsidized loans for the purchase of milk buffalo, they have little to gain; these loans are only profitable when women's labor is discounted (Sharma and Vanjani, 1993). In addition, milk consumption is decreasing in rural India while its production is increasing; poor households, in particular, are selling all the milk they produce to buy food and fodder and repay loans used to buy livestock (Sharma and Vanjani, 1993). In Pakistan, in contrast, one study in the rainfed Punjab indicates that 92% of the total milk production is consumed by the households in which it is produced (Parveen and Hussain, 1988).

Dairy development in Nepal has meant even more severe problems for women.[7] Although the programs for extending credit to farmers for the purchase of milk buffalo have resulted in households owning as many as five buffalo, a recent critique of these projects noted that they have "exacerbated the subordinate position of women" (Thomas-Slayter and Bhatt, 1994, p. 468). Women do almost all the work involved in caring for these buffalo, especially fodder collection, which is onerous because the animals are stall fed. Women of both castes in the area (Brahmin and Tamang) report that they work much longer hours than before. More significant, "They cite little or no personal gains from these activities," and "none reported extra spending money or an increase in personal assets" (Thomas-Slayter and Bhatt, 1994, pp. 485–486). In addition, school registers show that girls are dropping out of school to help their mothers. Because men dominate the market aspects of this activity, they get the loans and they get the profits.[8] Women merely provide the labor.

Another, more general, critique of women's development in Nepal (Acharya, 1995), is that 20 years of efforts to develop rural women have resulted in women losing ground to men. In particular, rural women's contribution to household decision-making processes actually declined from 1978 to 1992 (Acharya, 1995).[9]

Finally, the critiques of the dairy projects reviewed above all mention one additional problem: fodder shortages. Fodder shortages clearly constrain the development of livestock production; they also increase its negative impacts on people, especially females and the poor. Women must go farther to collect fodder, increasing their work hours and physical stress. Girls must help their mothers, which keeps them out of school. At least in rainfed Pakistan, fodder shortages cause men too to invest significant additional time. Agricultural resources may be diverted from food to fodder crops. Common property resources, key sources of fodder (especially for land-poor households), continue to shrink and degenerate in quality, producing less and inferior kinds of fodder (Jodha, 1995). Farming households (especially the poorest) are increasingly forced to buy fodder, which thus becomes an expensive input. Finally, it may be very difficult to improve the fodder productivity of common property resources (Robinson, 1985). Unfortunately, the shortage of fodder seems to be just as invisible to development planners[10] as the labor necessary to collect it.

Writing about Pakistan's development experience, Kazi reported that rural women continue to be treated as consumers in need of welfare rather than producers in need of technical assistance, credit, and markets (Kazi, 1995; Kazi and Raza, 1992). The history of dairy projects in India and Nepal suggests that women can actually do worse than being offered only welfare-type projects: they can degenerate from invisible producers to visible but exploited laborers (see Carpenter, 1991). The technical assistance rarely reaches them because extension agents in South Asia are almost all male; and even when it does, women's traditional knowledge about livestock and fodder is not respected (Sharma and Vanjani, 1993).[11] Where credit for livestock purchase is available only to men and men market the milk, as in Nepal, women get no return for their additional labor. Where credit is given to women and they can market the milk themselves, as in India, the high cost of inputs still keeps their profit so marginal that they and their families can no longer afford to consume any of the milk they produce. Hence the real profits are made by the male-dominated dairy industry. In short, developing the dairy sector has made women's lives harder, and, at least in Nepal, it is lowering their status in their households as well. Rural women in Pakistan, still excluded from such development, may actually be better off than their "developed" counterparts in India and Nepal.

The failure of dairy projects to help (let alone empower) women is no different from the well-documented failure of development projects generally to raise up other underprivileged categories of people.[12] Development is not neutral; it is generated by and thus favors established relations of power.[13] As Sharma and Vanjani (1993) wrote, the real purpose of the "white revolution" is to extract milk from the rural areas so that the dairy industry can market it in the urban areas. Explicit project goals—the goals stated in the project papers—may call for the alleviation of poverty or the empowerment of women, but goals implicit in the structure and the process of development enrich the wealthy and strengthen the powerful. The exhaustion of women and the parallel exhaustion of common property resources are the inevitable consequences of these implicit goals.

Additional Notes

1. I would argue that ecofeminists have exaggerated, or at least overgeneralized, the importance of women's economic activities for the environment. On ecofeminism, see Shiva (1989); for a good critique, see Jackson (1993).

2. More such research is needed, however. These studies' findings are based on interviews with one woman per household; men should have been interviewed too, as well as women who had different roles in the household. Furthermore, survey research undoubtedly misses much that long-term participant observation would reveal: for instance, women in strongly patriarchal societies informally influence men's decisions even in matters about which they have no decision-making authority. Finally, the Bahar (1987) study on crop decisions differentiates (but the Haque, 1986, study on livestock decisions does not) between decisions made by different age groups as well as genders in the household. This is important because the senior women in the household may have considerable decision-making authority and the daughters and daughters-in-law very little.

3. For an apt criticism of this assumption, see Jackson (1993).

4. Moreover, even these projects have been largely unsuccessful because planners have failed to make use of existing research (Kazi, 1995).

5. These include the Baluchistan Rural Support Programme, the Sarhad Rural Support Corporation in the North West Frontier Province, and the National Rural Support Programme (Kazi, 1995).

6. These programs also have problems finding women with technical training in agriculture and forestry to act as trainers.

7. For a detailed examination of the interaction between women and forests in Nepal, see Denholm (1990).

8. And the women say the men spend the profits on alcohol and gambling (Thomas-Slayter and Bhatt, 1994).

9. The decisions involve, among other things, farm management, domestic expenditure, and disposal of household products and capital transactions. Acharya (1995) blamed this decline primarily on the commercialization of the rural economy.

10. Sharma and Vanjani (1993) wrote that the extent of the fodder shortage "came as a shock" to dairy officials trying to talk women into taking out loans for milk buffalo.

11. Women's indigenous knowledge about livestock and fodder is a topic much in need of research.
12. See, for example, Dove (1993) and Ferguson (1990).
13. See, for example, Escobar (1991).

Additional References

Acharya, M. 1995. Twenty years of WID and rural women of Nepal. *Asia-Pacific Journal of Rural Development,* 5(1):59–77.

Bahar, S. 1987. *Women in crop production and management decisions in Barani Punjab: Implications for extension.* Occasional paper, Social Sciences Division, Parkistan Agricultural Research Council, Islamabad, Pakistan.

Carpenter, C. 1991. *Invisible women, invisible production: Economic functions of Purdah for rural households in Pakistan.* Paper presented at the American Association of Anthropology.

Denholm, J. 1990. *Reaching Out to Forest Users: Strategies for Involving Women.* MPE Series No. 3, March, pp. 14–30. Kathmandu, Nepal: ICIMOD.

Dove, M. R. 1993. A revisionist view of tropical deforestation and development. *Environmental Conservation,* 20(1):17–56.

Escobar, A. 1991. Anthropology and the development encounter: The making and marketing of development anthropology. *American Ethnologist,* 18(4):658–682.

Ferguson, J. 1990. *The Antipolitics Machine: "Development," Depoliticization and Bureaucratic Power in Lesotho.* Cambridge: Cambridge University Press.

Haque, H. 1986. *Role of women in livestock production and management decisions: implications for extension.* Paper presented at Parkistan Agricultural Research Council (PARC), Islamabad.

Ibraz, T. S. 1993. The cultural context of women's productive invisibility: A case study of a Pakistani village. *The Pakistan Development Review,* 32(1):101–125.

Jackson, C. 1993. Doing what comes naturally? Women and environment in development. *World Development,* 21(12):1947–1963.

Jodha, N. S. 1995. Common property resources and the dynamics of rural poverty in India's dry regions. *Unasylva 180,* 46:23–29.

Kazi, S. 1995. Rural women, poverty and development in Pakistan. *Asia-Pacific Journal of Rural Development,* 5(1):78–92.

Kazi, S., and B. Raza. 1992. Women, development planning and government policies in Pakistan. *The Pakistan Development Review,* 31(4):609–620.

Parveen, S., and S. Hussain. 1988. *Livestock on the Farming System of the Barani Punjab.* Tarnab, Peshawar: Agriculture Research Institute.

Robinson, P. J. 1985. *The role of forestry in farming systems with particular reference to forest-grazing interactions.* Ph.D. diss., University of Edinburgh.

Sharma, M., and U. Vanjani. 1993. When more means less: Assessing the impact of dairy "development" on the lives and health of women in rural Rajasthan (India). *Social Science Medicine,* 37(11):1377–1389.

Shiva, V. 1989. *Staying Alive.* London: Zed Books.

Thomas-Slayter, B., and N. Bhatt. 1994. Land, livestock, and livelihoods: Changing dynamics of gender, caste, and ethnicity in a Nepalese village. *Human Ecology,* 22(4):467–494.

Chapter 11

Gender, Seeds, and Biodiversity

CAROLYN E. SACHS

Department of Agricultural Economics and Rural Sociology
Penn State University
University Park, PA 16802
USA

KISHOR GAJUREL

Department of Agricultural Economics and Rural Sociology
Penn State University
University Park, PA 16802
USA

MARIELA BIANCO

Department of Agricultural Economics and Rural Sociology
Penn State University
University Park, PA 16802
USA

Abstract *Crops have been domesticated and improved by millions of men and women throughout the world for centuries. During the late 20th century, however, the pool of crops being cultivated has declined dramatically. Crop diversity has been reduced as a result of the introduction of new crop varieties, replacement of traditional crops by cash crops, and devaluation of local knowledge. Yet, small farmers and gardeners in many regions of the world continue their efforts to conserve crop diversity in their fields as a strategy to provide for family needs and ensure food security. Gender division of labor and access to resources result in men and women having different relationships with the plants they grow and the seeds they maintain. In two studies of seed savers in Pennsylvania (USA) and the Peruvian Andes, we found that women assume the main responsibility for keeping seeds and maintaining genetic variability. Despite enormous differences in environment, socioeconomic conditions, and cultural traditions, women in both regions share similar motivations for maintaining diversity. These farmers and gardeners save seeds primarily for their own use and exchange. They are concerned with improving the nutrition and health of their families, ensuring safe and adequate food supplies, and maintaining family and cultural traditions.*

Keywords Gender, crop diversity, seeds, seed saving.

All over the world, declining biodiversity threatens people's livelihoods, cultures, and standards of living. Degradation of the environment, destruction of natural habitats, and changes in cultural strategies for survival contribute to the increasing loss of biodiversity and also to the impoverishment of women (Abramovitz, 1994; Shiva, 1995). Declines in biological resources often result in declining standards of living for many people in the world, especially women and the poor (Abramovitz, 1994). Women, in many cultural contexts, rely on diverse biological resources to provide food, clothing, housing, and other needs for their families. As access to these resources declines through environmental degradation or inequitable distribution of resources between men and women, women's workloads often increase and their ability to provide food for their families decreases. As a result of gender divisions of labor, women and men have different knowledge about plants and other biological resources (Sachs, 1996). Efforts to preserve biodiversity have generally neglected women's work and knowledge about crops and other natural resources. This chapter focuses on women's knowledge and efforts to maintain crop diversity. First, we discuss reasons for the decline in crop genetic diversity; then, we focus on two studies of seed saving in the United States and the Peruvian Andes.

Biodiversity and Crop Diversity

Biological diversity is generally defined as the variety and variability among living organisms and the ecological systems in which they exist (Office of Technology Assessment, 1987). Biodiversity occurs at three levels: ecosystem diversity, species diversity, and genetic diversity. Species diversity is striking. Approximately 1.4 million species of living organisms have been described: 750,000 insects, 41,000 vertebrates, and 250,000 plants. Biologists estimate that the actual number of species may actually be between 5 and 30 million (Wilson, 1988). Each species contains an enormous amount of genetic information and no two members of the same species are genetically identical. For example, many flowering plants have more than 400,000 genes. Preserving biodiversity is not merely saving endangered species from extinction, but also maintaining genetic diversity within species. Despite the enormous biodiversity, humans directly depend on very few species for food.

Much of the international attention on preserving biodiversity focuses on conservation of tropical forests and preservation of endangered species and their habitats. Although forest and species conservation activities are crucial given the rapid destruction of tropical forests and the irreversibility of species extinction, another area that must receive more attention from environmentalists is the continual decline in crop diversity. Mankind's dependence on a narrow pool of crops and the genetic uniformity within crops threaten the long-term survival of major food crops. Crop genetic diversity provides security for farmers and contributes to the stability of the food supply.

Many biologists have been concerned that only a very few plants feed the world. In the 1970s, various scientists estimated that between 7 and 30 plants supply the majority of human food. Mangelsdorf, as cited in Harlan (1975), listed 15 crops that provide 90% of humans' food. The National Academy of Sciences (1975) listed 20, and Harlan (1975) listed 26. These scientists expressed concern that the number of plants that feed the world are limited and their numbers are declining. Their findings were used to illustrate human vulnerability to starvation as a consequence of dependence on such a small pool of crops for the bulk of its food. However, since the 1970s, other authors have challenged scientific claims that humans rely on a small number of crops for their food. Scientists used global production data to estimate the number of crops that "feed the world," thus masking differences due to high levels of data aggregation. Using national-level supply data, Prescott-Allen and Prescott-Allen (1990) found that the number of plant species that provide 90% of the world's food supply is 103, much higher than earlier studies suggested. Prescott-Allen and Prescott-Allen also suggest that the figure of 103 species feeding the world probably underestimates the importance of many other species. For example, they explain that their data does not take into account production from home gardens or local markets; economic, ethnic, and regional differences in diets; and plant ingredients that may be used in small amounts that are indispensable for a particular cuisine, such as lemon grass, ginger, and tamarind. Thus, they refute the argument that humans are vulnerable to starvation due to reliance on a few food plants; rather, they suggest that vulnerability to starvation is due more to reduction of the diversity of plant species, by not maintaining intraspecies genetic diversity, as well as political, social, and economic forces that inequitably distribute access to food (Prescott-Allen and Prescott-Allen, 1990).

The crops humans rely on today have been domesticated by millions of people through a long, slow process. Both men and women domesticated and improved crops for generations as they adapted them to a wide diversity of ecological, climatic, and social circumstances. Farmers have created diverse agroecological systems, crops, and varieties. Crop diversity provides security for farmers in the face of pests, diseases, climatic changes, and market fluctuations (Amanor et al., 1993). However, the wide pool of crops that people have cultivated and maintained has declined dramatically in the late 20th century. Women farmers, gardeners, and plant collectors have contributed enormous efforts in the domestication of plants. Unfortunately, the neglect of women's knowledge and work with crops in preserving biodiversity has contributed to the demise of genetic resources.

The decline in crop diversity is a result of several factors, including the introduction of new varieties, replacement of traditional crops with cash crops, and habitat destruction. By far, the introduction of new varieties, especially high-yielding, hybrid varieties, is the primary cause of the decline in crop diversity. Women's work in seed saving has pushed against this trend toward decreasing crop diversity. Although women have played a key role in this effort to conserve crop genetic resources, interested scientists and researchers have often neglected women's contributions.

Seed Saving and Gender

Due to gender divisions of labor and access to resources, women and men, in most locations, have different relationships with plants and their seeds. In most situations, women have the major responsibility for reproductive labor, including feeding the family. In agricultural households, these activities generally include the majority of tasks involving subsistence crops after they leave the fields. Seed storage is often one of the least visible of these postharvest activities, which include grain storage, food processing, and food preparation. As grains or other crops come from the fields, women often decide what will be stored, processed, and saved for next year's crops. In making these decisions, women concentrate on providing adequate and nutritious food for their families throughout the year. Because they cook, women must consider cooking qualities of the food for particular meals they will prepare. They consider taste and texture of the foods according to what meals will be prepared and if the food will be fed to children, adults, older people, or particular animals. They consider multiple uses of crops. For example, in Swaziland, people eat maize green (fresh), dried and ground into meal for porridge or brewed into beer. In selecting seeds to plant each season, women choose different varieties that are suitable for each of these uses. In Ethiopia, women save different varieties of wheat for specialized uses. For example, they use *Triticum aestivum* for making bread, *T. turgidum conv. durum* and *aethiopicum* for making macaroni, pastries, local breads, and porridge, *T. turgidum conv. dioccon* for soup women eat during pregnancy and weaning, and a number of different varieties for various alcoholic drinks (Worede and Mekbib, 1993). Women in India use rice grains for cooking and feed rice straw to their animals. Thus, saving seeds involves numerous considerations such as taste, cooking quality, seasonality, multiple uses, nutrition, quantity, and storage potential.

Women also have particular knowledge of plants due to differential access to resources such as land and capital. Access to land is limited for many people in poor agricultural households, but many households have access to small plots of land surrounding their houses. In many areas of the world, women's gardens provide essential food for the nutrition and health of their families. For example, women in northwest Ethiopia intercrop more than six crops in their backyards, including maize, fava beans, sorghum, cabbage, tomatoes, potatoes, pumpkin, and gourds (Worede and Mekbib, 1993). Women also gather wild plants from their farms and from nearby boundaries, wastelands, and forests. Gathering of wild roots, plants, and fruits requires little capital, but can significantly contribute to the food supply of the family. As Ireson's study in this volume shows, Laotian women gather 37 plant food items, 68 medicinal products, and 18 items for household use.

Scientists and development planners consistently undervalue women's knowledge of plants and seeds. Jiggins (1986) pointed out the difficulties in convincing plant breeders at the international agricultural research centers the importance of men's and women's different skills, knowledge, and experience with seeds. However, perhaps even more problematic is what Abramovitz (1994) refers to as the erosion of the "library of indigenous knowledge." Increasing legal restrictions on the use of common land prohibit people from collecting wild plants. There is no guarantee that women will maintain access to plant resources if class and gender struggles over these resources escalate. Abramovitz (1994) also questions if women will continue to pass on their particular knowledge to younger generations. As older people with knowledge of plants and animals die, their wisdom goes with them (Rocheleau, 1991). Many young people no longer learn about plants from their elders. The young are increasingly reliant on off-farm income, spending more time on other activities, and—perhaps consequently—devaluing traditional knowledge. Devaluing of traditional knowledge is furthered by the introduction of new crop varieties and new crops.

New Crop Varieties

Scientific developments, such as green revolution technologies and biotechnology, have changed and continue to transform crop production throughout the world. Agricultural science and development efforts have successfully increased agricultural production, but simultaneously they have threatened crop diversity. Green revolution technologies developed by the international agricultural research centers in the 1960s promised to lead the way out of the Malthusian dilemma through increasing yields of grain crops such as wheat, maize, and rice. Newly developed hybrid varieties known as HYVs (high-yielding varieties) became the "solution" and rallying cry at the heart of the green revolution.

These new varieties produced higher grain yields and were quickly adopted by many farmers. The spread of these new varieties was phenomenal. In 1976, HYVs covered 44% of all land in wheat and 27% of all land in rice in developing countries (Fowler and Mooney, 1990). The success of these new varieties depended on access to inputs such as irrigation, fertilizer, insecticides, herbicides, and fungicides. Unlike seeds from traditional varieties, which can be saved from year to year, seeds from hybrid varieties do not reproduce themselves and therefore must be purchased. Thus, although farmers may obtain better yields with these new varieties, poorer farmers, including many women farmers, often cannot afford the seeds or inputs for hybrids. Benefits proved uneven; wealthier farmers with more capital to purchase seeds, fertilizer, and pesticides have benefited disproportionately from improved varieties. Unfortunately, although the green revolution successfully increased grain yields, the proponents accomplished little in their efforts to improve the welfare of small farmers or to decrease malnutrition and hunger.

In developing new varieties, plant breeders selected characteristics that often did not coincide with what small farmers needed. Plant breeders selected new varieties for high grain yields and fertilizer responsiveness. Although high grain yields are important, small farmers, including many women farmers, need varieties with additional characteristics, including early maturation to increase food supplies early in the season, particular cooking qualities and textures, ability to grow well when intercropped with other species, and suitability for multiple uses, such as fodder for animals and grain for people. van Oosterhout's (1993) study of sorghum growing in Zimbabwe provides an example of the needs of small farmers in Zimbabwe. Farmers grew between 8 and 13 varieties of sorghum. They used multiple criteria for choosing varieties. In order of importance, their criteria were gastronomic, early maturity, and agronomic. Their complex gastronomic descriptions included "threshability, ease of winnowing, processing and milling, good taste for beer and sadza (a traditional staple), colour of resulting food products, time required in cooking, keeping quality of the cooked grain, texture of endosperm and suitability for use in multiple food products, and storage quality" (van Oosterhout, 1993, p. 91). By contrast, scientists carefully selected the newly introduced sorghum varieties primarily for high grain yields, and have consequently reduced the variability of sorghum varieties. The new rice varieties provide another useful example of plant breeders' and farmers' different criteria. These new rice varieties have short stems to prevent plants from falling over (lodging) prior to harvest. Although these varieties produce more grain, they produce less straw or biomass, which farmers use for livestock fodder or other purposes. Women, who are often responsible for obtaining fodder for animals in rice-based economies in Asia, must search longer and harder for fodder as the amount of biomass available from rice straw decreases. Plant breeders did not recognize that the parts of a plant that are "unwanted" depend on the class and gender of the user (Shiva, 1995). Thus, HYVs increased production, but failed to benefit many small farmers and resulted in less diversity within species. In addition to the declining diversity within species, many new crops have replaced farmers' traditional crops.

New Crops

The exchange of germplasm and introduction and disappearance of crops from particular farming systems has occurred for centuries. The "exchange" of plants and seeds occurred under colonialism, usually to the benefit of the colonial powers (Fowler, 1995). Columbus's voyage to the Americas signaled the beginning of what has become known as the "Columbian exchange." Columbus and others after him brought maize, potatoes, peanuts, beans, and squash from the Americas to Europe. These new crops increased the food supply in Europe and strengthened a newly industrializing Europe. Botanical gardens and scientific institutions facilitated the exchange of plants beginning in the 17th century and continued their search for valuable plants throughout the 19th and early 20th centuries (Brockway, 1979). Europeans controlled and profited from the exchange of plants that could only be grown in the tropics, such as sugar cane, coffee, sisal, and rubber. England, France, Germany, the Netherlands, and Belgium controlled the exchange of these tropical plants between hemispheres, transferring plants indigenous to Latin America to plantations in Asia and Africa to support European industrialization (Brockway, 1988). Crops, such as rubber, provided major sources of wealth for colonial empires.

The long-term struggle for control of crop resources reflects an important fact: genetic resources are not equally distributed across the globe. In the 1930s, Nikolai Vavilov, a Soviet scientist, noticed that some regions had hundreds of varieties of crops and identified centers with the greatest diversity of crops. He identified eight centers of crop genetic diversity: China, India, Central Asia, the Near East, the Mediterranean, Ethiopia, Southern Mexico and Central America, and the Andean region. Subsequent researchers have identified additional centers of diversity and insist that centers of diversity are not necessarily centers of origin for particular crops. However, few question Vavilov's realization that centers of genetic diversity were located in what is now the Third World. Most of the important food crops have origins in the Third World. Wheat originates in Asia, rice in China and Southeast Asia, maize and beans in Mexico, potatoes in the Andes, and sorghum and millets in Ethiopia (Genetic Resources Action International [GRAIN], 1992). By contrast, Northern Europe and North America were genetically poor in providing food crops and have long been dependent on Southern countries to provide crop diversity. This long-term, often uneven, exchange of crop genetic resources between different regions has resulted in the current distribution of crops and control of genetic resources throughout the world.

Seed Saving: Gene Banks or Farmers' Fields?

The question of how to conserve crop genetic resources engenders heated discussion at the global level. The scientific community, international organizations, and national governments have generally supported gene banks (ex situ) as the most effective strategy for long-term preservation of genetic resources. However, others have argued that preservation in farmers' fields (in situ) is more effective and more equitable (GRAIN, 1992). The international agricultural research community began to show concern for genetic erosion in the 1960s. In 1974, the International Board for Plant Genetic Resources (IBPGR) was established as part of the Consultative Group on International Agricultural Research (CGIAR). The purpose of IBPGR was "to establish a global network of activities to further the collection, conservation, documentation, and use of germplasm for crop species" (Williams, 1988, p. 241). Initially, the IBPGR focused its efforts on establishing ex situ collections in gene banks. However, IBPGR's centralized decision-making and gene banks came under increasing scrutiny for not serving the interests of developing countries or small farmers (Kloppenburg, 1988). Many of the initial efforts of IBPGR did not link conservation of genetic resources in gene banks to use in farmers' fields. Also, scientists engaged in national breeding programs rarely used the genetic resources in the international gene banks. Researchers lacked the capacity to grow many of the varieties they saved. They grew plants in a limited range of agroecological conditions. Therefore, the varieties no longer responded to changing environmental and social conditions. Currently, scientists have little information about the varieties stored in gene banks; consequently, plant breeders infrequently use many varieties. In addition to these problems of seed conservation, the major controversy over gene banks involves political control over germplasm (GRAIN, 1992). For several centuries, explorers, botanists, and other scientists freely collected varieties as the "common heritage" of humans, providing little advantage to the people or nations donating the plant resources. Now, with the legalization of the Plant Variety Protection Act, many Northern companies or breeders can claim property rights to the resources that small farmers maintained for centuries. In addition, biotechnology and recombinant DNA research rely on acquisition and control over genes and genetic resources. These genetic resources, which were once the "common heritage" of humans, can now be commodified and controlled by industry.

Problems associated with ex situ conservation in gene banks as well as struggles for control over germplasm point to the role of farmers in conserving genetic resources. Presently, farmers' efforts to preserve genetic diversity, in situ conservation, are gaining increasing visibility and credibility. Small farmers in many regions of the world have maintained heterogeneous populations of seeds for generations. Resource-poor farmers in low-input marginal environments maintain species and genetic diversity as a strategy to ensure adequate food (Montecinos and Altieri, 1992). In addition, farmers often grow these varieties in close proximity to wild relatives, allowing for exchange between wild and cultivated varieties. As Worede (1992) explains, in many developing countries farmers still hold the bulk of genetic resources; their fields are active gene banks.

Nongovernmental organizations have led the way in recognizing the importance of farmers' conservation of genetic resources and the need for local control and participation in seed saving. Organizations such as GRAIN, Rural Advancement Fund International (RAFI), and Seed Savers Exchange promote grassroots approaches for genetic conservation. For example, in the late 1980s, RAFI developed a strategy for community seed banks and held workshops for local nongovernmental organizations in Ethiopia, Indonesia, and Chile (Mooney, 1992). Both GRAIN and RAFI emphasize that genetic diversity cannot be preserved without diversity in farmers' fields and cultural diversity in farming communities. In the United States, Seed Savers Exchange promotes use and exchange of seeds between growers. It publishes yearly books of the seed varieties maintained by their members for exchange with other members. In 1996, its 1,011 members offered 11,572 unique varieties of plants for exchange with other growers (Seed Savers Exchange, 1996).

More recently, agricultural scientists are becoming convinced about the importance of farmers' knowledge and in situ conservation (Montecinos and Altieri, 1992). The International Board for Plant Genetic Resources now emphasizes in situ conservation as well as gene banks. Perhaps, as Kloppenburg (1991) suggests, dialogues between agricultural scientists and local farmers can promote sustainable farming systems and transform agricultural science to more adequately meet the needs of small farmers. Two case studies of seed saving in the United States and Peru illustrate efforts by local people to conserve crop diversity.

Seed Saving in the United States

Crops in the United States are strikingly uniform. Although it is difficult to document how much crop diversity has been lost in the United States, RAFI has attempted to document the extent of genetic erosion, comparing U.S. Department of Agriculture (USDA) lists of vegetable varieties offered by commercial seed companies in 1903 and 1983 (Fowler and Mooney, 1990). They found that 97% of the varieties from 1903 are no longer available. In addition, of those crop varieties that continue to exist, only a small number of varieties account for a large percentage of the acreage planted (Kloppenburg, 1988). Genetic uniformity and monocultural practices make these crops particularly vulnerable to pest and disease outbreaks.

In recent years, North Americans have displayed increasing interest in reintroducing genetic diversity and indigenous species of plants and animals. Articles in newspapers, magazines, and popular media show evidence of renewed interest in heirloom crops. Nongovernmental organizations and small seed companies have promoted interest in crop diversity and heirloom varieties. Seed Savers Exchange in Decorah, Iowa, successfully facilitates exchanges of seeds between hundreds of seed savers in the United States and other growers. Many of these seed savers are gardeners, and some of them specialize in particular crops such as tomatoes, peppers, or onions. Other organizations specialize in crops from particular regions and cultural heritages. Native Seeds in New Mexico encourages saving and planting of Native American varieties of plants such as maize, beans, peppers, and squash. Landis Valley in Southeastern Pennsylvania saves and distributes seeds saved by German American farmers.

Despite this growing interest in seed saving, little knowledge exists concerning who is saving seeds and why. We conducted a study of seed savers in Pennsylvania to investigate their reasons for seed saving and any problems they encountered saving seeds. To find seed savers, we contacted organizations including Seed Savers Exchange, the cooperative extension's Master Gardener program, the Pennsylvania Association of Sustainable Agriculture, individuals involved in seed saving, and gardening magazines. We compiled a list of 62 seed savers, but located only 21 seed savers, with whom we conducted personal or telephone interviews. The seed savers we interviewed were overwhelmingly women (90%). They ranged in age from 29 to 79 years, with an average age of 45. Although some seed savers had not completed high school, 48% had graduated from college, and some had graduate education.

The seed savers saved vegetable and flower seeds, but no grain crops. Slightly less than half (44%) of those we interviewed saved only vegetables and 39% saved both vegetables and flowers. Table 1 shows the vegetables and number of varieties grown by

Table 1
Vegetable Seeds Saved and Number of Varieties

Vegetable	% Growing	Mean # of Varieties	Range
Tomatoes	66	7	1–20
Beans	62	3	1–10
Squash	43	2	1–7
Peas	33	1	1–3
Lettuce	29	3	1–9
Peppers	19	5	1–9
Cucumbers	14	1	1–2
Potatoes	14	1	1
Carrots	14	1	1
Kale	10	1	1
Spinach	10	1	1
Okra	10	1	1
Radishes	5	1	1
Turnips	5	1	1
Onions	5	1	1
Garlic	5	1	1
Pumpkin	5	1	1

the respondents. Seed savers were most likely to save tomatoes, beans, squash, peas, and lettuce. They saved the largest number of varieties of tomatoes, followed by peppers. Several people who were saving varieties of tomatoes explained:

> I like the tomatoes. There are so many unusual types: striped ones, yellow ones, fuzzy ones, really strange kinds of varieties. Garden peach (fuzzy one), Cherokee purple, Persimmon, Golden Pearl (little tiny yellow, so sweet!), Egyptian Tomb (the seeds were found in a tomb and have since been passed down). I have also been corresponding with a woman in Czechoslovakia, so I grow what she sends me. A big yellow radish.

Another woman explained that she is now saving tomatoes for drying rather than canning or freezing, so she is experimenting with new varieties:

> My major project now are paste tomatoes—I'm into heirloom varieties. I am trying 40 varieties this year and try 10–20 seeds out of different packets. I'm particularly interested in paste tomatoes for drying so I am experimenting—I've tried at least 100 varieties from a tomato supply catalog. I have selected a wonderful pink, pear-shaped variety that is excellent for drying. Most people choose tomatoes for canning, but I prefer to dry my tomatoes.

Most of the respondents were saving only a small quantity of seeds sufficient for their own use. However, some people saved seeds to exchange through Seed Savers Exchange or with other gardeners or friends. When asked about the quantity of seeds they saved, most people described the amounts using household measurements such as tablespoon or cup, whereas a few used weights such as ounces and pounds. For example, one woman explained,

> With regard to quantity, it depends on how the crop produces and how adamant I was about getting the seeds. Maybe a tablespoon of seeds from each variety. I don't save massive quantities, but I have more than enough for my family, and I send them to friends who are interested in some unusual varieties.

Another woman who listed her seeds in Seed Savers Exchange responded,

> I don't save very much—no more than a cup, maybe a half cup. Just a few plants make a lot of seeds. I don't get many orders from Seed Savers Exchange. Always have more than enough for myself.

Seed savers listed many reasons for saving seeds, including interest in gardening, continuing family tradition, saving money, and preserving varieties. As several respondents explained:

> I became interested in seed saving for many reasons. I didn't like that seed companies offer mostly hybrids—they force you to keep buying their seeds each year. I like to find varieties that do well in my areas. By saving seeds, they will be more acclimated to my area and each year I can select for qualities for my area.

> I started gardening when I was ten years old and dad allowed me to plant my own vegetables. I started saving seed about three years ago. I constantly read about both gardening and seed saving. I got involved in seed saving because I want to help seed survive, I want to be independent from the market, and want to save money.

In terms of continuing family tradition, one woman explained:

> My grandfather used to save seeds—sweet corn—the best I ever remember eating. But I don't have the seeds anymore. I have tried to find that variety, but I can't. Knowing that there is that good kind of corn that isn't available makes me want to save seeds.

Another woman saved different varieties because they tasted good or were unique:

> I save Mexican cave beans, cucumber seeds, red Russian kale, lettuce seed, squashes: old fashion pumpkin or moshata, and guild golden; tomatoes: brandy wine, schimmeg creig, mortgage lifter, fargo yellow pears, yellow current, and black prince. I save those seeds because they are unique and their fruits are tasty.

The majority of seed savers regularly exchange seeds through seed saving networks or with friends. Three out of five of the seed savers exchanged seeds through organizations such as Seed Savers Exchange, Organic Gardening Seed Exchange, and cooperative extension Master Gardener programs. However, this figure may overestimate the number of seed savers involved in formal exchanges because we chose our sample of seed savers from these sources. Many of the seed savers obtained their seeds by mail through catalogs or seed savers exchanges. Others exchanged seeds with friends or other gardeners and mentioned that saving seeds provided a special connection to other people. Several people explained that when they grow plants from seeds given to them by friends or relatives, they remember the person who provided the seeds. As one woman expressed:

> Friends and other gardeners share seeds. People say I saved these precious seeds and when I give them away, it establishes a connection with the person. We share back and forth. It's a good way of being with people—to have their plants in your garden. Also, it's much better than spending money.

Seed savers gave mixed reviews of commercial seeds and seed companies. The majority of respondents considered their seeds superior to seeds from commercial sources. They preferred their seeds because of better germination, no chemical treatment, adoption to microclimate, and open pollination. As one seed saver explained:

> My seeds are more adaptive for my microclimate. Also, sometimes I save only the most vigorous plant, best yielding—not all the seed. I choose the hardiest and best. More disease resistance. Bettering the genetic makeup of the future plant. I know seed companies have to maintain standards, but you don't know to what extent.

Half of the respondents noted problems with seed companies. Their criticisms were as follows:

> Their seeds are too expensive; the varieties are poor.

> The trouble with the commercial seed places is that they are all getting into hybrids. I think we can develop seeds that are more acclimated to the environment they are being raised in.

> I don't think they look at the big picture. Their main concern is yield and if it will make them money.

> Not good, some varieties are lost because they stopped selling them.

> I seem to notice they are offering mostly hybrids. I think they are more concerned with making money. Only the smaller companies are concerned with the old varieties.

Many of the seed savers buy from and try to support small seed companies, especially companies that offer heirloom and open-pollinated varieties. The majority of seed savers regularly read and order from seed catalogs.

These seed savers are part of the growing movement in the United States to revive and maintain a diverse variety of vegetables, flowers, and herbs. The vast majority of the seed savers we interviewed were women, but men also save seeds. The gendered nature of seed saving is connected to the gendered nature of gardening. Seed savers primarily identify themselves as gardeners, although some are also farmers. They are saving seeds primarily for their own use and for exchange. Saving money, providing tastier food, improving nutrition and health, preserving genetic diversity, and maintaining family and cultural traditions motivate people to save seeds. Oddly enough, some of these same factors contribute to seed saving efforts in the Peruvian Andes.

Seed Saving in Peru

Farmers in the Peruvian Andes produce their animals and plants under diverse and difficult ecological and climatic conditions. They produce crops in mountainous environments ranging from 1,000 to 4,200 meters, with poor soils and often steep slopes. In this atmosphere of environmental variability and uncertainty, they produce and maintain a large and diverse pool of both indigenous and introduced crops. Vavilov identified the Andean region as one of his eight centers of diversity. Table 2 shows some of the indigenous Andean tubers, roots, grains, and legumes grown at a variety of altitudes. Scientists and others are concerned about the decline in the use of indigenous crops and declining genetic diversity within these crops (Fano and Benavides, 1992; National Research Council, 1989). Studies in the Andes have shown that women have the major responsibility for selection of seeds and use of different varieties (Tapia and de la Torre, 1993).

An interdisciplinary team of scholars from the United States and Peru is currently involved in a study of Andean root and tuber crops. Biologists, agronomists, economists, nutritionists, and sociologists are working together to understand farmers' practices and strategies for producing these crops and for maintaining genetic variability.[1]

Table 2
Andean Crops

Common Name	Scientific Name	Altitude (m)
Tubers		
Potato	*Solanum andigenum*	1,000–3,900
Bitter potato	*Solanum juzepczukii*	3,900–4,200
Oca	*Oxalis tuberosa*	1,000–4,000
Ulluco	*Ullucus tuberosum*	1,000–4,000
Roots		
Arrachacha	*Arracacia xantorrhiza*	1,000–2,500
Achira	*Canna edulis*	1,000–2,500
Ajiba	*Pachyrhizus tuberosus*	1,000–2,000
Yacon	*Polymnia sonchifolia*	1,000–2,000
Chago	*Mirabilis expansa*	1,000–2,500
Maca	*Lepidium meyenii*	3,900–4,100
Camote	*Ipomoea batata*	0–2,800
Grains		
Maize	*Zea mays*	0–3,000
Quinoa	*Chenopodium quinoa*	0–3,900
Kaniwa	*Chenopodium pallidicaule*	3,200–4,100
Amaranth	*Amaranthus caudatus*	0–3,000
Legumes		
Lupine	*Lupinus mutabilis*	500–3,800
Beans	*Phaseolus vulgaris*	100–3,500

Source: Tapia and de la Torre (1993).

During 1995 and 1996, we studied two communities in the Peruvian Andes to understand how the three tuber crops, oca, ulluco, and mashua, fit into the farming system of Andean people. In 1995, we collected data in Picol, a small community with 25 families located less than 20 kilometers from Cusco. In 1996, we collected data from the second community, San Jose de Arizona, a larger community with more than 100 households located in Ayacucho province. Students conducted interviews with 21 families in Picol and 40 families in San Jose de Arizona.

Farmers in the two communities grew a wide diversity of both indigenous and introduced crops. In Picol, farmers grew potatoes, oca, ulluco, mashua, lupine, barley, onions, lettuce, and cabbages, with potatoes as their principal crop. Community members used potatoes primarily for subsistence but they also used it as a cash crop. Onions grown on irrigated land provided the major cash crop of farmers. All of the farmers we interviewed in 1995 grew at least some minor tuber crops: oca, ulluco, and mashua. Farmers intercropped the minor tubers or grew them in separate plots. The diversity of tubers varied between household; economically better off households reported growing more varieties than poorer households (Bianco, 1996). In comparison to potatoes, minor tubers were cultivated in small amounts. Farmers used oca, ulluco, and mashua primarily for subsistence, although some women bartered small amounts in exchange for maize (Trivelli, 1996). During our later work in Picol in 1996, we found some farmers did not grow any minor tubers. Through close observation of the harvesting and storing of tubers, M. Ramirez (personal communication, 1996) and the farmers identified more morphological types of tubers than were identified in 1995.

In San Jose de Arizona, a community with greater variation in altitude, more land, and no irrigation, farmers grew maize, beans, peas, potatoes, bitter potatoes, quinoa, amaranth, barley, wheat, oca, ulluco, and mashua. They grew potatoes as the major cash crop as well as the major subsistence crop. In contrast to Picol, this community did not have irrigation and therefore relied solely on rainfed production. Although the farmers did not grow irrigated vegetables, they had more ecological diversity and were thus able to grow a greater diversity of crops on dry land than farmers in Picol, including maize, potatoes, barley, quinoa, squash, wheat, fava beans, peas, lupines, oca, ulluco, and mashua.

The majority of the farmers grew oca, ulluco, and mashua in small amounts. Farmers grew between 1 and 6 varieties of oca, 1 to 4 varieties of ulluco, and 1 to 5 varieties of mashua. All family members participated in the harvest, with women assuming more responsibility for sorting, storing, trading, processing, and cooking the tubers. The farmers grew minor tubers for consumption. As in Picol, farmers preferred a white, sweet variety of oca because it is especially good for eating. Unfortunately, the same variety is also particularly susceptible to insect damage. The majority of farmers in San Jose reported that they grow fewer varieties now than they did ten years ago, primarily because of problems obtaining seeds. The majority of farmers obtain seeds by saving them, although a few farmers buy seeds in markets.

Thus, in both communities, farmers grew a wide diversity of crops for subsistence, barter, and the market. In both communities, production has changed in response to market conditions, access to irrigation and inputs, political circumstances, and availability of seeds. Farmers value oca, ulluco, and mashua and they continue to grow small amounts of different varieties primarily for subsistence. Because Andean people value these crops, many of which grow only in particular ecological conditions, people barter and exchange minor tuber crops for other crops—such as maize and fruits—to improve their diets and nutrition. In both locations, women assume primary responsibility for marketing and barter of crops. Thus, the exchange of seeds occurs between women at the markets. Also, nongovernmental organizations in Peru sponsor seed fairs to promote exchange of diverse varieties between farmers (Tapia and Alcides, 1993). Women, the major participants at seed fairs, maintain enormous diversity and varieties of crops. For example, one woman grew 28 varieties of oca and another woman had 56 varieties of potatoes (Tapia and de la Torre, 1993).

Our research in Peru indicates that farmers grow a wide diversity of both indigenous and introduced crops. Although farmers continue to grow a wide diversity of tuber crops—oca, ulluco, and mashua—the amount of species diversity seems to be declining due to lack of availability of seeds. Whereas both men and women know and select different varieties of the minor tuber crops, women have the major responsibility for saving, processing, preparing, bartering, and selling oca, ulluco, and mashua. Our research will continue to explore farmers' strategies and problems in maintaining crop diversity and genetic diversity within these crops.

Conclusion

Preservation of crop diversity and diversity within varieties represents a major challenge for maintaining food security. Although initial efforts by international and national organizations, focused on preservation of seeds in gene banks, have preserved genetic diversity of crops, current efforts suggest that preservation of diversity within crops must occur both in gene banks and in farmers' fields. Small farmers and gardeners in many parts of the world continue to save a wide variety of seeds to meet their subsistence needs and ensure food security.

Seed saving in most regions of the world is gendered. In our studies in the United States and Peru, both women and men participate in seed saving, although women clearly predominate. Farmers and gardeners value diversity for cooking, eating, ensuring adequate food supplies, and adaptation to ecological environments, climatic, and market conditions. In Peru, women save seeds in their households, exchange seeds with other members of their communities, and buy seeds in the local markets. In the United States, the majority of seed savers we interviewed were women gardeners who saved seeds primarily for their families' subsistence needs and to exchange with other growers, friends, or relatives. Scientists and development planners must increasingly recognize the importance of both women's and men's knowledge, activities, and needs related to seed saving for the purpose of preserving genetic diversity of crops.

The renewed interest in preserving genetic diversity comes at a time when diversity is declining due to the increasing prevalence of hybrid varieties and a focus on production for the market. With the switch to cash crop production, farmers become less concerned with diversity of their crops and more concerned with the amount of production. Preservation of crop diversity requires preservation of farming communities and cultural diversity as well. In the United States, some people are attempting to reclaim diversity, as the renewed interest in seeds of Native Americans, German Americans, and others reveals. In Peru, people in the Andes have struggled against a long history of colonialism and marginalization to maintain their culture, preserve their way of life, and save their indigenous crops. These crops provide subsistence and nourishment and also symbolize attachment to family, community, and cultural traditions.

Note

1. A number of people have contributed to this project. Carolina Trivelli, Mariela Bianco, Robert Torres, Ramiro Ortega, Marleni Ramirez, and Steve Smith were instrumental in data collection in the village in Cusco province in 1995. David Dominquez, Flavio Cahuana, Uriel Sulcate Acuna, Andrea Mayer, Terre Flores, Steve Smith, Marleni Ramirez, and Carlos Perez collected data in the village near Ayacucho in 1996.

References

Abramovitz, J. N. 1994. Biodiversity and gender issues: Recognizing common ground. In *Feminist Perspectives on Sustainable Development*, ed. W. Harcourt, pp. 198–212. London: Zed Books.

Amanor, K., K. Wellard, W. de Boef, and A. Bebbington. 1993. Introduction. In *Cultivating Knowledge: Genetic Diversity, Farmer Experimentation and Crop Research*, eds. W. de Boef, K. Amanor, and K. Wellard, with A. Bebbington, pp. 1–13. London: Intermediate Technologies.

Bianco, M. 1996. *Farming systems and indigenous knowledge: The case of minor Andean tubers in the Peruvian highlands*. Master's thesis, Pennsylvania State University.

Brockway, L. H. 1979. *Science and Colonial Expansion: The Role of the British Royal Botanic Gardens*. New York: Academic Press.

Brockway, L. H. 1988. Plant science and colonial expansion: The botanical chess game. In *Seeds and Sovereignty: The Use and Control of Plant Genetic Resources*, ed. J. Kloppenburg, pp. 49–66. Durham, NC: Duke University Press.

Fano, H., and M. Benavides. 1992. *Los Cultivos Andinos en Perspectiva: Produccion y Utilizacion en el Cusco.* Lima, Peru: Centro de Estudios Regionales Andinos, Bartolome de las Cases—Centro Internacional de la Papa.

Fowler, C. 1995. Biotechnology, patents and the Third World. In *Biopolitics*, eds. V. Shiva and I. Moser, pp. 214–225. London: Zed Books.

Fowler, C., and P. Mooney. 1990. *Shattering: Food, Politics, and Loss of Genetic Diversity.* Tucson: University of Arizona Press.

Genetic Resources Action International (GRAIN). 1992. Why farmer-based conservation and improvement of plant genetic resources? In *Growing Diversity: Genetic Resources and Local Food Security*, eds. D. Cooper, R. Vellve, and H. Hobbelink, pp. 1–16. London: Intermediate Technologies.

Harlan, J. 1975. *Crops and Man.* Madison, WI: American Society of Agronomy and Crop Science Society of America.

Jiggins, J. 1986. *Gender-Related Impacts and the Work of the International Agricultural Research Centers.* Washington, DC: World Bank.

Kloppenburg, J. 1988. *First the Seed: The Political Economy of Plant Biotechnology, 1492–2000.* Cambridge: Cambridge University Press.

Kloppenburg, J. 1991. Social theory and the de/reconstruction of agricultural science: Local knowledge for an alternative agriculture. *Rural Sociology*, 56:519–548.

Montecinos, C., and M. Altieri. 1992. Grassroots conservation efforts in Latin America. In *Growing Diversity: Genetic Resources and Local Food Security*, eds. D. Cooper, R. Vellve, and H. Hobbelink, pp. 106–116. London: Intermediate Technologies.

Mooney, P. R. 1992. Towards a folk revolution. In *Growing Diversity: Genetic Resources and Local Food Security*, eds. D. Cooper, R. Vellve, and H. Hobbelink, pp. 125–145. London: Intermediate Technologies.

National Academy of Sciences. 1975. *Underexploited Tropical Plants with Promising Economic Value.* Washington, DC: Author.

National Research Council. 1989. *Lost Crops of the Incas: Little Known Plants of the Andes with Promise for Worldwide Cultivation.* Washington, DC: National Academy Press.

Office of Technology Assessment. 1987. *Technologies to Maintain Biological Diversity.* Washington, DC: U.S. Government Printing Office.

Prescott-Allen, R., and C. Prescott-Allen. 1990. How many plants feed the world? *Conservation Biology*, 4(4):365–374.

Rocheleau, D. 1991. Gender, Ecology, and the Science of Survival. *Agriculture and Human Values* 8(112):156–165.

Sachs, C. 1996. *Gendered Fields: Rural Women, Agriculture, and Environment.* Boulder, CO: Westview Press.

Seed Savers Exchange. 1996. *Seed Savers 1996 Yearbook.* Decorah, IA: Author.

Shiva, V. 1995. Biotechnological development and the conservation of biodiversity. In *Biopolitics*, eds. V. Shiva and I. Moser, pp. 193–213. London: Zed Books.

Tapia, M. E., and R. Alcides. 1993. Seed fairs in the Andes: A strategy for local conservation of plant genetic resources. In *Cultivating Knowledge: Genetic Diversity, Farmer Experimentation and Crop Research*, eds. W. de Boef, K. Amanor, and K. Wellard, with A. Bebbington, pp. 111–118. London: Intermediate Technologies.

Tapia, M. E., and A. de la Torre. 1993. *La Mujer Campesina y Las Semillas Andinas.* Lima, Peru: Food and Agriculture Organization.

Trivelli, C. 1996. *Secondary crops in peasant economies: Minor tubers in the Peruvian Andes.* Master's thesis, Pennsylvania State University.

Van Oosterhout, S. 1993. Sorghum genetic resources of small-scale farmers in Zimbabwe. In *Cultivating Knowledge: Genetic Diversity, Farmer Experimentation and Crop Research*, eds. W. de Boef, K. Amanor, and K. Wellard, with A. Bebbington, pp. 89–95. London: Intermediate Technologies.

Williams, J. T. 1988. Identifying and protecting the origins of our food plants. In *Biodiversity*, ed. E. O. Wilson, pp. 240–247. Washington, DC: National Academy Press.

Wilson, E. O. 1988. *Biodiversity*. Washington, DC: National Academy Press.
Worede, M. 1992. Ethiopia: A genebank working with farmers. In *Cultivating Knowledge: Genetic Diversity, Farmer Experimentation and Crop Research*, eds. W. de Boef, K. Amanor, and K. Wellard, with A. Bebbington, pp. 78–96. London: Intermediate Technologies.
Worede, M., and H. Mekbib. 1993. Linking genetic resource conservation to farmers in Ethiopia. In *Cultivating Knowledge: Genetic Diversity, Farmer Experimentation and Crop Research*, eds. W. de Boef, K. Amanor, and K. Wellard, with A. Bebbington, pp. 78–84. London: Intermediate Technologies.

Chapter 12

Women and Agroforestry: Four Myths and Three Case Studies

LOUISE FORTMANN

Department of Forestry and Natural Resources
145 Mulford Hall
University of California
Berkeley, CA 94720
USA

DIANNE ROCHELEAU

The Rockefeller Foundation and International Council
　for Research in Agroforestry
P.O. Box 30677
Nairobi, Kenya

Abstract *Women are traditionally important participants in both the agricultural and forestry components of agroforestry production. Women are frequently ignored in the design of agroforestry projects because of commonly held myths about their participation in both production activities and in public life. The involvement of women in agroforestry projects and activities are examined in case studies from the Dominican Republic, India, and Kenya. Considerations for including women in agroforestry projects are discussed.*

Introduction

Over a decade ago Boserup [10] documented the role of women in development. Nonetheless, many development projects continue to be designed without consideration of their effect on women or of the role of women in their implementation. Forestry and agroforestry projects are no exception. A study of 43 World Bank forestry projects found that only eight made specific reference to women [47].

In part this state of affairs is due to the masculine images conjured up by the word forestry. The reality is, in fact, often the opposite of the image. This article reviews the literature on women's involvement in agroforestry and presents three case studies on the results of including or excluding women from agroforestry projects.

Myths and Realities

Women have traditionally played important roles in agricultural production and in the use and management of trees. The importance of these roles is, however, often obscured by the prevailing myths held both by donors and by the local personnel in ministries about the roles and status of women:

Myth 1: Women are housewives and are not heavily involved in agricultural production
Myth 2: Women are not significantly involved in tree production and use
Myth 3: Every woman has a husband or is part of a male-headed household
Myth 4: Women are not influential or active in public affairs.

These are the myths. Here are the realities.

Agricultural Production Is a Traditional Role of Women

Women are not just housewives. In many, if not most, societies they are, in fact, farmers often bearing the major or sole responsibility for food production.

Region by region, country by country, ethnic group by ethnic group, detailed studies have documented Boserup's point that women's labor and women's decision making are crucial to agricultural production and development. (CF. 19).

A study of 95 developing countries found that women comprised from 17.5% (in Central and South America) to 46.2% (Sub-Saharan Africa) of the total agricultural labor force in 1970 (18). If only food production were to be considered, these figures could be expected to be far higher. The predominant role of women in both plow and hoe agriculture has been demonstrated for Sub-Saharan Africa [11]. The involvement of women, particularly poor women, in agricultural production in India has been reported [49]. Women have been demonstrated to play a major role in agricultural work and agricultural decision making in Nepal [1, 7, 40, 42, 43]. Mazumdar [36] states that in Asia "rural women constitute the single largest group engaged in agriculture and the production of food." Colfer [15] documents the central role of women in swidden agriculture in East Kalimantan. Women in Latin America have been shown to be actively involved in agricultural production [3, 16]. Women are frequently responsible for smallstock husbandry and for the feeding of larger livestock, particularly milk cows and calves. Thus agroforestry projects which involve fodder trees, the servicing of crops by trees or intercropping of crops and trees must include women, since it is often women who grow the crops or care for the livestock which will be involved.

The Gathering and Use of Forest Products Is a Traditional Role for Women

A second myth is that only men are users of and responsible for trees. In fact women are often the prime users of forestry products such as fuelwood, wild foods and fodder. Wood et al. [58] noted that women "are primarily responsible for wood collection and utilization and often the initial establishment and tending of the wood stock around the village." Bennett [7] found in Nepal that 78% of the fuel collection was done by women and 84% by women and girls combined. The 1980 report of the Expert Group of Women and Forest Industries of UNESCAP noted that as much as two-thirds of the time spent collecting fuelwood was that of women. In the Sahel it is women who collect fuelwood [33].

Women may make different uses of forest products than men do. Hoskins [30] contrasts the interest of men and women in forest resources. Men are more likely to be interested in forest products for commercial sale and in the use of products farther from home. Women collect fuelwood and both human and animal food from forests and individual trees. They are knowledgeable about the burning characteristics of various species and about species which have food values. Both men and women make medicine from forest products but for different purposes. Women also use forest products for such purposes as basket making and dyeing. It might be expected that women would have a more detailed knowledge of trees and their uses. And indeed, Hoskins [28] found in Sierra Leone that women could list 31 products which were harvested or produced in nearby bushes and trees while men could list only eight. Similarly, the priorities of men and women in species preference for Indian afforestation programs often conflict [48].

Women's close involvement with forest products and agricultural production often results in their having a greater awareness of environmental problems than men. For example, the implementers of a soil and water conservation project in Mali discovered that local women had already undertaken conservation efforts which would have been destroyed by the project [29].

Wiff [57] describes a reforestation and soil conservation project in which women were deliberately denied access to credit and technical support. Nevertheless in those places where they could work without credit, it was the women who led the conservation movement.

Thus, the presumption that it is only men who are involved in forestry is often totally wrong. In the case of fuelwood and minor forest products it is often *only* women who are involved. It is women who know what is needed for these uses. It is women who know which trees are suitable and which are not. And it is women who will use the final product.

Women as Heads of Households Are Assuming Nontraditional Roles in Agriculture and Forestry

Increasing numbers of women head their own household, sometimes by choice, sometimes as a result of personal events such as death of a spouse, divorce, desertion, abandonment, or of social trends such as male out-migration.

The woman-headed household is an increasingly common social unit found in substantial numbers in every region of the world. It has been estimated that between 25 and 33% of all households in the *world* (emphasis in the original) are de facto headed by women [12]. In a study of 73 developing countries, the lowest percentage of women-headed households was 10.1% in Kuwait. The highest was Panama with 40%. Thirty-seven percent of the countries had 10–14% women-headed households; 43% had 15–19%; 23% had 20–24%; and 7% of the countries had over 25% women-headed households [12].

As a result of heading their households, women have assumed new roles. An historian process of women undertaking "male" tasks and working in "male" sectors in the absence of men has been reported for a number of societies [4, 14, 34]. The lesson for agroforestry projects is that "male" roles and "male" sector jobs are not fixed but are increasingly being undertaken by women household heads as well as by other women.

Women and Women's Groups Are Important in Community Organization and Mobilization

March and Taqqu [35] have documented the important influence of women's informal associations in both the private and public sphere. Within the domestic sphere women exercise influence on public events through their information links based on lineage ties and through their ability to withhold goods necessary for men's public participation [35].

Women's solidarity groups may take a very active stance in defending their own interests. In the Nigerian "Women's War" of 1929, women, incensed over the rumor that an on-going census meant that they were about to be taxed, refused to allow census enumerators to count them on their property. Women organized confrontations with the colonial administration which at times led to the use of armed force on the part of the latter [35].

Women gain leverage in the economic sphere through rotating work and credit associations [35]. They may also influence events through their roles as healers or religious figures [35].

Women also participate in leadership roles in formal institutions. For example, in Botswana women are generally the mainstay of social welfare organizations and even hold positions of authority in "male" organizations such as farmers committees [22].

The influence of women is demonstrated by a project in Cameroon where a protective nursery fence was torn down by the men fearing that the government was trying to seize their land by planting trees. The women who recognized the need for fuelwood, persuaded the men to rebuild it [29].

Women, then, both individually and in groups have private influence on public action by men and undertake public action themselves. The potential for women's public action in areas such as reforestation, soil and water conservation and the like is especially high because they are the principal sufferers from environmental degradation [23]. It is they who must walk farther for water, fuelwood, and fodder. It is they who must produce subsistence on increasingly degraded soils. It is they who often are both able and likely to organize the community for action.

The Case Studies

The following three case studies examine women's involvement in different aspects of agroforestry. The first is a study of women's involvement in the implementation and benefits of an agroforestry project. The second examines women's importance in mobilizing a community. The third looks at some of the techniques necessary to involve women in agroforestry projects.

Plan Sierra Development Project, Dominican Republic

Plan Sierra is an integrated rural development project with strong agroforestry and reforestation programs. During its first three years (1979–1981) Plan Sierra developed innovative approaches to agroforestry, soil conservation and forestry training and extension. While women were consciously included in some aspects of these programs, nonetheless their interests in "invisible" subsistence activities were overlooked. Securing women's participation and serving women's interests were sometimes confounded. Both the successes and the failures of Plan Sierra's early years exemplify many of the key issues for women in agroforestry development projects.

Production Systems in the Sierra. The Sierra is a rugged, relatively isolated region in the Central Mountains of the Dominican Republic. Household income, health status, and educational level of residents are well below national averages. The economy has suffered from boom-bust cycles in mining and lumbering. Out-migration has been high, particularly among men 20–40 years of age. The area has been largely deforested through commercial timber exploitation and the practice of shifting cultivation, which continues on the forest fringes. Soil degradation and erosion are widespread and the region's hydrologic balance has been severely disrupted [5, 46].

Most agricultural production takes place within agroforestry systems based on mixtures of field crops, coffee, pasture, and forest. These are combined in simultaneous intercropping mixtures as well as in rotations over time. Most farm families manage such diverse holdings for both subsistence and commercial ends.

Coffee and cattle are the major commercial enterprises among large landowners, while smallholders sell coffee or annual crops for cash. Most smallholder households are also heavily dependent on off-farm employment. Subsistence and cash crops are intercropped. Pastures are usually studded with multipurpose palm trees. Local cottage industries include furniture making, food processing, and production of palm fiber containers for sale to the tobacco industry and for local use with pack animals [44].

Women's Participation in Production. Women share with men the harvesting of annual crops and the coffee harvest (as owners and/or hired farmworkers). Women raise the small animals (hogs and chickens) for meat and egg production, they usually milk the cows (for home consumption), and they tend home gardens with vegetables, bananas, and herbs. Responsibility for fuelwood and water gathering falls mostly on the women with some help from the children.

Cheese, candy, and cassava processing are almost entirely women's enterprises. Palm fiber containers are produced and sold by women with help from children and elderly family members. Women artisans also weave the seats and backs onto locally manufactured wooden chair frames. The weaving is subcontracted as piecework from men's woodworking shops.

Plan Sierra Programs. Plan Sierra was an integrated rural development project designed to serve a 2,500 km^2 area within the Sierra. It included strong agricultural, reforestation and soil conservation components from the outset. Agroforestry initiatives emphasized coffee systems. Farmers (mostly smallholders) received subsidized credit and intensive training courses to facilitate establishment of multipurpose shaded coffee stands. Credit and technical assistance were also provided for the establishment or improvement of fruit orchards, often intercropped with annual food crops [46]. Aside from coffee and fruit, tree planting on small farms was not treated separately from reforestation of large-scale state and private holdings. Reforestation efforts focused on indigenous and exotic pine trees for watershed management or timber. Most small farmers were unwilling to plant these trees on their own property because in the Sierra the prohibition against cutting trees (regardless of land title) has been most strongly enforced in the case of pines and other timber species.

Support services for agroforestry and reforestation included soil conservation, nursery development, employee training, and community education programs. Hundreds of local men were hired to construct nurseries and to engage in soil conservation/forestry training, extension and construction activities throughout the 2,500 km^2 project area. Agroforestry and reforestation, were promoted through teacher training courses in local ecology, reforestation and sustainable production systems. The teachers were later enlisted as participants/promoters in community tree-planting campaigns.

Women's Participation in Agroforestry and Reforestation. As in many similar projects in Latin America [57] local women's participation in Plan Sierra was initially limited to health services/home hygiene/home economics [45]. Women professionals were concentrated primarily in Health, Education and Rural Organization units (M. Fernandez, personal communication).

The maturation of Plan Sierra programs and their respective inclusion/exclusion of women provides some insight into issues that need to be addressed in future agroforestry/development projects in Latin America. The key concerns include access to employment, training, credit, land, and appropriate technical assistance.

Women's Involvement in Employment and Training. Local women were initially hired as home-economists, secretaries, cooks, and cleaners. Eventually some of the nurseries hired women to water seedlings and to fill polyethylene bags with potting soil. Male nursery supervisors considered women to be more efficient and patient at this tedious task. Some women who originally took this non-traditional work with reservations later acquired an active interest in plant propagation techniques. Plan Sierra administrators and some technical staff encouraged this trend. They trained a group of paratechnical women horticulturists who became known as the "budders and grafters." The job attained a high status and was accepted as a woman's task, setting a precedent for inclusion of more women in technical nursery work. As in many other similar projects sexual harassment by a few coworkers and managers made work difficult for some women employees, but administrators eventually ousted the offenders. Women began to specifically request assignment to the nurseries and a few expressed interest in conducting forestry or soil conservation field work with visiting women researchers. Some of the "budders and grafters" talked of investing part of their earnings in land and/or small citrus groves, an option they would not have considered previously.

Women's Involvement in Technical Assistance/Training. Women are actively sought as volunteer laborers and promoters for community reforestation. They also become involved in technical training and project implementation through the teachers' training courses. About half the teachers in the Sierra were women, and all teachers attended the same training sessions regardless of individual specialization or prior training. But it was the project not the women themselves who benefited from these activities. This was not by design but by default. Since they were not consulted in project design, their concerns and needs were not adequately addressed in the technical programs. During a mid-project evaluation of women's needs, women requested, among other things, assistance with home gardens and cottage industries (weaving and food processing), all within the domain of agroforestry [13, 44, 45].

Some women who produced woven containers complained of lack of access and/or dwindling, insecure supplies of palm fiber (E. Georges, personal communication). The palms (*Sabal umberculifera* Martius and others) are also important sources of wood, food, and animal feed, and they are often located in pastures or fallow land owned by neighbors or relatives. The women have free access (an apparent advantage) but no guarantee of future access and no control over cutting and replacement. The palm frond supply is both free and unreliable. Supplies in some instances were threatened by the felling of palms for cheap construction material. Many local men decided to use the trees for construction or cash income after their hog-feed value was undermined by a swine fever epidemic and a subsequent embargo on hog-raising.

The management and improvement of this disrupted multiple use agroforestry system were not integrated into either the pasture management or rural industry programs. The marketing of the finished products was addressed by Plan Sierra (Marketing and Rural Industry Unit), nearly doubling the farm gate price of containers. Neglect of the technical and management aspects of this enterprise may have been conditioned, in part, by the low cash income (and low priority) relative to such enterprises as coffee. The importance of this cottage industry to rural women lies in the fact that it requires neither land nor capital and work can be performed by women, children, and the elderly.

The fuelwood shortage also hit hardest in small holdings. Some women had closed down their cassava bread processing operations due to lack of fuelwood nearby and the high cost of purchased fuelwood. Others commented on the increasing time and effort (and trespass) necessary to secure the same quantity of fuel for home use.

Gathering, selection, consumption, and the potential for subsistence production of fuelwood in rural farms or towns were not directly addressed in any of the programs. Research on commercial charcoal production in dry forests had been conducted in the region but these studies were not utilized during this period [32, 38]. Some trials of exotic species including fuelwood species were initiated in 1981 (V. Montero, unpublished) but no experiments or farm trials were conducted with proven indigenous fuelwood trees.

The fuelwood problem was not initially recognized as a high priority issue. When the issue did arise there was a lack of training and experience among technical staff in the choice, propagation, management, and promotion of fuelwood species, particularly for planting on farm. Nor were there any women foresters (or agroforesters) available within the country. As such, a major opportunity to involve and serve local women in agroforestry and reforestation was lost.

Women were more likely to benefit directly from the technical assistance if their cash income or food sources coincided with those most often managed by men. For example those who had (or wished to establish) coffee holdings (as heads-of-households or as spouses of owners/managers) were invited to attend training courses for improved agroforestry systems based on coffee. Women were consciously included although they were not considered to be the main clients of this training/extension/credit program. They were in effect given equal access to training and equal formal commitments for technical assistance in a field considered to be a man's domain. Obtaining credit proved to be more difficult depending on marital status.

Women's Access to Credit and Land. Both of these policy aspects were somewhat beyond the ability of Plan Sierra to change. Some selected subsidized credit and land reform programs were included within the overall project but lack of access to credit and land constituted serious obstacles to implementation of on-farm or community tree-planting projects by and for women, particularly those settled in small towns, on non-tilled land or on property of absentee husbands.

Lessons from Plan Sierra. Plan Sierra illustrates both the possibility for employment and training of women and the need to utilize experience in project definition, technology design, and extension/implementation. This example demonstrated that even in regions where women do not traditionally till the soil, they can and should be offered employment and training in nursery and horticultural techniques, some of which can be identified as women's occupations.

The experience of Plan Sierra also indicates the need for prior consultation with (women) clients of agroforestry development projects about issues of immediate concern to them and about potential action to solve problems or otherwise improve their lot. This would, in many cases, imply a reordering of priorities in project identification, technology design and species selection criteria to better meet the needs of rural women.

Employment and training of women in fields already recognized as important (but not exclusively defined as men's work) could then be extended to training of women personnel for more flexible roles in agroforestry extension programs for rural families, including subsistence farmers and smallholders.

An NGO—The Chipko Movement

The history of the Chipko Movement in the Uttarakhand region of India illustrates the active role of women in mobilizing for forest protection and reforestation. In some cases the divergent interests of men and women served as a key factor leading the women to act as a group both to save existing forests and to choose useful species for reforestation and agroforestry projects.

The Production System of the Himalayan Foothills. The Uttarakhand region encompasses a variety of ethnic groups and a diversity of ecological zones with some features common throughout. Most of the people live in rugged formerly-forested hill country in small, isolated villages. Relative to the rest of India the hill regions are poor and underdeveloped [8]. Large commercial enterprises (mining and lumbering) have been extractive, economically unstable and ecologically unsound [9]. Out-migration has been substantial, particularly among young men [8].

The people of Uttarakhand rely on their immediate surroundings (including the forests) for production of most necessities. Smallholder agriculture is based on subsistence food production (grains and pulses) with supplementary cash crops. Larger landowners rely more on commercial crops including fruit, grains, and other annuals. Animals are kept by both small- and largeholders for milk production, draught power, hides, wool, and sometimes meat.

Cropland and homesteads are privately owned while communally owned forests and grazing lands are managed by village *Panchayats* (councils). Communal forests are sources of fuelwood, food, herbs, medicine, and construction poles, as well as fiber and wood for handicrafts. Forests may also be grazed directly or may serve as sources of "cut and carry" fodder for stall-fed animals. Many of the remaining large tracts of forest held by the state are also used for these purposes by nearby villagers [9].

Role of Women in Production, Uttarakhand. Women plant, weed, and harvest the crops after the men prepare the soil. The burden of fuelwood and water gathering (a 1–15-km daily journey in rough terrain) also falls on the women. Women usually tend the small animals and must often collect fodder for larger stock. Child care, food preparation, and housekeeping are also women's responsibilities. As such, women have a clear interest in the management of cropland as well as in the multiple products of the existing forests [2].

Women's domain tends to be subsistence production, while men concern themselves more with wage labour and/or cash crops [48]. Women's labor, however, is often required by male heads of household for their cash crops enterprises. Men's and women's priorities are largely differentiated along cash and subsistence lines.

The Chipko Movement. The movement began with confrontations between hill people and outsiders over rights to develop or harvest the forests. In one of the first incidents, in 1972, the people of Uttarakashi, a village in Chamoli district, adopted the practice of tree-protection demonstrations, hugging trees to prevent their being cut down. These actions were responses to large scale deforestation of the Himalayan Foothills. Clear-cutting had denuded the mountains and resulted in widespread environmental damage (erosion and floods) as well as the loss of a productive resource for nearby villages [6].

Chipko has gradually developed from a protective movement to prevent commercial clear-cutting, to a broad "movement for the ecological and human rights of the hill people and for adherence to a conservation ethnic" [3]. Since 1975, afforestation has been increasingly emphasized. There are now two distinct sectors within Chipko, one focusing more on protection of existing forests (led by Bahuguna) and the other promoting afforestation and development of sustainable village production systems based on forests and agroforestry. This latter group (led by Chandi Prasad Bhatt of Chamoli District) has joined the Ministry of Environment and others to conduct "eco-development" camps for massive tree-planting campaigns [2, 9]. Through local participation these voluntary afforestation efforts have achieved 85–90% survival rates.

The potential of Chipko in promoting sustainable agroforestry systems for the Himalayas has been widely recognized in non-government circles [2, 56]. The potential role of Chipko and the hill women has also been recognized by government officials. In a letter to the Chief Secretary of the Government of Uttah Pradesh, Swaminathan [53] urged the forest department to follow the lead of Chipko and to support and train the hill women for the task they had already begun.

Women's Participation in the Chipko Movement. The success of Chipko has been largely due to the continued initiative and support of the hill women [2]. While some of the earlier demonstrations focused more on men's demands for local control and the development of timber production, by 1978 the women (and their concerns) had assumed a more prominent role. They emphasized the multiple use/subsistence value of the forests for Himalayan villages [31].

Reports from Himalayan villages indicate that women are prepared to confront their own communities, as well as outsiders to protect Panchayat forests (village commons) from encroachment. One of the key incidents occurred in 1974 when 27 women of Reni village (near the Tibetan border) successfully protected the Panchayat forest against 60 men (some armed) from a neighboring village while their own men were away, engaging in a protest against the contractor. In a more recent case the women of Dungari Paitoli (Chamoli District, Uttar Pradesh) who had not even heard of Chipko, reacted spontaneously to a move by their own men to sell the Panchayat forest in exchange for a potato cultivation project. Loss of this forest stand would have added 5 km per day to their fuelwood collection journeys. With their own efforts and subsequent help from Chipko the women defeated the project but also incurred the wrath of the men. The incident pointed out the need to include women in village decision making and sparked demands for election of women to Forest Panchayats in Chamoli District [2, 32].

Women have also influenced the species selection criteria and general orientation of the tree-planting projects conducted by Chipko, the Ministry of Environment, and other groups [48]. They have requested fuelwood, fodder, and food-producing trees for home use in contrast to the men's preference for fruit or timber trees for sale.

Lessons from the Chipko Movement. The first lesson to be learned from Chipko is that there may be considerable divergence between the interests and priorities of the local community and those of the encompassing state or national system. Similarly, there may be equally strong divergence between the interests of men and women in the local community. Implicit in both is the potential conflict in the needs of the cash and subsistence sectors.

The second lesson is the clear demonstration of women's strong and conscious interest in the management of trees at both farm and village levels. Because they are more dependent on resources gathered off-farm (water, fuel, fodder), women may also be the first to recognize the need for balanced management (or rehabilitation) of local ecosystems. In order for women to bring this about, they must have access to land and/or decision making about land at the farm and village levels and they must receive technical training in the management of existing forests, in reforestation, and in the integration of trees into village and farm grazing lands, croplands, boundaries, and roadsides.

The third lesson is the potential impact of women and women's groups even in the face of superior physical force and the opposition of men in their own families. The dramatic confrontations at Reni and Dungari Paitoli, mentioned above, were based on small women's mutual aid groups [39]. Although none of the Chipko literature analyzes this or other women's organizations involved in the movement, it is not unlikely that much of the women's ability to organize and sustain action flows from women's informal solidarity networks and associations. The similarity to the Women's War described above is striking. The clear lesson for agroforestry projects is the need to include women in decision making about forest use and management in order to insure both that their interests are included in the project and that their very considerable organizational energies are on the side of the project.

Tree Planting and Agroforestry Workshops in Kenya

KENGO (Kenya Energy Non-Government Organizations Association) Tree-Planting and Agroforestry Workshops are conducted in cooperation with the Kenya Ministries of Energy, Environment and Agriculture. These week-long meetings combine training and consultation of rural constituents and implementers of grassroots forestry and agroforestry projects relevant to fuelwood production and use. Because of the nature of the workshops, questions and issues surface which might normally take years to manifest themselves clearly within projects or government programs. The Kisii district workshop typifies many of the women's issues facing agroforestry in Kenya. Additional information was also drawn from proceedings of a prior meeting in Kitui [37]. These events serve not only as sources of information per se but as concrete examples of how to set about (begin) consulting and involving rural people in decision making and training. Women and women's concerns are explicitly included in this process.

Production Systems in Kisii District. Kisii is a densely populated, intensively cultivated, high rainfall district within Nyanza Province. The hilly terrain is almost entirely devoted to privately owned crop and pasture land, in equal proportions. Smallholder farms of similar size predominate. Crop production systems are a mix of cash and substance enterprises with emphasis on commercial products (tea, coffee, pyrethrum, and sugar cane). Maize and bananas are the subsistence staple crops [52]. Animal production is a supplementary enterprise that supplies milk for home consumption and sale.

Exotic fuelwood and timber trees (*Eucalyptus saligna, Acacia mearnsii* and *Cupressus spp*) are closely fit into the farmlands [52]. These exotic trees have displaced but have not completely replaced indigenous vegetation in the region. They are planted along fencerows, in ravines, in small woodlots near boundaries and are sometimes intercropped with cash and food crops. Building poles grown on-farm are a source of cash for many households. Fuelwood has also become a commercial good which families purchase regularly from those who manage to produce a surplus.

Hedgerows between farms consist of a diverse species mix, including herbaceous and woody plants used for fuelwood, cattle fodder, medicines and fiber. Fallow plots are often grazed or may be planted to soil-improving multipurpose herbaceous and woody species. Scattered swamps and hilltop woodlands also provide fuelwood, a wide selection of fibers, and medicinal plants. Communal swamp and hilltop lands are managed by the county councils.

Off-farm labour is a major source of income in Kisii. Both men and women pick tea for piecework wages and many men work in the towns or in urban areas outside the district.

Role of Women in Kisii Production Systems. Women who head households are the main decision makers and perform or manage most of the farm work, including soil tillage, maintenance of buildings, and other tasks considered men's work. In all cases women collect the water and fuelwood; even purchased fuelwood must be transported, a service performed (for a fee) by women's groups. Women also collect fodder for cattle, grass for thatching, fiber for handicrafts, and a wide variety of medicinal plants. While the men are considered to be the main crop managers in Kisii, women always participate in planting and harvesting. Food preparation, child care, and housekeeping are also women's work. Most women weave baskets and mats with fibers gathered from woody shrubs in the swamps. These may be used in the home and can also be an important source of cash income.

Summary of the Kisii Workshop. Approximately 40 people attended the Kisii workshop, including five representatives of women's groups (local and district levels) and a fairly even mix of men representing church groups, farmers, and local officials (chiefs, sub-chiefs). Two District Forest Officers (Busia and Kisii) and the District Agricultural Officer were key participants, as government representatives directly responsible for nursery management, extension programs, and implementation of special projects. Ministry of Energy and Ministry of Agriculture personnel from the Farmer's Training Centre nurseries served as tutors. Also present were ten technical resource people from outside the region, five of whom were women.

The five-day workshop itinerary included presentations by resource people, group discussions, and group participation in practical nursery exercises and cookstove demonstrations. A field trip combined observation and discussion of existing farming systems with visits to small nurseries and to forestry and agricultural project sites. Small group discussions of extant agroforestry systems produced a list of existing practices and useful species. The groups also identified potentials and constraints for agroforestry improvement/development in the region. A separate informal meeting/consultation was held for women participants, resource people, and FTC employees.

Women's Participation in the Kisii Meetings. Although the women were included throughout, their most valuable contributions and their strongest expression of interest were confined largely to the informal small-group activities, such as the nursery practicum, identification and description of plants and practices during the field trip, and open exchange and discussion of information during the informal women's meeting. During the frequent and even heated exchanges of questions, answers, criticisms, and suggestions among the men in the larger sessions, the women were silent—even when topics directly relevant to them, such as improved stoves or intercropping food, fuel, and cash crops, were under discussion. In subsequent interviews the women requested that they be involved more with separate discussions of fuelwood, fiber, fodder, and cookstoves, and that "agroforestry" be left to the men. The decisions about cropping systems, they contended, would not be theirs to make. The men echoed similar sentiments about division of interests (though not about spheres of control) when they asked why they were being subjected to "irrelevant" discussions of cooking, cookstoves, and fuelwood.

During the women's meeting, both existing production systems and potential improvements were discussed. Information was freely offered as to the amount of fuelwood used, the time and/or money spent to collect and deliver it, the role of women's groups in fuelwood delivery, the degree of dependence on off-farm sources, and the species used, in order of preference. The group also touched on fodder collection, since most women in the area were said to collect fodder off-farm. Another issue that surfaced, both in this context and on the field trip, was importance of woody shrubs and herbaceous plants found in hedgerows, along roads, and in unoccupied bottomlands. Some of the women identified more than 20 indigenous species highly valued and frequently used by them. The forestry and agricultural personnel were unfamiliar with most of these, especially the non-commercial shrubs and herbs. These plants are often multipurpose and may provide fuel, fiber (two or more types), fodder, and medicinal ingredients. This is a critical consideration to incorporate into decisions about the future development of bottomlands in the area [52].

The interview-discussion with women participants raised issues about the kinds of fuelwood required, the type of multipurpose trees/shrubs that might be introduced, indigenous trees and shrubs that could be propagated/managed more intensively, the potential for incorporating more of these plants on-farm, and constraints. Among the constraints cited were control of farm land and labor and negative attitude of extension agents (and husbands) towards trees in the cropland. The separate meeting was necessary, but not sufficient, to define the social specifications for agroforestry systems to meet the women's needs. With their information it was possible to pose more pertinent questions to the men in order to identify points of convergence, or conflicts of interest. Fencerows were suggested by both men and women as a potential niche for fuelwood trees. Other possible niches are small fuel/fodder lots in improved fallow plots (short rotation), and interplanted rows of fuel and polewood trees, the latter being more of interest to the men, who could provide land and labor for clearing. The use of separate species for fuel and poles would avoid problems with allocation of a given plant to two conflicting uses by different parties

Lessons from the Kisii Agroforestry Workshop. The contrast in the apparent interest and knowledge exhibited by the women in formal and informal contexts was striking. The mere format of the mixed group evidently imposed restrictions on the level of participation by women. It would be even less likely to elicit comments between men and women in the distribution and management of farm resources. Issues of control and decision making can hardly be raised, let alone worked out, unless there is a chance to consider these as a separate group.

The experience from this workshop indicates the need not only to include women and issues of concern to them, but to provide a flexible format for them to explore and express those concerns. Some provision for separate discussion by women is critical, for reasons of both form and content of participation.

Another issue critical to the development and implementation of agroforestry technologies is the nature of available training, from district level supervisors to farm household members. Agroforestry is not generally well understood, and the aspects most relevant to women are rarely included in the training of technical personnel. Further local knowledge about indigenous plants (including shrubs and herbs) needs to be recorded, disseminated, and incorporated into improved technologies and training programs.

Existing knowledge and technology need to be taught to both men and women who are expected to apply it on the farm. The district women's group representative asked the resource people: "What can you give me to show a group of people how to choose the right plants, where to get seeds or seedlings and how to grow them? I need something to work with." The answer is at present there is almost no basic training and promotional material on agroforestry, indigenous plants, seed collection/distribution [54], multipurpose trees, and preparation techniques. In addition, training personnel and documents need to be quite distinct for men and women in Kisii, given their differences in language and literacy, and divergent perspectives on agroforestry technologies and plant species.

The Kisii workshop raised several issues but perhaps strongest among them is the need for programs, personnel, and training materials geared specifically for women, taking into account the complete range of roles in rural production (existing and potential) and in community affairs. Within such programs (whether in forestry, agriculture, home economics, or rural development institutions) the sexual division of labor, interests, control, and benefits on-farm and within-community must be recognized and addressed. This applies to both form and content of any program that is to succeed in involving and serving rural women.

Lessons for the Future

We can learn a number of lessons about women's involvement in agroforestry from these case studies and the literature reviewed.

Participation in Implementation and Benefits Are Not the Same

The women who worked as "builders and grafters" in Plan Sierra benefited in terms of wages and knowledge which they could transfer to their own private economic activity. The number of such women was limited by the size of the project. Women also participated in tree planting. However, the project did not address two of the issues most important to women as a whole: fuelwood and fiber for handicrafts. Unless care is taken, women's participation in a project in the form of their labor may give them relatively little in return.

Men's and Women's Priorities for Agroforestry May Differ

The division of labor in agriculture has often resulted in men's involvement with export cash crops and women's involvement in the subsistence sector. Plan Sierra focused on the men's export crop of coffee while the women's non-export cash enterprises received lower priorities. In Chipko, women's subsistence needs conflicted with men's interest in wage employment. Similarly, in Kisii women's desire for fuel and fodder trees conflicted with the male extension worker's image of good farming practices. There is a clear need to establish who uses what trees for what purposes before planning a project.

Men and Women Have Differential Access to Resources

The women in Plan Sierra suffered from lack of access to credit and land. This is a problem experienced by women, especially women heads of household, the world over [1, 3, 17, 20, 21, 26, 27, 36]. The women of Kisii were ignored by their agricultural extension agent. Lack of extension contact is also a problem typically faced by women farmers [3, 20, 32, 38, 51]. The rights of men and women over trees may also differ [41].

Project design must take into consideration what resources women actually have to work with. Either resource constraints must be alleviated or the project must be geared to the resources women actually control. Failure to do this will result in an exclusion of women from project benefits. In terms of extension contact, the case has been made repeatedly for female extension workers to work with women. This strategy not only avoids restrictions on male-female contact and interaction, but it is likely to facilitate communication between the female farmer and the extension worker. Reluctance on the part of male forestry officials or extension staff to work with women may have to be overcome by requiring them to report their visits with women [5].

Not All Women Are Alike

The nature of women's participation in agriculture has been shown to differ by social class and control of resources [16, 20, 21, 49]. Just as men and women's priorities may differ, the priorities of rich and poor may differ.

This point is especially important to remember when women's organizations (formal or informal) are to be involved in promoting/utilizing agroforestry. It is often the case that only wealthier, better educated women have the leisure time to spend in formal organizations. It is also often the case that such women capture the leadership of both formal and informal organizations. While there is a great deal to be said for utilizing the organizational capacity of women's organizations and associations, care must be taken that the poor are not excluded by this means. Associations should be identified which are "structured so as to redistribute introduced resources equitably" and in which "all members participate equally in or have equal access to group decision-making procedures and avenues of redress" [35].

Different women and different women's groups and associations will have distinct interests and capacities. This should be viewed as a programming strength, not a difficulty.

Special Arrangements May Need to be Made for Women's Participation

The Kisii meeting demonstrated the chilling effect the presence of men may have on women's participation. In some cases it may be necessary to approach women separately in order to learn what they know and what they want. But at other times, it is necessary to be sure that men and women are involved together in order to alleviate any fears that men may have about their wives' activities, and to coordinate family participation.

Summary

Women are traditionally active participants in both the agricultural and forestry components of agroforestry production systems. They are also private and public participants in community life and decision making. The failure to include women in agroforestry projects has several detrimental effects. It excludes the increasing proportion of rural households which are headed by women from project benefits. It may prevent project designers from benefiting from women's special knowledge. It may exclude (or even harm) activities and products such as fuelwood, basket making and minor forest products which are part of women's economic sphere.

The inclusion of women is essential for the success of agroforestry projects but it may require change in both approach and personnel of forestry and extension departments. This would include diagnosing the existing agricultural and resource management systems to determine what women presently use and what they need; analyzing the constraints imposed on women's agroforestry by social institutions such as land tenure or property laws or the division of labor by gender; arranging a format for discussions with women which enables them to express their concerns and questions freely; and by hiring women extension workers and technical personnel who can work easily with both men and women.

References

1. Acharya, M., and Bennett, L. (1981). The rural women of Nepal Volume II Part 9, Kathmandu, Centre for Economic Development and Administration.
2. Agarwal, A., Chopta, R., and Sharm, K. (1982). The state of India's environment 1982, a citizen's report. New Delhi, Centre for Science and Environment, pp. 33–55.
3. Alberti, A. (n.d.). Some observations of the productive role of women and development efforts in the Andes. Unpublished manuscript.
4. Allan, W. (1965). The African husbandman. Edinburgh, Oliver and Boyd.
5. Antonini, G., Ewel, K., and Tupper, H. (1975). Population and energy: a system analysis of resource utilization in the Dominican Republic. Gainesville, University of Florida Press.
6. Bahuguna, S. (1982). Walking with the Chipko message. Tehri-Garhwal, Chipko Information Centre.
7. Bennett, L. (1981). The Parbatiya women of Bakundal. The status of women in Nepal. Volume 11 Part 7. Kathmandu, Centre for Economic Development and Administration.
8. Berreman, G. D. (1972). Hindus of the Himalayas: ethnography and change. Berkeley, University of California Press.
9. Bhatt, C. P. (1980). Ecosystem of the Central Himalayas and Chipko movement, determination of hill people to save their forests, Gopeshwar (UP), Dashauli Gram Swarajya Sangh, pp. 7–31.
10. Boserup, E. (1970). Women's role in economic development. London, George Allen and Unwin.
11. Bryson, J. C. (1981). Women and agriculture in Sub-Saharan Africa: implications for development (an exploratory study). J. Develop Stud 17(3):28–45.
12. Buvinic, M., and Youssef, N. H. (1978). Women-headed households: the ignored factor in developing planning. Washington, DC, International Centre for Research on Women.
13. Chaney, E., and Lewis, M. (1980). Planning a family food production program, some alternatives and suggestions for Plan Sierra, San Jose de las Matas, Plan Sierra.
14. Cliffe, L. (1975). Labor migration and peasant differentiation: Zambian experiences. J Peasant Stud 5(3):326–346.
15. Coffer, C. J. P. (1981). Women, men and time in the forests of East Kalimantan. Borneo Research Bulletin 1981 (September): 75–85.
16. Deere, C. D. (1982). The division of labor by sex in agriculture: a Peruvian case study. Econ Dev Cu 30(4):795–811.
17. Dietel, E. (1982). A profile of small-scale traders in Western Kenya: an alternative credit approach for women. Cornell University: Unpublished Masters Thesis.

18. Dixon R. (1983). Land, labour and the sex composition of the agricultural labour force: an international comparison. Develop Cha 14(3):347–372.
19. FAO Committee on Agriculture. (1982). Follow-up to WCARRD: the role of women in agricultural production. Rome. FAO.
20. Fortmann, L. (1979). Women and agricultural development. In Kim, K. S., Mabele, R., and Schultheis, M. J., eds. Papers on the political economy of Tanzania. Nairobi, Heinemann Educational Books, Ltd.
21. Fortmann, L. (1984). Economic status and women's participation in agriculture: a Botswana case study. Rur Soc 49(2):452–464.
22. Fortmann, L. (1983). The role of local institutions in communal area development. Gaborone, Applied Research Unit, Ministry of Local Government and Lands.
23. Fouad, I. (1982). The role of women peasants in the process of desertification in Western Sundan. Geojournal 6(1):25–30.
24. Gupta, R. K. (1980). Alternate strategies for rural development in Garhwal Himalaya. In: Singh, T., and Kaur, J., eds. Studies in Himalayan ecology and development strategies. New Delhi, The English Book Store 218–228.
25. Gupta, R. K., Singh, G., Katiyar, V. S., Bhardwaj, S. P., Puri, D. N., Ram, Babu, and Tejwani, K. G. (1979). Watershed management—a tool for integrated rural development and flood control. Paper presented at the National Symposium on 'Soil Conservation and Water Management in the 1980's'. Dehra Dun, Central Soil and Water Conservation Research and Training Institute.
26. Hammer, T. (1977). Wood for fuel: energy crising implying desertification. The case of Bara, the Sudan. University of Bergen.
27. Hammer, T. (1982). Reforestation and community development in the Sudan. Bergen, Development Research and Action Programme, The Chr. Michelsen Institute.
28. Hoskins, M. (1982a). Observations on indigenous and modern agroforestry activities in West Africa. Paper presented at the United Nations University Workshop 'Problems of Agroforestry,' University of Freiburg.
29. Hoskins, M. (1982b). Social Forestry in West Africa: myths and realities. Paper presented at the annual meeting of the American Association for the Advancement of Science, Washington, DC.
30. Hoskins, M. (1983). Rural women, forest outputs and forestry projects. Rome, FAO.
31. Ioshi, G. (1982). Men propose, women oppose the destruction of forests. New Delhi, Information Service on Science and Society-Related Issues, Centre for Science and Environment 1–5.
32. Jennings, P., and Ferreiras, B. (1979). Recursos energeticos de bosques secos en la Republica Dominicana. Santiago R D, Centro de Investigaciones Economicas y Alimenticias, Institute Superior de Agricultura.
33. Ki-zerbo, J. (1981). Women and the energy crisis in the Sahel. Unasylva 33(133):5–10.
34. Kuper, H., Hughes, A. J. B., and van Velson, J. (1954). The Shona and Ndebele of Southern Rhodesia. London, International African Institute.
35. March, K., and Taqqu, R. (1982). Women's informal associations and the organizational capacity for development. Ithaca, Cornell University Rural Development Committee.
36. Mazumdar, V. (1982). Another development with women: a view from Asia. Development Dialogue 1982 (1-2):65–73.
37. Mengech, A., and Aworry, A. (1982). Proceedings of the Kengo Workshop held in Kitui in October, 1982. Nairobi: Kenya Energy Non-governmental Organizations Association.
38. Mercedes, J. (1980). Estudios Para el diseño de un sistema de manejo integrado en areas de bosque seco en La Republica Dominicana. Ag Eng Thesis, Santiago, Instituto Superior de Agricultura Universidad Catolica Madre y Maestra.
39. Mishra, A., and Tripathi, S. (1978). Chipko Movement, Uttarakhand women's bid to save forest wealth. New Delhi, People's Action.
40. Molnar, A. (1981). The Kham Magar women of Thabang. The status of women in Nepal Volume II Part 2. Kathmandu, Centre for Economic Development and Administration.

41. Obi, S. N. C. (1963). The Ibo law of property. London, Butterworth.
42. Pradhan, B. (1981). The Newar Women of Bulu. The status of women in Nepal Volume 11 Part 6. Kathmandu, Centre for Economic Development and Administration.
43. Rajaure, D. (1981). The Tharu women of Sukhwar. The status of women in Nepal Volume 11 Part 3. Kathmandu, Centre for Economic Development and Administration.
44. Rocheleau, D. (1983). An ecological analysis of soil and water conservation in hill-slope farming systems: Plan Sierra, Dominican Republic. Gainesville, University of Florida, Department of Geography.
45. Safa, H., and Gladwin, C. (1981). Designing a women's component for Plan Sierra. Gainesville, Centre for Latin American Studies, University of Florida.
46. Santos, B. (1981). El Plan Sierra: una experiencia de desarrollo rural en las montanas de la Republica Dominicana. In Novoa, A., and Posner, J., eds. Agricultura de ladera en America Tropical. Turrialba, C.A.T.I.E.
47. Scott, G. (1980). Forestry projects and women. Washington, DC, The World Bank.
48. Sharma, R. (1981). Greening the countryside. New Delhi, Information Service on Science and Society—Related Issues, Centre for Science and Environment: 1–4.
49. Sharma, U. (1980). Women, work and property in North-West India. London, Tavistock Publications.
50. Spring, A. (1983). Extension services in Malawi. Paper presented at the XIth international congress of anthropological and ethnographic sciences, Vancouver.
51. Staudt, K. (1975/76). Women farmers and inequities in agricultural services. Rur Afr 9:81–94.
52. Stroud, A. (1983). A vegetation assessment of Kisii District, Kenya. Nairobi, National Environment and Human Settlement Secretariat.
53. Swaminathan, M. S. (1980). Ecodevelopment of the Uttrakhand Region. In: Bhatt, C. P., Ecosystem of the Central Himalayas and Chipko Movement. Gopeshwar, Deshauli Gram Swarajya Sangh 3–6.
54. Teal, W. (1984). The Kenya public tree seed directory. Nairobi, Heinemann Educational Books, Ltd. (In press).
55. United Nations Economic and Social Commission for Asia and the Pacific. (1980). Report of the expert group meeting on women and forest industries. Bangko.
56. Varghese, B. G. (1978). Introduction. In: Mishra, A., and Tripathi, S., Chipko Movement, Uttarakhand women's bid to save forest wealth. New Delhi, People's Action 1–3.
57. Wiff, M. (1977). La mujer en el desarrollo agroforestal en America Central. Annex to the report on the FAO/SIDA seminar on the role of silviculture in rural development in Latin America. Rome, FAO.
58. Wood, D. H., et al. (1980). The socio-economic context of firewood use in small rural communities. Washington, DC, USAID.

Epilogue

Over the last ten years both the context for women and agroforestry and our own understanding of it have changed considerably. In 1984 we wrote from a professional context that was reluctantly receptive, at best, and actively hostile, at worst, to the prospect of women as practitioners, professionals, and clients in the domain of agroforestry technology design, development, and dissemination. Since then we have moved from women's roles in agroforestry (never our own choice, but an oft-requested topic) to women's interests at stake in agroforestry, to gender and agroforestry. Most recently we have engaged in feminist studies of gendered agroforests and the gendered science and practice of agroforestry. We have moved beyond the plots and farms focus of the early 1980s to the study of multiple sites of gendered political power over trees and people: from fields, forests, labs, research stations, and rural households, to the corridors of agrarian, environmental, and forest policy. And we have returned to the gendered fields of power on the ground, to the patchwork of people, trees, crops, and livestock living in relation in rural landscapes. Rather than "women's roles in agroforestry," we focus on agroforestry as both agent and outcome of gendered land, labor, and commodity markets, shaping and being shaped by gendered ecosystems and landscapes.

The four myths that we posited as driving agroforestry policy and practice in 1984 persist, although they are perhaps less pervasive. Many agroforestry research and development institutions pay more attention to women. Although relatively few institutions still buy into the first two myths—that women do not participate in agriculture or forestry—some have not gone beyond considering women as merely a source of labor to be exploited for agroforestry technology interventions. Institutions that have dispensed with the fourth myth that women are politically powerless have mobilized women's political support for conservation or production agroforestry programs. The third myth, that all women have husbands and live in male-headed households, is more entrenched. Some agroforestry programs have acknowledged the existence of female-headed households and have invited those women to participate in programs designed for men, as if they were men. Most agroforestry programs have yet to recognize that such households may have different priorities and constraints. Still unchallenged, even further beneath the surface, rests the corollary of the "women as subservient wives" myth: a widespread assumption that women in male-headed households share the interests, priorities, and opportunities of their husbands or the household as a unit. This often leads to policies based on a mistaken expectation that agroforestry project benefits will "trickle down" from male participants to the women and children in their families.

As agroforestry projects have become increasingly environmental in their focus and have moved beyond the plot and farm, some of these old myths have been reworked. The unitary, homogeneous community has replaced the unified household in the new mythology, which continues to obscure the specific interests of women in general, the poor, and less powerful ethnic, religious, or occupational groups.

There has been, however, a groundswell of alternative efforts to address both women and men within a framework of multiple land users with legitimate claims on the time, attention, and material resources of agroforestry policy, technical assistance, and research institutions. These innovations and demands come from technocratic and academic circles as well as from the ranks of grassroots activists, exploring questions of gendered knowledge, gendered places and property, and gendered organizations.

Since we wrote, the academic and policy conversation about women and agroforestry has expanded in a number of ways. Both the field of agroforestry as a whole and gender and agroforestry field studies have broadened conceptually and geographically from a "Third World" to a global/local focus that includes North America and Europe, spanning urban, suburban, and rural contexts. Having introduced women and gender onto the international agroforestry agenda, many feminist scholars and practitioners have entered into a series of debates within feminism, based on questions about the origin and "ranking" of gender inequities relative to class, race, nationality, sexuality, and a host of other differences among women. Conversely, the field studies and development innovations on gender and agroforestry have provided important insights for feminist scholars as well as for political ecologists and cultural ecologists in academic and policy circles, and have contributed to the formulation of feminist political ecology as a field of study.

Five major areas promise to contribute theory, policy, and practice over the next few years:

1. The nature of gendered agroforests on a local and regional scale within gendered ecosystems and economies
2. Gendered ecosystems and biodiversity
3. Gendered landscapes, resource tenure, and property
4. Gendered resource mapping and countermapping
5. Gender, communities, and conflicts over trees, forests, and their production, mapping, and resolution

Over the past decade we have participated as both fieldworkers and as visiting researchers in the gendered practice and politics of agroforestry research and development and have reflected as academics on the current trend. What looms largest for us now is to build a useful and ethical coalition of theory and practice to address the concerns of women and men in the gendered agroforests of the planet, from city streets to dryland and farms to ancient rainforests.

PART FOUR

THE GENDERING OF ENVIRONMENTAL AND SOCIAL MOVEMENTS

Photograph courtesy of Carolyn Sachs.

Chapter 13

Men, Women, and the Environment: An Examination of the Gender Gap in Environmental Concern and Activism

PAUL MOHAI

School of Natural Resources
University of Michigan
Ann Arbor, MI 48109

Abstract *Relatively little information yet exists regarding gender differences in environmental concern and activism. What information is available has so far provided a mixed picture, with some studies indicating men to be more concerned than women, others indicating women to be more concerned, and still others finding no significant differences. This study provides additional evidence from national survey data. From these data, women were found to express greater concern for the environment than men before and after applying multivariate controls for age, education, labor force/homemaker status, and other variables. However, the magnitude of the differences was not great. Gender differences in environmental activism provided an ironic contrast. Even though women indicated somewhat greater concern, rates of environmental activism for women were substantially lower than for men. Furthermore, these differences were greater than differences in rates of general political participation and persisted in spite of multivariate controls for socioeconomic status, homemaker status, and other variables. That the environmental activity of women appears to be constrained by factors in addition to those constraining general political activity is similar to earlier findings regarding the environmental activity of blacks. Common threads in these findings are explored.*

Keywords Environmental attitudes, environmental movement, gender, political participation.

Introduction

Studies of gender differences in concern for the environment have been relatively few, especially ones national in scope. Much of the information that currently exists about such differences is from studies that have examined concerns about *local* environmental issues and problems that pose a potential threat to community health and safety (Blocker and Eckberg, 1989; Brody, 1984; George and Southwell, 1986; Hamilton, 1985a, 1985b; Nelkins, 1981; Passino and Lounsbury, 1976; Solomon, Tomaskovic-Devey, and Risman, 1989). These studies have consistently shown women to be significantly more concerned about such issues than men.

Studies that have compared men and women on attitudes about *general* environmental issues, that is, environmental issues not specifically limited to those in the neighborhood or community (Arcury and Christianson, 1990; Arcury, Scollay, and Johnson, 1987; Blocker and Eckberg, 1989; Lowe and Pinhey, 1982; McStay and Dunlap, 1983;

Mitchell, 1979; Van Liere and Dunlap, 1980), have been less clear. Differences have tended to be modest. Moreover, the direction of difference (i.e., women more concerned versus men more concerned) has varied from study to study. As a result, no firm conclusions can be drawn about the effects of gender on concern about general environmental issues, and more analysis and explanation clearly need to be done in this area.

Although much of the interest in examining gender differences in environmental concern comes from an interest in understanding the impact of women on the environmental movement (see, for example, McStay and Dunlap, 1983), surprisingly little has been done to examine the extent of environmental activity of women and factors related to it. Ironically, what information exists has tended to show that even though women may be somewhat more concerned about the environment than men, they are less politically active on these issues (McStay and Dunlap, 1983; Mitchell, 1979). Why women's concerns about the environment should not translate proportionately into activism is unknown.

Given the less than complete understanding of the influence of gender on environmental concern and activism, further examination of its effects is made in this paper. Specifically, the purpose is to evaluate the extent of gender differences in environmental concern from national survey data and to evaluate alternative explanations of such differences. An additional purpose is to examine the links between gender and environmental *activism*. As there are yet no explanations of these linkages, hypotheses are derived from the literature on the general political participation of women and tested.

Gender Differences in Environmental Concern

Theoretical explanations of gender differences in environmental concern lead to the expectation that women are more concerned about such issues (local or otherwise) than men (Arcury et al., 1987; Blocker and Eckberg, 1989; George and Southwell, 1986; McStay and Dunlap, 1983). This expectation is based on the argument that from childhood on women are socialized to be family nurturers and caregivers, that is, to develop a "motherhood mentality." The nurturing attitudes that result from this socialization carry over to concerns about a wide range of social issues such as poverty, homelessness, problems of the elderly, and racial discrimination. Similarly, these attitudes translate into attitudes toward nature and the environment that are more protective than those of men. Moreover, it has been hypothesized that the attitudes obtained from this socialization are reinforced by the roles that women assume in their adult lives as homemakers and mothers. In contrast, men are socialized to be family breadwinners and economic providers. The socialization of men results in a "marketplace mentality" that gives priority to economic growth and development and that may portray environmental pollution as a necessary tradeoff for growth (Arcury et al., 1987; Blocker and Eckberg, 1989; George and Southwell, 1986; McStay and Dunlap, 1983). As in the case of women, the attitudes acquired through socialization may be further reinforced by the roles men assume in adulthood, namely, as employees in the workforce and as family providers.

Whether women, in reality, are more concerned about the environment than men has not been determined conclusively by empirical studies. The clearest and strongest evidence for gender differences has come from studies examining concerns about local environmental issues such as nuclear power (Brody, 1984; George and Southwell, 1986; Nelkins, 1981; Passino and Lounsbury, 1976; Solomon et al., 1989), toxic waste contamination of the local water supply (Hamilton, 1985a, 1985b), local energy development (Stout-Wiegand and Trent, 1983), and local air and water pollution (Blocker and Eckberg, 1989), with women tending to express greater concern than men. It has been shown that concern about such local hazards involves fears about threats to health and safety (Brody, 1984; George and Southwell, 1986; Nelkins, 1981; Passino and Lounsbury, 1976).

In contrast, evidence regarding gender differences in concern about general environmental issues has been slight and often contradictory. Blocker and Eckberg (1989) found no statistically significant differences between men and women in the Indianapolis area regarding concerns about general environmental issues. From their statewide survey of Washington, McStay and Dunlap (1983) found gender differences to be statistically significant but concluded that the differences were modest. From national surveys, Lowe and Pinhey (1982) and Mitchell (1979) also found such differences to be modest. Although it was not statistically significant, Arcury et al. (1987) found men to be more concerned than women about acid rain in a statewide survey in Kentucky. Likewise, Arcury and Christianson (1990) found men to score higher on their "environmental worldview" scale. Van Liere and Dunlap (1980) too observed, in their review of studies conducted in the 1960s and 1970s, that gender differences regarding concern about general environmental issues have tended to be slight and often contrary, with men showing more concern in some studies and women showing more concern in others.

Evidence supporting the family nurturer and economic provider explanations has been particularly limited. Tests of the effects of parental, homemaker, and workplace roles on gender differences in concern have provided some evidence about these explanations, but the results have often been mixed. For example, Hamilton (1985a, 1985b) found in his studies of several New England communities that women with children were significantly more concerned about toxic waste contamination of the local water supplies than women without children, or than men (with and without children). In contrast, Blocker and Eckberg (1989) found that parenthood had no effect on the concerns of either men or women in the Indianapolis area regarding local air and water pollution. Similarly, in a study of San Luis Obisbo County residents, George and Southwell (1986) found that parenthood had no effect on women's concerns about the startup of a local nuclear power plant, although men with children were found to be more in favor than men without. In examining the effects of full-time employment and homemaker status, Blocker and Eckberg (1989) found that neither was related to concerns about local air and water pollution.

So far, Blocker and Eckberg's (1989) study has been the only one to examine the effects of family and workforce roles on concern about general environmental issues. Interestingly, whereas they found these roles to be unrelated to concerns about local environmental issues, they did find them to be related to concerns about general environmental issues—but in the unanticipated direction. Contrary to the economic provider explanation, they found that full-time employment in the labor force was positively rather than negatively related to one of two general environmental concern scores. Furthermore, contrary to both family nurturer and economic provider explanations, women who identified themselves as homemakers were found to rank lower, not higher, on the two scores than women employed full time in the labor force. In the case of parenthood, Blocker and Eckberg obtained mixed results. They found fathers to rank lower on one of their environmental concern scores than men without children, while mothers ranked higher on this score than other women.

Taking these results together, the lack of consistent evidence for the effects of family and workplace roles casts doubt on family nurturer and economic provider explanations.

At the least, to the extent that these explanations are valid, gender differences may be more dependent on socialization differences than on the roles men and women come to occupy.

In sum, evidence to date regarding gender differences in concern for the environment is mixed. Differences appear to be greatest and most consistently found in regard to concerns about local environmental hazards, hazards that may pose immediate threats to family health and safety. Differences in concern regarding general environmental issues are less clear, with most studies to date revealing rather slight, contrary, or nonexistent differences between men and women. Likewise, evidence regarding the effects of family and workplace roles on concern has been mixed, raising doubt about family nurturer and economic provider explanations. Given the inconclusive nature of the evidence, a purpose of this paper is to provide additional information from national survey data. Specifically, from these data the extent of gender differences in concern about general environmental issues in evaluated. Also, the effects of homemaker and labor force roles are examined and an explanation of their effects is elaborated. Although additional information about the effects of parenthood is also desirable, such information was not available from the national survey data used in this study. In any event, Blocker and Eckberg's results suggest that homemaker status and labor force status may have a more important effect on concern about general environmental issues than parental status.

Gender Differences in Environmental Activism

If women are more concerned than men about the environment, then one might expect that higher levels of concern should translate into higher levels of environmental activity. Although little evidence exists regarding gender differences in environmental activism, what data are available indicate that this is not the case. Apparently, even though women may be somewhat more concerned about general environmental issues, they are somewhat less politically active on these issues than men. For example, from a national survey Mitchell (1979) found that 7% of all women, compared to 9% of all men, claimed they belonged to either a national or local environmental organization. From a statewide survey in Washington, McStay and Dunlap (1983) found that men were more likely than women to indicate that they have engaged in activities such as contacting a government official, contacting a business or industry, and writing a letter to a newspaper or magazine regarding their concerns about an environmental issue. On the other hand, they found no significant differences between men and women on activities such as attending a public meeting or hearing, signing a petition, and donating money or time to an organization. Why have the environmental concerns of women not translated into activism proportionately to that of men? McStay and Dunlap theorize that the lower levels of environmental activity of women may be a consequence of women's lower level of political activity generally. In other words, whatever barriers inhibit the general political activity of women may similarly inhibit environmental activity.

The literature on gender differences in general political participation thus may provide clues to explaining gender differences in environmental activism. Evidence regarding gender differences in general political participation is extensive and appears to point to a consistent pattern. On the whole, women appear to be less politically active than men, although the differences tend to be slight (Andersen, 1975; Andersen and Cook, 1985; Bourque and Grossholtz, 1974; Lansing, 1974; McDonagh, 1982; Milbrath,

1965; Smith et al., 1980; Verba and Nie, 1972; Welch, 1977; Welch and Secret, 1981). Three explanations to account for gender differences in political participation have emerged. Welch (1977) has referred to these as (1) the political socialization explanation, (2) the situational explanation, and (3) the structural explanation.

The *political socialization* explanation takes the view that women are socialized from childhood on to be politically passive while men are socialized to be politically active (Andersen, 1975; Andersen and Cook, 1985; Campbell et al., 1960; Smith et al., 1980; Welch, 1977). That is, from very early on boys and girls are socialized to view politics as a "man's game." Furthermore, as women assume mother and homemaker roles in adulthood, politically passive attitudes may be reinforced as women's attention is drawn toward home and family life and away from "outside world" events and politics (Andersen, 1975).

Focusing more on the constraints imposed on women, rather than on their political attitudes, the *situational* explanation takes the view that the roles of mother and homemaker tie women down with family and home responsibilities, leaving them with little time and opportunity to take part in activities outside the home, such as in the political arena (Orum et al., 1974; Welch, 1977; Welch and Secret, 1981). Also, being tied to the home, homemakers do not have the opportunity to get involved in political discussion that would stimulate their interest and involvement in politics.

The *structural* explanation views gender differences in political participation as resulting from differences in socioeconomic characteristics such as educational attainment, income, and occupational status (Welch, 1977; Welch and Secret, 1981). It has been shown that these variables, especially education, are strong predictors of political participation (Milbrath, 1965; Smith et al., 1980; Verba and Nie, 1972). Generally, socioeconomic factors are believed to be related to factors that facilitate political participation. These include the availability of resources (such as time, money, or knowledge of the political system), organizational affiliations, sense of political (personal) efficacy, and other factors (Mohai, 1985; Smith et al., 1980; Verba and Nie, 1972). Since women, on the whole, historically have had lower education, income, and occupational status than men, their level of political activity can in turn be expected to be lower (Welch, 1977).

Of the three explanations, the socialization explanation appears to be the least supported (Bourque and Grossholtz, 1974; McDonagh, 1982; Orum et al., 1974; Welch, 1977). Instead, recent studies have tended to show that differences in political participation between men and women are mostly related to differences in homemaker/labor force status and socioeconomic characteristics (Andersen, 1975; Andersen and Cook, 1985; Lansing, 1974; McDonagh, 1982; Welch, 1977; Welch and Secret, 1981), tending to support situational and structural explanations. In some studies, controlling for socioeconomic characteristics has accounted for virtually all of the gender differences in political participation (see, for example, McDonagh, 1982, and Welch and Secret, 1981).

Whichever explanation is correct, it is clear that all lead to the expectation that women, on the whole, are less politically active than men. Also, it is logical to expect that the factors inhibiting general political activity likewise inhibit environmental political activity. But if women are more concerned about the environment than men, to what extent does that compensate for the factors inhibiting activism? Whatever that may be, the results of Mitchell's (1979) and McStay and Dunlap's (1983) studies suggest that the environmental concerns of women do not completely transcend these barriers. Furthermore, although it would be logical to expect women's environmental activism to equal, if not exceed, that of men once these factors have been controlled, women may be constrained by an additional factor. This has to do with the fact that, as a group, women are faced with more problems and issues than men. These include women's concerns about social, economic, and political equality with men. Because of the greater number of issues confronting women, the time, money, and other resources available for women to devote to any particular issue, such as the environment, may be proportionately less than for men. Thus, even if women are more concerned about the environment, and all other factors inhibiting the general political participation of women are discounted, we may find women to be less environmentally active. In other words, the *relative* amount of resources available for environmental activity rather than the *absolute* amount available for general political activity may be a key in understanding gender differences in environmental activity.

The notion of relative resource availability has also been suggested as an explanation of why blacks are less environmentally active than whites, even when the environmental concerns of both groups are equated and factors inhibiting general political participation are discounted (see Mohai, 1990). This explanation may be equally relevant in explaining gender differences in environmental activism, since the situation of women is similar to that of blacks: Both groups are faced with a larger array of problems and issues than their counterparts, which may affect the relative amount of resources available to them for environmental activity. In sum, the literature on the political participation of women suggests several reasons why women may be less environmentally active than men, even if women are as concerned as or more concerned than men about general environmental issues. The factors that limit the general political activity of women in all likelihood similarly affect environmental activity. In addition, the relative amount of resources available to women for environmental activity may be less. Since little direct information about gender differences in environmental activism is currently available, a purpose of this paper is to provide such information from national survey data. Specifically, the extent of gender differences in environmental activism is assessed, and explanations of these differences are examined.

Data and Method

Data used to examine gender differences in environmental concern and activism were obtained from the Survey of the Public's Attitudes toward Soil, Water, and Renewable Resources Conservation Policy (Fischer et al., 1980), a major national stratified sample survey sponsored by the U.S. Department of Agriculture (USDA) and conducted by Louis Harris, Inc. The survey consisted of 7010 face-to-face interviews, resulting in a sample of 3255 men and 3755 women. Approximately 30 items dealing with environmental and conservation issues are included in this survey. These reflect general environmental concerns, as none of the items relate to concerns about local environmental issues. Factor analysis was applied to these items in order to identify those that could be used to construct reliable (Cronbach's alpha of at least .60) indicators of concern. Three such indicators involving 13 items resulted.

One indicator of concern, dealing with perceptions about the seriousness of a range of environmental problems, was constructed from averaging responses to questions such as "How serious a problem do you feel water pollution is today?" and "How serious a problem do you feel getting rid of chemical and nuclear wastes is today?" Table 1 contains a complete list of the items. The range of responses was from 4, very serious

Table 1
Differences in Means of Individual Environmental Concern Items Between All Men and Women

Environmental Concern Item	All Men (N = 3255) Mean	SD	All Women (N = 3755) Mean	SD	Difference
How serious a problem is:					
a. Presence of chemicals in food	3.07	1.09	3.36	0.93	−0.29***
b. Misuse of soil and water	3.22	0.97	3.28	0.89	−0.06**
c. Safely getting rid of chemical and nuclear wastes	3.47	0.86	3.52	0.79	−0.05*
d. Loss of wildlife habitat	3.14	1.03	3.21	0.93	−0.07**
e. Water pollution	3.42	0.87	3.53	0.76	−0.11***
f. Shortage of fresh water	2.90	1.13	3.05	1.06	−0.15***
How likely in the future are shortages of:					
a. Scenic landscapes	3.07	1.49	3.39	1.46	−0.32***
b. Lakes and rivers suitable for recreation	3.09	1.48	3.39	1.43	−0.30***
c. Unspoiled places for fish and wildlife	3.43	1.44	3.76	1.32	−0.33***
How many acres of land out of 100 should be allocated to outdoor recreation	14.62	12.37	13.27	9.64	1.35***
How many gallons of water out of 100 should be allocated to protect fish and wildlife	17.26	12.07	16.89	9.09	0.37
How many tax dollars out of 100 should be allocated to:					
a. Clean up streams, lakes, and rivers	14.67	9.98	14.20	8.23	0.47*
b. Provide places for fish and wildlife	12.96	9.33	13.03	8.06	−0.07

*$p < .05$.
**$p < .01$.
***$p < .001$.

problem; 3, somewhat serious problem; 2, not sure; to 1, hardly serious problem at all. Cronbach's alpha of the resulting index was .72.

A second indicator of concern, dealing with perceptions about probable shortages in the future of a range of environmental amenities, was constructed from averaging responses to questions such as "How likely do you think it is that there will be shortages of pleasant views of scenic landscapes in this country 10 years from now?" and "How likely do you think it is that there will be shortages of natural unspoiled places for fish and wildlife to live?" (see Table 1). The range of responses was from 5, very likely to be shortages; 4, somewhat likely; 3, not sure; 2, somewhat unlikely; to 1, very unlikely. Cronbach's alpha of the resulting index was .79.

A final indicator, assessing the relative importance placed by the respondent on environmental protection, was constructed from items that asked how he or she would allocate limited resources to alternate uses such as household uses, industrial uses, agricultural uses, energy development, and environmental protection. The indicator was constructed by averaging responses to such items as "How many gallons of water out of 100 to you think should be allocated to fish and wildlife?" and "How many tax dollars out of 100 do you think should be allocated for cleaning up streams, lakes, and rivers?" (see Table 1 for complete list). These questions forced respondents to reveal the relative importance they placed on environmental protection as compared to economic and other objectives because, for example, each gallon of water allocated to industrial or household uses was one less that could be allocated to fish and wildlife and vice versa. The Cronbach's alpha of the resulting index was .65.

These items are not exhaustive in their coverage of opinions about all environmental issues. Nevertheless, they cover an array of nature protection, resource conservation, and health-related issues and are consistent with the broad range of indicators of environmental concern reported in the literature. Their breadth and scope are such that it is reasonable to assume that individuals concerned about environmental quality issues will score high on these items. Also, as will be seen in the analysis below, the resulting environmental concern indicators are significantly related to indicators of environmental activism.

Several items in the questionnaire indicated various forms of political participation. First, respondents were asked whether they had "ever personally done anything, either on your own or as part of a group, such as write a letter, go to a public meeting, go to a demonstration, or talk to an official to try to influence any policy or program of the federal government?" Next they were asked: "Have you ever personally done anything, either on your own or as part of a group, to try to influence any government policy or program having to do with the conservation of land, water, forests, or places for fish and wildlife to live? . . . things like writing a letter, going to a public meeting, going to a demonstration, or talking to a government official." Those answering the second question affirmatively (20.2% of the sample) were classified as "environmental" contactors [the term "contactor" is used in the same sense as employed by Mohai (1990) and Olsen (1970)]. Those answering the first affirmatively but the second negatively were classified as "other" political contactors (18.4%). Those answering both negatively were classified as "noncontactors" (61.4%).

Respondents were also asked about organization membership (although the list of organizations was limited): "Here is a list of various kinds of organizations. Will you please tell me all the types of groups in which you have participated by attending at least one meeting or event within the past year?" The list included (1) a group or organization concerned about protecting fish or wildlife or the environment, (2) a farm organization or group, (3) a group or organization specifically concerned about conserving soil and water, (4) a garden club, (5) an organization or club for outdoor sports or recreation, and (6) none. Those indicating membership in the first category were classified as "environmental" organization members (8.5%). Although the list is clearly not comprehensive, those who did not indicate membership in the first category but indicated membership in one or more of the others were classified as "other" organization members (20.5%). Those indicating "none" were classified as "nonmembers" (71%).

As is the case with the concern items, the two environmental activism indicators are not exhaustive. Both contacting and organization membership questions are more closely related to nature and resource protection issues than to health-related issues. Also, the two items do not reveal whether respondents' activities were focused at the local or national levels. Finally, the indicators classify individuals as activists or nonactivists, etc., rather than measuring the degree or frequency of activity.

Socioeconomic and demographic variables included age, income, education, occupational status, current and past residence, full-time employment, and homemaker status. In the questionnaire, age was recorded in one of twelve categories ranging from "18–19" to "75 and over." The midpoint of each range was used to estimate respondents' age. Seventy-nine years was used for the final open-ended category. Likewise, family income was recorded in one of 10 categories ranging from "under $2500" to "more than $50,000." As before, the midpoint of each range was used to estimate the income of respondents, with $60,000 used for the final open-ended category. Educational attainment was recorded in one of eight categories ranging from "none" to "at least 1 year of graduate school completed." Midpoints were used to estimate years of education completed, with 17 years used for the final open-ended category.

Ranks following Hollingshead (1957) were assigned to the occupational categories provided in the questionnaire: 1, unskilled and farm labor; 2, service and private household worker; 3, skilled labor; 4, clerical and sales; 5, farm manager; 6, manager; 7, professional. Place of past and place of current residence were ranked as 5, large metro (city of 250,000 or more plus suburbs); 4, small metro (city of 50,000 or more plus suburbs); 3, urban (city of under 50,000, not in suburbs); 2, rural (not farm or ranch); or 1, farm or ranch. For gender, "1" was assigned to females and "0" to males. Similarly, "1" was assigned to respondents who indicated full-time employment ("0" to all others), and "1" was assigned to those indicating themselves to be homemakers (originally worded as "housewife" in the questionnaire) and "0" to all others.

Finally, since resource availability (in the absolute, not relative, sense), including knowledge of government, is an expected link between socioeconomic factors and political participation (see above), a "knowledge of government" variable was included in the analysis. This was derived from the following item in the questionnaire: "If you wanted to influence a decision made by departments and agencies of the executive branch of the federal government, do you feel you fully understand how to do this, mostly understand, understand only a little, or hardly understand at all how to do this?" Values of 4, 3, 2, and 1 were assigned respectively to the four possible responses.

Results

Composite indices of environmental concern were used in the analyses to provide reliable measures of environmental concern that could be used efficiently in the multivariate analyses. However, item-by-item comparisons were also made to determine whether gender differences varied with specific environmental issues. Little variation was found. Women tended to rate higher on most items, regardless of the issue (see Table 1). In fact, women rated higher on all the items dealing with perceived seriousness of environmental problems and perceived probable future shortages of environmental amenities. Variation from this pattern was found only for the forced-choice items. For two of these items, "gallons of water allocated to fish and wildlife" and "dollars allocated to fish and wildlife," there were no statistically significant differences between men and women. For the remaining two, "acres allocated to outdoor recreation" and "dollars allocated to cleaning up streams, lakes, and rivers," men rated higher than women. That this does not necessarily indicate that men are more concerned about streams, lakes, rivers, and recreation is demonstrated by the fact that men were less likely than women to feel that there will be future shortages of lakes and rivers suitable for recreation and were less likely to feel that water pollution is a serious problem (Table 1).

Why women did not score higher on the forced-choice items may be explained by the fact that women tended to allocate more resources to the alternatives dealing with family and household purposes. For example, in contrast to men, women on average allocated more acres of land "for homes for people and families," more gallons of water "for household use," and more tax dollars "for increasing the availability of water." If instead women had allocated the differences to the environmental alternatives, women would have scored higher than men on all of them. (Although these results are not displayed in the tables, they can be obtained from the author on request.) That women should give higher priority to family and household purposes than men is not surprising in light of the premises of the family nurturer explanation. However, use of forced-choice questions where family and household uses is one of the options clearly results in fewer resources left for the environmental items.

Since differences between men and women do not appear to vary by issue, the following composite indices were used as three separate measures of general environmental concern: perceived seriousness of environmental problems, perceived future shortages of environmental amenities, and importance of allocating resources to environmental protection. Multiclassification analyses were performed on each of these in order to determine whether gender differences in concern are independent of other socioeconomic and demographic variables that have been previously hypothesized to account for some of the differences (McStay and Dunlap, 1983). These included age, education, income, and place of past and current residence.

Results of the analyses are given in Table 2. As can be seen, women scored higher than men on two of the indices ("perceived seriousness" and "perceived shortages"), while men scored higher on the third ("importance of allocating resources to environmental protection"), before and after applying multivariate controls. Although the differences are statistically significant, eta and beta values (the measures of association between gender and the environmental indices before and after applying controls) indicate that the effect of gender is rather modest, tending to confirm McStay and Dunlap's (1983) and Blocker and Eckberg's (1989) findings. Eta and beta values are especially low for the third index, in which men scored higher than women. Although modest, the effect of gender nevertheless appears to be independent of the other socioeconomic and demographic variables, as beta values are both statistically significant and not appreciably reduced from the etas as the result of applying controls. Taking the results together, women on the whole appear to be somewhat more concerned about the environment than men.

To assess whether the gender differences are accounted for by homemaker status or full-time employment in the labor force (the "family nurturer" and "economic provider" explanations), comparisons between men and women in the sample were made so that the unique effects of each factor (i.e., homemaker status and full-time employment) could be identified. First, in order to determine whether homemaker status accounted for the differences, an assessment was made of whether the differences were effectively eliminated by dropping homemakers from the analysis (this was a useful approach since the status of homemaker is virtually nonexistent for men; only 8 out of 3255 men in the survey indicated they "kept house"). Thus, the multiclassification analysis was repeated

Table 2
Differences in Means of Environmental Concern Indicators Between Men and Women (All Women, Women Outside the Home, Full-Time Employed Women, and Homemakers)

Group	Valid n^a	Unadjusted Mean	SD	Eta	Adjusted Mean^b	Beta
Perceived Seriousness of Environmental Problems						
All men	2964	3.21	0.64		3.21	
All women	3370	3.34	0.58	.11	3.34	.10***
All men	2964	3.21	0.64		3.21	
Women outside the home	2253	3.35	0.58	.12	3.35	.11***
Full-time employed men	1913	3.22	0.62		3.23	
Full-time employed women	1047	3.38	0.56	.12	3.37	.11***
Women outside the home	2253	3.35	0.58		3.34	
Homemakers	1117	3.29	0.59	.05	3.31	.03
Perceived Shortages of Environmental Amenities						
All men	3026	3.20	1.23		3.21	
All women	3471	3.52	1.19	.13	3.51	.12***
All men	3026	3.20	1.23		3.22	
Women outside the home	2309	3.57	1.17	.15	3.55	.13***
Full-time employed men	1939	3.23	1.24		3.25	
Full-time employed women	1066	3.62	1.17	.15	3.59	.13***
Women outside the home	2309	3.57	1.17		3.55	
Homemakers	1162	3.43	1.20	.05	3.46	.04*
Importance of Allocating Resources to Environmental Protection						
All men	2836	59.66	30.69		59.80	
All women	3205	57.64	23.59	.04	57.52	.04***
All men	2836	59.67	30.69		60.09	
Women outside the home	2140	58.79	24.30	.02	58.23	.03*
Full-time employed men	1851	61.97	31.44		62.38	
Full-time employed women	1014	60.30	24.05	.03	59.56	.05*
Women outside the home	2140	58.79	24.30		58.21	
Homemakers	1065	55.34	21.94	.07	56.49	.03*

^{a}The sample size resulting from excluding cases with missing values.
^{b}Means have been adjusted for age, education, income, and current and past residence.
*$p < .05$.
**$p < .01$.
***$p < .001$.

in which only men and "women outside the home" (i.e., women excluding homemakers) were compared. As the results in Table 2 indicate, differences remain statistically significant. In fact, the magnitudes of the differences are similar to the case where men and women are compared overall. These results therefore indicate that gender differences in concern are independent of the homemaker status of women. In fact, contrary to expectations derived from the family nurturer explanation [and consistent with Blocker and Eckberg's (1989) findings], homemakers score lower, not higher, than women outside the home across all three concern indicators (Table 2). However, eta and beta values indicate the differences are slight, and only in two of three cases are they statistically significant.

To assess whether differences in concern between men and women are accounted for by differences in full-time employment in the labor force, full-time employed men were compared with full-time employed women. Restricting the analysis to this group allows for an assessment of whether gender differences in concern persist after the effects of full-time employment are controlled. Results in Table 2 indicate that they do. The magnitude and direction of the differences in concern between full-time employed men and women are very similar to those found in the case where men and women are compared overall. The results thus indicate that the gender differences are independent of full-time employment in the labor force. Furthermore, in contrast with the economic provider explanation (and once again consistent with Blocker and Eckberg's findings), concern scores for full-time employed men and women are as high as or higher than those for men and women overall.

Gender differences in environmental activism were next examined. First, the extent of overall differences between men and women were assessed. Results in Table 3 indicate that men are significantly more likely to be environmentally active than women. Specifically, men (all) are more likely to be environmental contactors than women (all) by a ratio of nearly 3 to 2 and more likely to be environmental organization members by a ratio of nearly 2 to 1.

To account for these differences, structural and situational explanations were examined. To test the structural explanation, rates of environmental contacting and organization membership between men and women were compared after adjusting for education and income. The rates were also adjusted for differences in environmental concern and "knowledge of government" (the latter is a resource affecting the likelihood of political activity and is related to socioeconomic status). The adjusted differences in rates of environmental contacting and organization membership were estimated from logit regression equations (see Table 4 for a complete specification of these equations). Environmental contacting and organization membership were coded as dichotomous variables ("1" for contacting, "0" for not; "1" for organization membership, "0" for none). Gender, income, education, knowledge of government, and environmental concern were the independent variables in both cases.

As can be seen in Table 3, although the adjusted differences in rates of environmental contacting and organizational membership of men (all) and women (all) are reduced somewhat, the differences remain statistically significant. These results thus indicate that differences in socioeconomic characteristics, environmental concern, and knowledge of government are not sufficient by themselves to explain the differences in rates of environmental contacting and organization membership between men and women.

The effects of homemaker status and full-time employment on gender differences in environmental activity were examined to test the situational explanation. First, men were

Table 3
Differences in Likelihood of Environmental Contacting and Organization Membership Between Men and Women (All Women, Women Outside the Home, Full-Time Employed, and Homemakers)

	All Men (%)	All Women (%)	Difference Unadjusted (Adjusted[a])	All Men (%)	Women Outside the Home (%)	Difference Unadjusted (Adjusted[b])
Environmental contactor	24.2	16.7	7.5*** (4.0***)	24.2	18.5	5.7*** (3.3**)
Environmental organization member	11.1	6.3	4.8*** (4.2***)	11.2	7.6	3.6*** (4.0***)
Other political contactor	19.2	17.7	1.5 (−1.0)	19.2	18.5	0.7 (−1.0)
Other organization member	21.6	19.5	2.1* (0.0)	21.6	20.5	1.1 (0.0)

	Full-Time Employed Men (%)	Full-Time Employed Women (%)	Difference Unadjusted (Adjusted[b])	Women Outside the Home (%)	Homemakers (%)	Difference Unadjusted (Adjusted[a])
Environmental contactor	25.8	19.0	6.8*** (4.7**)	18.5	13.3	5.2*** (1.6)
Environmental organization member	13.8	9.3	4.5*** (4.8***)	7.6	3.8	3.8*** (1.4)
Other political contactor	20.9	20.7	0.2 (−1.6)	18.5	16.2	2.3 (−0.4)
Other organization member	23.7	21.0	2.7 (1.4)	20.5	17.5	3.0* (1.3)

[a]Differences in percentages have been adjusted for education, income, and knowledge of government using the appropriate logit regression equations indicated in Table 4. In the cases of environmental contacting and organization membership, differences have also been adjusted for environmental concern.

[b]In addition to the above, differences in percentages have been adjusted for occupational status.

*$p < .05$.
**$p < .01$.
***$p < .001$.

compared with women outside the home to isolate the effect of homemaker status. As can be seen in Table 3, the differences in environmental contacting and organization membership between men and women remain statistically significant when homemakers are excluded from the analyses. And as before, the differences are statistically significant after they are adjusted for socioeconomic characteristics (including occupational status), knowledge of government, and environmental concern. Thus, homemaker status does not appear to account for the gender differences in environmental activism. That it has no significant independent effect on environmental activism is made further apparent when women outside the home are compared with homemakers. Before adjusting for socioeconomic and other variables, differences in environmental contacting and organi-

Table 4
Results of Logit Models Predicting Environmental Contactors, Environmental Organization Members, Other Political Contactors, and Other Organization Members

Variable	Environmental Contactor		Environmental Organization Member		Other Political Contactor		Other Organization Member	
	b	SE (b)	b	SE (b)	b	SE (b)	b	SE (b)
All men and all women								
Seriousness of problems	.27020***	.06410	.37320***	.09368				
Shortages of amenities	.00754	.03030	.10218*	.04276				
Importance of environmental protection	.00562***	.00120	.01066***	.00146				
Income[a]	.00974***	.00260	.00844*	.00352	.00654**	.00252	.00884***	.00238
Education[a]	.14186***	.01314	.17636***	.01866	.11184***	.01216	.09376***	.01124
Knowledge of government	.47564***	.03468	.39746***	.04742	.27616***	.03372	.15370***	.03220
Gender	−.26480***	.06950	−.53242***	.09758	.06850	.06666	−.00284	.06280
Constant	−5.47848***	.27080	−7.69546***	.40458	−3.65484***	.16052	−3.01164***	.14500
All men and women outside the home								
Seriousness of problems	.25386***	.07062	.33020***	.09998				
Shortages of amenities	−.01740	.03386	.11032*	.04644				
Importance of environmental protection	.00542***	.00130	.01008***	.00154				
Income	.00830**	.00298	.00754	.00392	.00426	.00292	.00760**	.00280
Education	.11490***	.01628	.14286***	.02288	.08344***	.01520	.08312***	.01426
Knowledge of government	.50434***	.03854	.40620***	.05128	.21354***	.03776	.13734***	.03612
Occupational status	−.03528	.02132	−.04778	.02892	−.08254***	.02106	−.01300	.01984
Gender	−.20590**	.07966	−.46768***	.10978	.06598	.07624	.00042	.07240
Constant	−4.86726***	.35006	−6.88486***	.50360	−2.76370***	.25670	−2.75474***	.24122

Full-time employed men and women

Variable	Coef	SE	Coef	SE	Coef	SE	Coef	SE
Seriousness of problems	.21162*	.08882	.27550*	.11634				
Shortages of amenities	−.03784	.04190	.07814	.05352				
Importance of environmental protection	.00610***	.00158	.01046***	.00182				
Income	.00872**	.00378	.00572	.00472	.00502	.00366	.00296	.00356
Education	.10992***	.02122	.12288***	.02760	.08214***	.02046	.05980**	.01886
Knowledge of government	.47162***	.04830	.39890***	.06066	.21254***	.04740	.10850*	.04548
Occupational status	−.05020	.02664	−.06428	.03402	−.09354***	.02664	−.04166	.02488
Gender	−.29158**	.10486	−.49282***	.13736	.09956	.09876	−.07820	.09464
Constant	−4.58058***	.46590	−6.15012***	.61498	−2.73020***	.35366	−2.13010***	.32570

Women outside the home and homemakers

Variable	Coef	SE	Coef	SE	Coef	SE	Coef	SE
Seriousness of problems	.44452***	.10454	.50534**	.17068				
Shortages of amenities	.06588	.04538	.11130	.07124				
Importance of environmental protection	.00668**	.00204	.01356***	.00274				
Income	.00844*	.00388	.00384	.00578	.00584	.00360	.01062**	.00338
Education	.15438***	.02132	.26678***	.03346	.13470***	.01862	.11406***	.01718
Knowledge of government	.51402***	.05142	.44254***	.07594	.38754***	.04790	.16768***	.04626
Homemaker	−.12804	.11318	−.33554	.18824	.02842	.10114	−.09156	.09554
Constant	−6.81114***	.45506	−10.12144***	.76352	−4.11302***	.24044	−3.30914***	.21644

[a] The variables "income" and "education" were divided by 1000 and 10, respectively, before being entered into the regressions.

*$p < .05$.
**$p < .01$.
***$p < .001$.

zation membership are statistically significant. When the differences are adjusted, they are no longer so.

To assess the effect of full-time employment on gender differences in environmental activism, full-time employed men were compared with full-time employed women. Their rates of environmental contacting and organization membership were compared to assess whether gender differences persist. Results in Table 3 indicate that they do, as the differences in rates of environmental contacting and organization membership are statistically significant. Furthermore, the differences remain statistically significant after they are adjusted for socioeconomic, knowledge of government, and environmental concern variables. Therefore, as in the case of homemaker status, full-time employment fails to account for the gender differences in environmental contacting and organization membership.

The above results indicate that neither structural nor situational explanations fully account for the gender differences in environmental activity. Women are significantly less likely to be environmental contactors or environmental organization members, even when socioeconomic factors, homemaker status, and full-time employment are taken into account. Controlling for such variables has been enough to eliminate differences in the rates of general political participation between men and women in a number of recent studies. Thus, a question remaining is whether the factor or factors inhibiting the environmental activity of women are unique to environmental activity or whether such factors reflect constraints on the general political participation of women.

To answer that question, rates of "other" political contacting and "other" organization membership were examined. Results in Table 3 reveal that these rates are only slightly higher for men (all) than for women (all) and that adjusting for socioeconomic factors and knowledge of government effectively eliminates the differences. Similar results are obtained when other intergroup comparisons are made. In none of the cases are differences in the rates of "other" political contacting and organization membership statistically significant once they have been adjusted for the socioeconomic variables. Such an outcome was clearly not the case for environmental contacting and organization membership, suggesting that the factors inhibiting the environmental activity of women are unique to the area of environmental activity and not simply an extension of lower general political activity. Identifying the factors that uniquely inhibit the environmental activity of women is beyond the scope of data used in this study. However, as suggested above, one possibility may be that the fraction of resources available to women for environmental activity may be less than that available to men, since women as a group face more problems and issues than men.

Discussion and Conclusions

The results of this study using national survey data indicate that women are somewhat more concerned about the environment than men. However, the differences are modest. This finding is in line with the findings of some earlier, nonnational studies that have examined gender differences in concern about general (i.e., nonlocal) environmental issues. That other studies have found men to be more concerned and still others have found no statistical differences may be confirmation that gender differences are indeed modest.

Although family nurturer and economic provider explanations have been offered to account for gender differences in concern, little evidence to support these explanations so far exists. Tests of the effects of homemaker status, full-time employment in the labor force, and parental status have been attempted, but so far most studies, including this one, have failed to demonstrate that workplace and family roles have a bearing on gender differences in concern. It may be that gender differences are strictly the result of differences in socialization and that the roles men and women come to occupy have little bearing on environmental attitudes. Or, simply, family nurturer and economic provider explanations may be incorrect. In any case, more research needs to be done to adequately explain gender differences in environmental concern (if indeed the magnitudes are worth explaining).

Less prior information exists regarding gender differences in environmental activism. Results of this study show the ironic result that although women may be somewhat more concerned about the environment than men, they are substantially less likely to be environmentally active. No explanation of this gap currently exists. Although it has been hypothesized that the lower level of environmental activity of women may be related to lower levels of political activity generally, the results of this study show that this is not the case. The gap between men and women in environmental activity is greater than the gap in general political activity. Furthermore, when variables thought to account for gender differences in political participation (such as full-time employment in the labor force, homemaker status, and socioeconomic factors) are controlled, gender differences in general political participation are eliminated while differences in environmental activism remain. These results suggest that although the factors that constrain the political participation of women may also constrain women's environmental activity, additional factors further constrain the latter.

Although identification of these additional factors is beyond the scope of the data used in this study, one explanation may lie in the observation that women as a group face more issues (principally involving concerns about social, economic, and political equality with men) than men as a group. As a result, women may be forced to divide their attention and resources in more ways than men. Thus, even when the absolute amounts of resources available to women for political participation are equated with those of men and other constraints on political participation are discounted, the fraction of resources available to women for environmental activity may nonetheless be significantly lower. This explanation has been similarly offered to account for why blacks are less environmentally active than whites even after differences in environmental concern and constraints on general political participation have been taken into account (Mohai, 1990). Clearly, future research needs to verify whether this explanation accounts for the gender (as well as racial) differences in environmental activism, or to identify other plausible explanations.

Also important for future research is further specification of the environmental issues of particular importance to women. Research to date appears to indicate that these are local environmental issues involving health and safety concerns. Similarly, women may be more politically involved in these issues than is revealed by the more general activism indicators used in this study. If women are more concerned than men about local versus general environmental issues, are they also proportionately more active on the former? Does a gender gap persist regarding political involvement in local issues? What factors constrain the environmental activity of women? Information regarding these questions would be useful in furthering our understanding of environmentalism as a social phenomenon, generally. Such information would also help in providing better understanding of the current and potential influence of women on the environmental movement and its future direction.

References

Andersen, Kristi. 1975. Working women and political participation, 1952-1972. *American Journal of Political Science* 19(3):439-453.

Andersen, Kristi, and Elizabeth A. Cook. 1985. Women, work, and political attitudes. *American Journal of Political Science* 29(3):606-625.

Arcury, Thomas A., and Eric H. Christianson. 1990. Environmental worldview in response to environmental problems: Kentucky 1984 and 1988 compared. *Environment and Behavior* 22(3):378-407.

Arcury, Thomas A., Susan J. Scollay, and Timothy P. Johnson. 1987. Sex differences in environmental concern and knowledge: The case of acid rain. *Sex Roles* 16(9/10):463-472.

Blocker, T. Jean, and Douglas Lee Eckberg. 1989. Environmental issues as women's issues: General concerns and local hazards. *Social Science Quarterly* 70(3):586-593.

Bourque, Susan C., and Jean Grossholtz. 1974. Politics as unnatural practice: Political science looks at female participation. *Politics and Society* 4(2):225-266.

Brody, Charles J. 1984. Differences by sex in support for nuclear power. *Social Forces* 63(1):209-228.

Campbell, Angus, Philip E. Converse, Warren E. Miller, and Donald E. Stokes. 1960. *The American voter.* New York: John Wiley.

Fischer, Victor, John Boyle, Mark Schulman, and Michael Bucuvalas. 1980. *A survey of the public's attitudes toward soil, water, and renewable resources conservation policy.* Washington, DC: U.S. Government Printing Office.

George, David L., and Priscilla L. Southwell. 1986. Opinion on the Diablo Canyon nuclear power plant: The effects of situation and socialization. *Social Science Quarterly* 67:722-735.

Hamilton, Lawrence C. 1985a. Concerns about toxic wastes: Three demographic predictors. *Sociological Perspectives* 28(4):463-486.

Hamilton, Lawrence C. 1985b. Who cares about water pollution? Opinions in a small-town crisis. *Sociological Inquiry* 55(2):170-181.

Hollingshead, August B. 1957. A two-factor index of social position. Department of Sociology, Yale University (mimeo).

Lansing, Marjorie. 1974. The American woman: Voter and activist. In *Women in politics,* ed. Jane S. Jaquette, pp. 5-24. New York: Wiley.

Lowe, George D., and Thomas K. Pinhey. 1982. Rural-urban differences in support for environmental protection. *Rural Sociology* 47:114-128.

McDonagh, Eileen L. 1982. To work or not to work: The differential impact of achieved and derived status upon the political participation of women, 1956-1976. *American Journal of Political Science* 26(2):280-297.

McStay, Jan R., and Riley E. Dunlap. 1983. Male-female differences in concern for environmental quality. *International Journal of Women's Studies* 6(4):291-301.

Milbrath, Lester W. 1965. *Political participation.* Chicago: Rand McNally.

Mitchell, Robert Cameron. 1979. Silent springs/Solid majorities. *Public Opinion* 2:16-20, 55.

Mohai, Paul. 1985. Public concern and elite involvement in environmental-conservation issues. *Social Science Quarterly* 66:820-838.

Mohai, Paul. 1990. Black environmentalism. *Social Science Quarterly* 71:744-765.

Nelkins, Dorothy. 1981. Nuclear power as a feminist issue. *Environment* 23(1):14-39.

Olsen, Marvin E. 1970. Social and political participation of blacks. *American Sociological Review* 35(August):682-697.

Orum, Anthony M., Roberta S. Cohen, Sherri Grasmuck, and Amy W. Orum. 1974. Sex, socialization, and politics. *American Sociological Review* 39(April):197-209.

Passino, Emily M., and John W. Lounsbury. 1976. Sex differences in opposition to and support for construction of a proposed nuclear power plant. In *The behavioral basis of design,* eds. Lawrence M. Ward, Stanley Coren, Andrew Gruft, and John B. Collins, Book 1, pp. 1-5. Stroudsburg, PA: Dowden, Hutchinson, and Ross.

Smith, David Horton, Jacqueline Macaulay, and associates. 1980. *Participation in social and political activities*. San Francisco: Jossey-Bass.

Solomon, Lawrence S., Donald Tomaskovic-Devey, and Barbara J. Risman. 1989. The gender gap and nuclear power: Attitudes in a politicized environment. *Sex Roles* 21(5/6):401–414.

Stout-Wiegand, Nancy, and Roger B. Trent. 1983. Sex differences in attitudes toward new energy resource developments. *Rural Sociology* 48(4):637–646.

Van Liere, Kent D., and Riley E. Dunlap. 1980. The social bases of environmental concern: A review of hypotheses, explanations, and empirical evidence. *Public Opinion Quarterly* 44:181–197.

Verba, Sidney, and Norman H. Nie. 1972. *Political participation in America*. New York: Harper & Row.

Welch, Susan. 1977. Women as political animals? A test of some explanations for male–female political participation differences. *American Journal of Political Science* 21(4):711–730.

Welch, Susan, and Philip Secret. 1981. Sex, race, and political participation. *The Western Political Quarterly* 34(1):5–16.

Epilogue

An increasing number of studies and reviews have underscored the observation, first made by Blocker and Eckberg (1989), that when environmental issues are framed as local issues posing potential health and safety risks, gender differences in concern tend to be particularly pronounced (Blocker and Eckberg, 1995; Davidson and Freudenburg, 1996; Flynn, Slovic, and Mertz, 1994; Mohai, 1992). A review by Davidson and Freudenburg (1996) has suggested that these differences are not limited simply to local environmental hazards, but include non-site-specific environmental risks as well, particularly those associated with nuclear energy and waste. In contrast to studies examining gender differences in concerns about local environmental risks, mixed findings are obtained when environmental concern is framed more generally (Blocker and Eckberg, 1989, 1995; Davidson and Freudenburg, 1996; Mohai, 1992; Stern, Dietz, and Kalof, 1993).

I suggested in the earlier version of this article that future research needs to explore in greater detail the environmental issues that are of special concern to women. Although existing research suggests that such issues of concern to women are local issues involving health and safety, there may be other environmental issues of particular concern to women that have not yet been identified. Measures of general environmental attitudes may camouflage the issues about which women are particularly concerned, hence leading to the mixed findings of past studies. Indeed, distinguishing among various types of environmental issues might lead to a finding that there are some issues about which men are especially concerned.

The approach of distinguishing among different types or dimensions of environmental concern was taken in the University of Michigan's 1990 Detroit Area Study. An objective of that study was to determine whether racial differences exist for some types of environmental issues but not for others, and whether for some issues African Americans are more concerned than whites (see Mohai and Bryant, 1996). The data can be similarly analyzed to determine more exactly the environmental issues that are of particular concern to women, as well as those that are of particular concern to men. The 1990 Detroit Area Study yielded a response rate of 69% and included face-to-face interviews with 309 men and 482 women (see Mohai and Bryant, 1992, 1996, for the methodological details of that study).

Based on a historical examination of the evolution of the environmental movement and the substantive issues that have been its focus, five dimensions of environmental concern were identified (see Mohai and Bryant, 1996). These included resource conservation (concern about resource scarcities, such as energy and water), nature preservation (concern about harm to plants and animals, natural areas, etc.), pollution (concern about air and water pollution, hazardous wastes, and other environmental problems with implications for human health), neighborhood environmental problems (concern about pollution and other environmental problems specific to people's neighborhoods), and global environmental issues (concern about issues that transcend both human and ecosystem health impacts, such as global warming, ozone depletion, and acid rain).

In the survey, respondents were first asked an open-ended question: "In your opinion, what are the most important environmental problems facing this country?" The purpose of the open-ended question was to identify the environmental issues most important to the respondent without biasing responses with a list of issues suggested by the researcher. Respondents were prompted for up to three mentions. Each issue mentioned was coded and categorized under one of the five dimensions of environmental concern. Initially, "running out of landfill/need for recycling" was coded under the

Table E1
Percentage of Men and Women in the Detroit Area Indicating Specific
Environmental Problems as One of the Most Important Facing the Country

Problem	Men ($N = 309$)	Women ($N = 482$)	Diff.
Pollution	76%	69%	7%*
Air pollution	48	43	5
Water pollution	40	32	8*
Hazardous wastes	20	18	2
Other toxic substances	8	6	2
Other unspecified pollution	11	9	2
Nature preservation	31%	33%	−2%
Loss of or harm to trees/plants	7	10	−3
Loss of or harm to fish/wildlife	4	5	−1
Loss of or harm to lakes, rivers, streams	14	12	2
Oil spills	9	6	3
Loss of or harm to wetlands	3	2	1
Harm to shore/coast lines and oceans	1	1	0
Loss of or harm to parks/open space	1	1	0
Running out of landfill/need for recycling	35%	42%	−7%
Running out of landfill	22	21	1
Too much waste/garbage being produced	10	12	−2
Need for recycling	9	14	−5*
Resource conservation	4%	4%	0%
Need to conserve energy	0	1	−1
Need to conserve water supply	1	1	0
Other	2	2	0
Neighborhood environmental problems	8%	12%	−4%
Too much trash/litter in neighborhood	2	7	−5**
Too much noise	2	1	1
Too many abandoned houses	0	1	−1
Too many household or neighborhood pests (e.g., rats, roaches, other insects)	1	0	1
Too much growth/overcrowding	2	4	−2
Global environmental problems	17%	18%	−1%
Global warming	3	3	0
Acid rain	4	2	2
Depletion of ozone layer	10	13	−3
Destruction of rain forests	3	3	0
Other environmental problems	6%	5%	1%

*$p < .05$. **$p < .01$. ***$p < .001$.

"resource conservation" category, but because such an unexpectedly large number of respondents mentioned it, it was made a separate category. As can be seen in Table E1, few differences were found between men and women in their likelihood of mentioning specific environmental issues as being among the country's most important. A statistically significantly greater proportion of men than women mentioned a pollution issue, but the difference was relatively small. No other categories (dimensions) of environmental concern yielded statistically significant differences between men and women, although women were somewhat more likely to mention "need for recycling" and "too much trash/litter in neighborhood" as being specific issues among the country's most important.

The open-ended question was followed by a series of closed-ended items that asked respondents to rate more precisely the seriousness of specific environmental issues representing the various dimensions of environmental concerns (because of space limitations, questions regarding resource conservation concerns were omitted). It is striking that for nearly every issue under the "pollution," "nature preservation," and "global environmental problems" categories, women were significantly more likely than men to rate it as a serious problem (see Table E2). Particularly striking is that the differences between women and men are greatest for the nature preservation issues, and not the pollution issues, which involve greater human health risks. Even more surprising, in light of studies to the contrary, is the lack of statistically significant differences between men and women in rating the seriousness of various neighborhood environmental problems (Table E3). However, there are several notable exceptions. Men were statistically significantly more likely to mention noise in the neighborhood as a serious problem.

Table E2
Gender Differences in Rating the Seriousness of Environmental Problems

Problem	Men	Women	Diff.
Pollution issues overall	3.59	3.66	−.07*
Air pollution	3.57	3.65	−.08
Pollution of drinking water	3.43	3.49	−.06
Safely getting rid of hazardous wastes	3.77	3.86	−.09**
Nature preservation issues overall	3.37	3.53	−.16***
Oil spills	3.68	3.78	−.10**
The loss of natural places for fish and wildlife to live	3.42	3.63	−.21***
Loss of natural scenic areas	3.02	3.18	−.16**
Global environmental issues overall	3.29	3.39	−.10*
Acid rain	3.28	3.39	−.11*
Depletion of the ozone layer	3.43	3.57	−.14*
Global warming or the greenhouse effect	3.14	3.23	−.09
Environmental issues overall (average of all of the above)	3.42	3.54	−.12***

*$p < .05$. **$p < .01$. ***$p < .001$.
Note: The rating scale for the seriousness of environmental problems was 4 = very serious problem, 3 = somewhat serious problem, 2 = not a very serious problem, and 1 = not a problem at all.

Table E3
Gender Differences in Rating of Neighborhood Environmental Problems

Problem	Men	Women	Diff.
Rating of *seriousness* of neighborhood environmental problems			
The noise level in the neighborhood	2.25	2.11	.14*
Abandoned or boarded up houses	1.47	1.57	−.10
Litter or garbage in the neighborhood	1.77	1.83	−.06
Rats, mice, or roaches	1.47	1.55	−.08
Exposure to lead	1.39	1.36	.03
Traffic congestion	2.47	2.48	−.01
Too much new construction	1.71	1.74	−.03
Average of all of the above	1.79	1.81	−.02
Rating of *quality* of neighborhood environmental attributes			
The number of available recreation or play areas nearby	2.50	2.69	−.19*
The general upkeep of the neighborhood	2.13	2.15	−.02
The quality of the air	2.52	2.62	−.10
The quality of the drinking water	2.35	2.57	−.22***
Average of all of the above	2.39	2.51	−.12*

*p < .05. **p < .01. ***p < .001.

Note: The rating scale for the seriousness of neighborhood problems was 4 = very serious problem, 3 = somewhat serious problem, 2 = not a very serious problem, and 1 = not a problem. The rating for neighborhood attributes was 5 = very poor, 4 = poor, 3 = adequate, 2 = good, and 1 = excellent.

Women, on the other hand, were significantly more likely to give a lower rating to the quality of their drinking water as well as to the availability of nearby recreation and play areas. Although differences did not reach statistical significance, women were also somewhat more likely to see abandoned houses, rats, mice, and roaches, and the quality of the local air as problems.

Results in Table E4 express the differences between women and men as a correlation. Positive coefficients mean men are more concerned, negative coefficients mean women are more concerned. The results are consistent with those displayed in Tables E1 to E3, which compared men and women on an item-by-item basis: women express greater concern for the environment than men along all dimensions. As before, differences appear to be greater for nature preservation issues than for the pollution and neighborhood issues, which imply more direct health and safety consequences. Nevertheless, these correlations represent rather modest associations, a finding consistent with the results of the national survey employed in the earlier version of this article (Mohai, 1992). Moreover, controlling for background variables, including homemaker and parental (men and women with children 12 years old or younger) status, alters the associations very little. The magnitudes of the coefficients are barely affected, although only in the case of nature preservation does the correlation remain statistically significant.

What conclusions can be drawn from the results of these newer data? As concluded earlier, although gender differences are evident, they remain, at best, modest—even when

Table E4

Correlation Between Gender and Various Dimensions of Environmental Concern, Controlling for Age, Political Liberalism, Education, Income, Size of Place of Residence, and Homemaker and Parental Status

	Gender		
Pollution	−.09*[a]	−.08*[b]	−.07[c]
Nature preservation	−.17***	−.13***	−.15***
Global environmental problems	−.08*	−.07	−.07
Seriousness of neighborhood environmental problems	−.02	.05	.05
Rating of neighborhood environmental attributes	−.08*	.00	.01

*$p < .05$. **$p < .01$. ***$p < .001$.

[a]Bivariate (Pearson) correlation between gender and the indicated dimension of environmental concern.

[b]Standardized regression coefficient for gender in multiple linear regression where the dependent variable is the indicated dimension of environmental concern and other independent variables are age, political liberalism, education, income, and size of place of residence.

[c]Standardized regression coefficient for gender in multiple linear regression where the dependent variable is the indicated dimension of environmental concern and other independent variables are age, political liberalism, education, income, and size of residence plus homemaker and parental status.

different dimensions of environmental concern are distinguished. Furthermore, background characteristics, including homemaker and parental status, appear to have little, if any, effect on these differences. This suggests that, to the extent that gender differences in environmental concern do exist, the differing socialization experiences of men and women may account for the differences, rather than the roles they occupy or other structural factors. These conclusions were made in the earlier version of this article and are bolstered by other recent studies (Blocker and Eckberg, 1995; Flynn, Slovic, and Mertz, 1994).

Surprising, however, is the finding of a weak association between gender and general and neighborhood pollution dimensions of concern, as this appears to contradict prior studies indicating that women are especially concerned about environmental health and safety risks. A possible reason for the variance in findings is that at the time the 1990 Detroit Area Study was conducted, no especially salient local environmental problem or crisis existed. The studies that have tended to find the most striking differences between men and women have analyzed citizens' reactions to an immediate local environmental crisis, such as the discovery of contamination of the local water supply (Hamilton, 1985a, 1985b) or plans concerning the construction of a nuclear power plant or waste facility (Brody, 1984; George and Southwell, 1986; Nelkins, 1981; Passino and Lounsbury, 1976; Solomon et al., 1989). That it is concern about local and immediate environmental crises, rather than generalized concern about local pollution and environmental health issues, where the gender gap in environmentalism is most pronounced is a hypothesis that needs further exploration.

Additional References

Blocker, T. J., and D. L. Eckberg. 1995. *Gender and environmentalism: Results from the 1993 General Social Survey*. Paper presented at the 1995 Meeting of the Southern Sociological Society, Atlanta, Georgia.

Brody, C. J. 1984. Differences by sex in support for nuclear power. *Social Forces,* 63(1):209–228.

Davidson, D. J., and W. B. Freudenburg. 1996. Gender and environmental risk concerns: A review and analysis of available research. *Environment and Behavior*, 28(3):303–339.

Flynn, J., P. Slovic, C. K. Mertz. 1994. Gender, race, and perception of environmental health risks. *Risk Analysis*, 14(6):1101–1108.

George, D. L., and P. L. Southwell. 1986. Opinion on the Diablo Canyon nuclear power plant: The effects of situation and socialization. *Social Science Quarterly*, 67:722–735.

Mohai, P. 1992. Men, women and the environment: An examination of the gender gap in environmental concern and activism. *Society and Natural Resources*, 4(1):1–19.

Mohai, P., and B. Bryant. 1992. Environmental racism: Reviewing the evidence. In *Race and Incidence of Environmental Hazards: A Time for Discourse*, eds. B. Bryant and P. Mohai, pp. 163–176. Boulder, CO: Westview Press.

Mohai, P., and B. Bryant. 1996. *Is there a "race" effect on concern for environmental quality?* Paper presented at the 1996 Meeting of the Rural Sociological Society, Des Moines, Iowa.

Nelkins, D. 1981. Nuclear power as a feminist issue. *Environment*, 23(1):14–39.

Passino, E. M., and J. W. Lounsbury. 1976. Sex differences in opposition to and support for construction of an proposed nuclear power plant. In *The Behavioral Basis of Design*, eds. L. M. Ward, S. Coren, A. Gruft, and J. B. Collins, Book 1, pp. 1–5. Stroudsburg, PA: Dowden, Hutchinson, and Ross.

Soloman, L. S., et al. 1989. The gender gap and nuclear power: Attitudes in a politicized environment. *Sex Roles,* 21(5/6):401–414.

Stern, P. C., T. Dietz, and L. Kalof. 1993. Value orientations, gender, and environmental concern. *Environment and Behavior*, 25(3):322–348.

Chapter 14

"MAKING A BIG STINK": Women's Work, Women's Relationships, and Toxic Waste Activism

PHIL BROWN

Brown University
Box 1916
Providence, RI 02912
USA

FAITH I. T. FERGUSON

Brandeis University
415 South Street
Waltham, MA 02254
USA

Abstract *Women constitute the majority of both the leadership and the membership of local toxic waste activist organizations; yet, gender and the fight against toxic hazards are rarely analyzed together in studies on gender or on environmental issues. This absence of rigorous analysis of gender issues in toxic waste activism is particularly noticeable since many scholars already make note that women predominate in this movement. This article is an attempt to understand how women activists transcend private pain, fear, and disempowerment and become powerful forces for change by organizing against toxic waste in their communities. This article systematically looks at these connections by examining data from survey research and case studies. The authors are particularly interested in the transformation of self of these women, with an emphasis on "ways of knowing." They also examine the potential of existing social movement theories to explain women's activism against toxic waste.*

Introduction

Grassroots activists who organize around toxic waste issues have most often been women, led by women. This is confirmed by case studies of numerous sites (Brown and Mikkelsen, 1990; Cable, 1992; Edelstein, 1988; Garland, 1988; Hamilton 1990; Krauss, 1993; Levine, 1982) as well as by observations of and interviews with leaders of national organizations, especially the Citizens' Clearinghouse for Hazardous Wastes and the Environmental Health Network (EHN; Capek, 1993; Krauss, 1993).

Despite high levels of participation by women in toxic waste activism, gender and the fight against toxic hazards are rarely analyzed together in studies either on gender or on environmental issues. The absence of rigorous analysis of gender issues in toxic

This is a revised version of a paper presented at the 1992 annual meeting of the American Sociological Association in Pittsburgh, Pennsylvania. We are grateful to Lynn Davidman, Verta Taylor, and the Journal reviewers for helpful comments on earlier drafts.

waste activism is particularly noticeable, since many scholars of toxic waste activism often note in passing that women predominate in this movement. Little work to date systematizes the overall characteristics of women toxic waste activists despite several case studies of the organization and activities of women in local toxic waste activist groups (Cable, 1992; Garland, 1988; Hamilton, 1990; Krauss, 1993) and many accounts of these groups in the popular media. Some work on toxic waste site organizing includes substantial discussion of women's roles (Brown and Mikkelsen, 1990; Cable, 1992; Edelstein, 1988; Levine, 1982); moreover, scholarly work concerning gender and environment is focused mainly on ecofeminism and on economic development issues in the Third World and not on the consequences of class, gender, and race that characterize the experience of local toxic waste activist groups (Freudenberg and Steinsapir, 1992; Nelson, 1990; Taylor, 1989). This article analyzes women's toxic waste activism by examining case studies and attitude surveys and then tracing the sources of this activism through a "ways of knowing" perspective (Field Belenky et al., 1986). We look at how social movement theories can explain this activism. Social movement theory fits in two ways. At the level of cognitive psychology and self-concept, "frame analysis" complements the ways of knowing approach by showing people's personal transformations of frames of understanding. At the more macro level of social roles and social structures, the "political process" model makes a contribution by situating the toxic waste movement in a political-economic context. Elements of these social movement theories can be applied to a new *global movement perspective.*

In their toxic waste activism, these women challenge the political and economic power structure as well as the gendered boundaries of behavior in their communities and in their families. Most of these women activists are housewives, typically from working-class or lower middle-class backgrounds, and most had never been political activists until they discovered the threat of toxic contamination in their communities.

Although grassroots women activists have not necessarily seen themselves as descendants of prior movements, especially the women's movement (Cable, 1992), they follow in the steps of generations of women activists who fought for occupational health and safety concerns throughout this century and who more recently have become involved in the women's health movement. Bale (1990) has noted some clear connections between various forms of "women's toxic experience." For example, early labor organizing around health issues often stemmed from women's workplaces and often involved specific women's health hazards. Women Strike for Peace, starting in 1961, was central to the movement against nuclear testing. Women's health activism concerning drug and contraceptive side effects, another set of technological hazards akin to toxic waste, is also a significant predecessor to current toxic waste struggles. Minority women active in toxic struggles do sometimes come out of a civil rights movement orientation (Bullard, 1993; Taylor, 1993).

In their efforts to understand the hazards and to draw attention to the consequences of toxic exposure, these women activists come up against power and authority in scientific, corporate, and governmental unwillingness to consider their claims or address their concerns. As activist Cathy Hinds of Gray, Maine, said about her initial efforts to get her contaminated water supply tested, "It almost seemed as if they were angry with us—as if we had done something wrong, and how dare we inconvenience them this way. It was like talking to someone with no ears" (Garland, 1988, p. 94). Authorities typically deny the need for action, largely on the basis that as women, particularly as housewives, activists cannot possibly know or understand the issues.

Women activists have a different approach to experience and knowledge. We view their different, gendered experience as based on their roles as people who center their worldview more on relationships than on abstract rights and on their roles as the primary caretakers of the family. These roles lead women to be more aware of the real and potential health effects of toxic waste and to take a more skeptical view of traditional science. This article examines these women's transformation of self, with an emphasis on Field Belenky et al.'s (1986) concept of "ways of knowing." That perspective traces the ways that women come to know things, beginning with either silence or the acceptance of established authority, progressing to a trust in subjective knowledge, and then to a synthesis of external and subjective knowledge.

We define *toxic waste* as the residue of toxic substances that are human-made or human-generated and known or suspected to be injurious to health. These occur from mining, extraction, manufacturing, agricultural application, consumer uses, transportation, and disposal. Clearly, not all toxics are "wastes" but may be the result of intended uses, such as pesticide applications. In common parlance, many people use the term *toxic waste* to mean toxics in general. These include radioactive materials and chemicals known to be injurious to health that are emitted in air, ground, or water. Toxic waste sites include past deposition of such wastes as well as current and planned toxic waste facilities such as landfills, incinerators, and transport and disposal facilities.

We define *activists* as people who take some public action in legislative, judicial, political, or media arenas to cause prevention or remediation of known or suspected toxic waste hazards. Activists usually act at the local level, by which we mean individual or joint action in specific locations. This usually means organizing or participating in local organizations that are not branches of any national organization, although they may have loose affiliations. We use the term *toxic waste movement* as do many of its activists: a loosely connected or even unconnected set of local activists and organizations, along with a few national groups, explicitly dedicated to aiding these local groups. Structurally, the toxic waste movement is thoroughly decentralized and far less organized than is the larger environmental movement.

Toxic waste activism frequently takes the form of *popular epidemiology* (Brown and Mikkelsen, 1990) whereby laypeople gather scientific data and also marshal the knowledge and resources of experts to understand the epidemiology of disease. To some degree, popular epidemiology parallels scientific epidemiology, such as when laypeople conduct community health surveys; yet, popular epidemiology is more than public participation in traditional epidemiology, since it emphasizes social structural factors as part of the etiology. Furthermore, it involves social movements, uses political and judicial approaches to remedies, and challenges basic assumptions of traditional epidemiology, risk assessment, and public health regulation. Popular epidemiology is, as we shall illustrate, largely a women's effort.

Lay attempts to do science or to use science, as in the case of popular epidemiology, are frequently subject to dismissal—if not ridicule—by professional scientists, mainly because of the practitioners' lay status. Because of their gender, women's attempts to use science for their own goals, rather than for normal goals of the scientific mainstream, are further disparaged by scientists in the service of government or industry. For example, when community activists in Los Angeles organized to prevent construction of a toxic waste incinerator, "the contempt of the male experts was directed at professionals and the unemployed, and Whites and Blacks—all the women were castigated as irrational and uncompromising" (Hamilton, 1990, p. 220).

We synthesize here two streams of research—one on attitudinal surveys of environmental concern and activism, another on case studies of local activism. While there are some mixed results in attitudinal surveys, the case studies show us a more definite foundation for a gender-based analysis of toxic waste activism. Precisely because the case studies tell stories about the complete surroundings of toxic waste activism, they offer us glimpses of the many gender-specific issues.

Experience, Knowledge, and Gender in Women's Toxic Waste Activism

Gender Issues Found in Case Studies

In each case study we have found, there are remarkable similarities in the women activists' transformation from housewives to activists. Although there are differences in time, in region, in the particular circumstances of each woman's life and the cause of the toxic waste nearby, overall there is a consistency of theme and of experience. Each woman who becomes a toxic waste activist first suspects that there may be a health problem in her neighborhood when her children become ill. Outreach to her neighbors leads to the discovery of a pattern of illness in the community. Local public health authorities respond to the suspected problem with evasion or denial and only after an organized community presents its demands strongly and publicly is there an effort at cleanup or resolution by the government. When the original source of the pollution is industry, as is usually the case, the activists' efforts to hold the industrial interests accountable is difficult because of unequal resources. One of industry's main strategies to resist activists is the use of expert science to discount or minimize the communities' claims of risk and harm. Rather than working with the results or data from "housewife surveys" (Nelson, 1990), scientists in the service of polluting industry or the government challenge them on the basis of their right to even be called evidence.

This challenge is made using objectivity as the standard against which lay work is judged, and objectivity debates are central to the gender-and-science question. We view this phenomenon in light of an awareness that science is historically, culturally, and structurally shaped very strongly by embedded beliefs about gender and women's role in science as well as about lay forms of knowledge and the value of subjective everyday experience. The women toxic waste activists' struggle not only is about the material conditions that have led to their exposure, but is centrally about the uses of knowledge and the validity of claims to recognition and authority as knowers.

The leaders of toxic waste groups are often women. This is commonly reported by national toxics organizations, in many case studies, and in the personal observations of scholars in this area. EHN, a national group that works with several hundred local organizations, estimates that 70 percent of activists in local and statewide groups are women, in comparison to 30 percent in national groups (Price-King, 1994). In Krauss's (1994) extensive interviews with toxic waste activists in various parts of the country, the majority were women. Raw percentages of women members of toxic waste activist groups may, however, underestimate actual activism. In Cable's study of Yellow Creek, Kentucky, women made up half the members, but mainly because most members were married couples. Since women carried out most of the work in that group, the 50 percent figure undercounts women's involvement. Cable (1994) believes that this phenomenon is typical in toxic waste organizations.

The only evidence from a national sample comes from Freudenberg's (1984) survey of organizations involved in environmental health issues. Freudenberg drew a sample of 242 groups using inquiries to environmental, public interest, consumer, health, and citizen action groups and from reviews of three national newspapers and various environmental publications over a three-year period. Usable responses came from 110 organizations. Although Freudenberg did not ask the gender of group leaders, he did ask the occupation. The most common occupation for the leaders of these groups was homemaker (41 percent), which is typically a response given by women. We can assume that even if only 17 percent of those listing other occupations were women, women clearly make up a majority. Most likely, more than 17 percent of those listing other occupations are women, making it likely that women are a majority.

Case studies of women toxic waste activists support Sara Ruddick's assertion that women's work and perceptions tend to be rooted, at least initially, in the concrete and the everyday. Bale suggests that the upsurge in environmental action in the 1970s, and toxic waste activism in particular, was for women "an attempt [by women] to attach meaning to their fears and pain" (1990, p. 421) resulting from exposure to toxic waste. Among toxic waste activists, this assertion is reflected in the following quote from a local activist: "The real issues came down to the human level. What we have seen in this community is kids die. When that happens, go for it" (Brown and Masterson-Allen, 1994, p. 276).

The traits and experiences of women who become toxic waste activists are not theirs simply because they are women who live in proximity to toxic waste hazards; rather, they conceptualize their action, both for themselves and a wider public, out of the meaning of womanhood, and especially of motherhood, in our culture. Such a broader social meaning of gender is clearly articulated by Fernandez Kelly as somewhat more complex than a simple social distinction between the sexes:

> Gender refers to meshed economic, political and ideological relations. . . . Gender circumscribes the alternatives of individuals of different sexes in the area of paid employment. . . . Gender is political as it contributes to differential distributions of power and access to viral resources on the basis of sexual difference. Gender implicates the shaping of consciousness and the elaboration of collective discourses which alternatively explain, legitimate, or question the position of men and women as members of families and as workers. (1990, p. 184)

Despite the centrality of gender and emphasis on mothering in our analysis, as Morgen points out, much of the literature on women and community organizing tends to begin with the presupposition that a sexual division of labor determines that women activists work primarily out of their conventional private-sphere responsibilities (i.e., family service and motherhood) and that this presupposition limits our understanding of women's activist work within the community. She notes, however, that "women's community-based political activism is a conscious and collective way of expressing and acting on their interests as *women,* as *wives and mothers,* as *members of neighborhoods and communities,* and as *members of particular race, ethnic and class groups*" (1988, p. 111). Rather than making an essentialist argument about women's nature as the determinant of this particular kind of activism, it is in this sense of distinctive identity rooted in gendered experience, especially the encounter with scientific expertise and the activists' growing belief in their own knowledge as authoritative, that we approach the work of women toxic waste activists.

While women often show a higher level of environmental concern in attitudinal surveys, they also report lower rates of activism in the broader environmental movement

(Blocker and Eckberg, 1989; Mohai, 1992; Portney, 1991); yet, they are heavily represented in both the leadership and the membership of local toxic waste activist organizations. Comparison of attitudinal studies to case studies suggests that there are important differences between environmental activists who work on a national or global scale—especially those affiliated with the older, more established environmental and conservation organizations such as the Sierra Club or the Appalachian Mountain Club—and members of smaller groups who are fighting the presence of a specific local hazard and its consequences (Dunlap and Mertig, 1992; Freudenberg and Steinsapir, 1992). Toxic waste activists differ from environmental activists in that the former include more women, more people of color, older people, and people with less education (Hamilton, 1990).

Gender Differences in Attitude Surveys

Case studies most often present toxic waste activism in the context of local communities defined by class and (sometimes) gender interests. Broader attitudinal surveys, by contrast, have been concerned with national or global environmental concerns, such as rain forest depletion or the greenhouse effect. They often miss the class and gender issues that distinguish local activism from more national or global concerns. Those attitudinal studies that have looked at the effects of gender on environmental concerns and activism have yielded conflicting results. Some researchers have found that women are more concerned with environmental issues than are men (Hamilton, 1985a,b; Howe, 1990; McStay and Dunlap, 1983; Portney, 1991; Schahn and Holzer, 1990; Solomon, Tomaskovic-Devey, and Risman, 1987). Others have found no clear gender differences in level of environmental concern (Sherkat and Blocker, 1991; Stefanko and Horowitz, 1989; Wright, 1992). Mohai (1992) reports modest, though statistically significant, outcomes in which women show a consistently higher level of interest in environmental hazards, conservation, and resource allocation issues than do men. He also reports, however, that women show a consistently lower level of environmental activism, lower even than their overall rates of political participation, relative to men. Several studies have found that gender effects are modified by proximity to the source or location of the environmental hazard (Portney, 1991).

Women generally show higher levels of concern when the hazard is a local issue. Blocker and Eckberg (1989) also found that women are significantly more concerned than men about local issues, although they express the same level of concern about general issues affecting household or private behavior. Others have found that women express greater concern or involvement over private behavior than do men (McStay and Dunlap, 1983; Schahn and Holzer, 1990).

How the Gendered Position of Women Influences Toxic Waste Activism

When women who define themselves primarily as housewives become involved in activism and work against local toxic waste hazards, they must find the resources to organize their communities, challenge local political systems, and hold corporate interests accountable. They find or develop many of these resources within themselves as they struggle and succeed in learning about science (e.g., epidemiology, hydrogeology, medicine, engineering), about politics and influencing public opinion, and about community organizing. These women also learn to cultivate external resources: they gain mentors in both scientific and political processes and become skilled in media relations. The women activists transform their everyday experiences, most typically their own and their neighbors' children's illness, into knowledge that they can use in the struggle against toxic waste, and they insist on its validity as knowledge. Such validity is contested by scientific experts and professionals, whose cultural beliefs about women and science lead them to refuse to accept the women activists' claims about the consequences of toxic exposure.

Women toxic waste activists encounter deeply held beliefs about women and science on many levels. In light of this, it is critical to understand that science itself, our perceptions of it, how and where and by whom it gets done, and what kinds of problems get selected for study all are highly gendered social constructions (Harding, 1986; Merchant, 1980). The prevalence of Baconian metaphors in our culture's beliefs about science—female nature, male science—complicate responses to women's attempts to do science or to use it in pursuit of other goals. This association has a profound impact especially on nonexpert women, trying to use science to achieve social justice. As Fox Keller (1986) suggests, our culture takes for granted the association of science and objective rational thought with masculinity and masculine ways of thinking. She notes that scientific thought is commonly held to be a masculine quality, although objectivity itself is held by definition to be gender neutral. For Fox Keller, the development of objectivity is related to the psychological development of a capacity to distinguish the self from others. While this differentiation has certain positive features, it also removes people from their relational capacities. For us, it is precisely those relational capacities that are manifested in women's toxic waste activism, since the activists' claims arise from their experiences as people in relationships.

Popular epidemiology is the method of choice for women facing toxic hazards. Women are generally denied access to scientific information, and their attempts at gaining access are trivialized as subjective and antiscientific. They therefore turn to approaches they develop themselves, such as informal health surveys and "lay mapping" of disease clusters. In some situations where sympathetic professionals are available and where the data are sufficient, activists collaborate with professionals in rigorous health surveys. A good example of women practicing this approach is the annual conference of EHN. Leaders and members of local toxic groups spend several days sharing their experience in investigation and action with other activists, physicians, epidemiologists, social scientists, and lawyers. These lay epidemiologists share health questionnaires, methods of data gathering and analysis, and other scientific issues. EHN acts as a national clearinghouse to assist local groups, but not to turn them into organizational branches. EHN staff work to develop alternatives to long and costly epidemiology studies. For example, they are correlating disease clusters with existing, accessible EPA data on toxic releases at the zip code level.

While women often take the lead in organizing toxic protests, there are countless cases in which organizing does not occur. People often have no experience in any form of organizing and are unable to locate assistance from those who do. If the early lay detectives do not have the right connections, subsequent organizing may be unlikely. This would involve connections to national groups, if citizens needed help in building an organization; legal connections, if litigation were considered; state and federal agencies, if health studies were considered; and local, state, and federal legislators, if political action were considered.

Sometimes people may have heard of toxic sites for which it took more than a decade to reach firm research conclusions, litigation, or remediation, and they fear embarking on such a long and arduous task. It is also common that communities develop disputes over the definition of toxic hazards, and alternative organizations vie for definition of the situation, hence diminishing action (Kroll-Smith and Couch, 1990). Boosterism by powerful local elites in support of polluting companies may prevent action. Similar problems are seen in severely economically depressed rural communities that support toxic waste dumps.

Sometimes a few people develop a good cluster analysis, yet still are unable to generate enthusiasm from others. An interesting example is in Leominster, Massachusetts, the location of a childhood autism cluster where the disease occurs at four times the expected rate. Several factors combine to make this site remain basically unorganized, with one family carrying the entire burden. First, the city government supports the case, rather than the typical governmental resistance. Second, there has been a lot of out-migration of afflicted families. Third, there is likely a stigma of autism, based on a long history of blaming parents, especially mothers, for causing the disease.

A clearer understanding of what causes some communities to develop activism and others to fail will require a detailed comparative study of many groups and members.

Ways of Knowing

To understand what Bale terms the "evolution of women's consciousness of the toxic experience" (1990, p. 431), we have adapted the model developed in *Women's Ways of Knowing* (Field Belenky et al., 1986) as a framework for analyzing women toxic waste activists' epistemological development. Through transformations in their ways of knowing, women toxic waste activists come to terms both with the nature of the toxic contamination and with denial or evasion by public officials and industry. Women tend to take up the side of "cultural rationality" in opposition to "technical rationality." That is, they are centrally concerned with individual suffering, impaired relationships, ordinary daily experience, and direct perception of health effects. Just as in the larger case of popular epidemiology, approaching a problem from subjective experience rather than from a stance of perceived rational objectivity (i.e., conventional science) gives the activists' claims greater legitimacy in their own eyes.

Local toxic waste activists typically describe the discovery of the truth about their communities' contamination as a developmental process. Their convictions about their government, their communities, and their own abilities follow a characteristic series of changes. Their self-development in understanding and using knowledge of science in defense of their children and communities conforms in many key ways to feminist relational psychology's descriptions of women's ways of knowing. Among women toxic waste activists, ways of knowing strongly affect the capacity for effective action. Activists' knowledge evolves from an initial trust in larger institutions—assumptions that government and businesses know and do what is morally right—to the discovery in their own neighborhoods of, for example, common incidence of rare childhood illness, to the ability to act on these discoveries. The process of coming to understand themselves as knowers is an important means by which women toxic waste activists empower themselves to act as forces for change in their communities.

This perception of knowledge and its uses, and also of activists basing their claims in subjective experience, is associated with a type of moral and psychological development that involves an orientation to an ethic of specific care rather than abstract rights (Gilligan, 1982). Toxic waste activists argue that the quantitative risk assessment approach ignores personal and community experience in favor of global calculations of financial accounting, potential psychological response, and, most importantly, probability of hazards at toxic sites. The "rights" of corporations are thus placed on an (allegedly) objective plane, as contrasted with the subjective plane of local response.

The conflict between these two is seen clearly in Love Canal, New York, where Hooker Chemical Company gave land to the town for a school with the proviso that the company would never be liable for chemical injuries. While Hooker relied on a legal document to protect itself from a lawsuit, Lois Gibbs's efforts on behalf of her own family and her community to solve the problem for everyone were based on a more personal notion of responsibility (Levine, 1982). There is a tension between objective (public, governmental, corporate, rational, male) and subjective (private, familial, emotional, female) that resembles in many ways the opposition between women toxic waste activists as claimants to science and the authority of experts to judge those claims.

We are not arguing that women toxic waste activists pursue their goals merely or solely because, as women, they experience a specific psychological developmental path; rather, we suggest that given social and material constraints that largely stem from gender and class, these activists find creative and effective ways to generate change in their communities. These creative forms grow out of a self-articulated ethic of responsibility and connection. In addition to clearly voicing a call to action based on justice, women toxic waste activists give credence to their claims based on a belief in the necessity and importance of caring and a recognition of interdependence. They find the actions of nonresponsive polluters and agencies wrong and requiring redress not simply because these actions violate their rights as citizens and members of a larger polity, but also because these actions violate a moral imperative of caring and responsibility.

This ethic of caring over rights is central to the whole toxic waste movement and can be extended by looking at the role of emotions in women's political participation. Taylor (1992) points out that scholars of social movements have failed to examine the importance of emotions. Indeed, the long-dominant resource mobilization theory made it difficult to bring in emotions, by virtue of its reliance on instrumental rationality. Taylor urges us to reconceptualize social movements to "break down the artificial barrier that exists between concepts of organization, rationality, and choice, on the one hand, and affective bonds, emotions, and impulse, on the other." We cannot, Taylor emphasizes, understand why women take up protest activity unless we understand their feelings of anger at male domination; nor can we understand the vitality of women's politics without grasping the nature of what Hochschild terms their "emotion culture," which women use in

> (1) channeling the emotions tied to women's subordination into emotions conducive to protest; (2) redefining feeling and expression rules that apply to women to reflect more desirable identities or self-conceptions; and (3) advancing an "ethic of care" that promotes organizational structures and strategies consistent with female bonding. (Taylor, 1992)

While Taylor is focusing on specifically feminist organizations, her observations are applicable to women's toxic waste activism.

When women claim the responsibility for assessing their assumptions about knowledge, the attention and respect that they might once have awarded to the expert is transformed. They appreciate expertise but back away from designating someone an *expert* without reservation. Evaluation of experts becomes an important responsibility that they assume. From Hinds:

> I had learned early on not to trust officials—we had trouble with them at every level. I try to work within the system, but when there's trouble, I know now how to hold them accountable. With the McKin plans, we were

watching them all the time, on issues like the cleanup itself, as well as where the air monitors should go, how many there should be, how often they would be read. We nailed them whenever we saw something wrong. And if they didn't listen, we called a press conference right at the site, and made a big stink. (Garland, 1988, pp. 103–104)

The two stages that Field Belenky et al. label *silence* and *received knowing* are often merged in women toxic waste activists. Although they generally express a perspective characteristic of received knowers, in the particular case of the scientific issues, they more closely resemble silent knowers. Such a perspective is illustrated by a quote from a victim of the Velsicol dumping in Hardemann County, Tennessee: "I took a water sample to the health department; they said nothing's wrong with it. I thought they was good people, smarter than I was. But they wasn't" (Brown and Mikkelsen, 1990, p. 145). Many narratives about women toxic waste activists describe how they were "just plain" housewives at the start who did not have anything to say about toxic contamination or its associated issues. For example, Gibbs relates her initial reaction to a news story about her neighborhood: "The problem didn't affect me, so I wasn't going to do anything about it, and I certainly wasn't going to speak out about it" (1982, pp. 9–10).

Subjective knowing, the turn in attention to the inner voice, comes with recognition of the validity of self-determined truths. Gibbs recounts what she was thinking when she tried to start collecting signatures at the very beginning of her activism:

When I got there, I sat at the kitchen table with my petition in my hand, thinking "Wait, What if people do slam doors in your face? People may think you're crazy." But what's more important—what people think or your child's health? Either you're going to do something or you're going to have to admit you're a coward and not do it. (1982, p. 13)

Hearing an "inner voice" of "self-protection, self-assertion, and self-definition" (Field Belenky et al., 1986, p. 54) leads to the beginning of activism. This is the point at which Anne Anderson in Woburn, Massachusetts, Hinds in Gray, and Gibbs in Love Canal began to believe that something was wrong in their communities, despite the denials of local authorities. As Gibbs said, "I used to have a lot of faith in officials, especially doctors and experts. Now I was losing that faith—fast!" (1982, p. 23). Field Belenky et al. state that as women "began to think and to know, they began to act" (1986, p. 77). Many subjective knowers come to this position as a result of an encounter with "failed authority," usually male. In the lives of many women toxic waste activists, the disillusionment comes from repeated encounters with officials in which their assumptions about the value of human safety over profits or convenience are violated by the officials.

Further stages designated by Field Belenky et al. include *procedural knowing,* a reasoned attempt at resolving the conflicts between external and subjective knowledge, and *constructed knowing,* which emerges in a voice that integrates the preceding voices of reason, intuition, and expertise of others. Among women toxic waste activists, procedural knowing evolves often out of contacts with a mentor in the role of scientific expert, since these mentors are able to help the activists in learning ways to use scientific expertise for their own goals. Constructed knowing is often focused on an attempt to transcend local issues and address larger concerns. Both Gibbs and Hinds moved beyond their tight-knit communities to work on toxic waste issues at a national level. In Hinds's words,

[A young boy] looked at the EPA officials and said, "I don't want to die." That boy, and the loss of my own first baby son, and hearing all the stories of people like us around the country—all that's been a fuel to me. It makes me think that, damn it, this is America, this stuff shouldn't be happening. (Garland, 1988, p. 105)

Experience and Knowledge

Although scientific experts and government officials typically view mundane experience as insufficient or unhelpful in finding or interpreting unusual health patterns, often the toxic waste activists' knowledge derives directly from their daily experience and from their relational orientation within the community. Ruddick notes that

> the physical home and the social household are usually part of a larger sociophysical community. There is nothing romantic about the extension of mother's activity from keeping a safe home to making their neighborhoods safe. . . . If children are threatened, mothers join together, in all varieties of causes, to protect the neighborhoods they have made. (1989, p. 80)

Women toxic waste activists' motivation to force changes in their communities seems to be rooted initially in personal and familial experiences—in the subjective experience of the everyday world (Krauss, 1993). This aspect of their experience marks it not only as culturally female but also as maternal thinking and action.

Smith (1987) explains how experiences within a traditionally female domain—the *everyday*—are often excluded from the realm of scholarly inquiry, because of the specific expression of gender inequality in academic disciplines. She argues that women do maintenance work—they perform the vast majority of housekeeping, child care, and other tasks of daily living—and that such work is devalued because of its mundane nature. But this work enables male professionals to work as if they are detached from material reality and the constraints of daily tasks, a structural expression of the platonic duality between mind and body. Men's work is accorded a higher rank and considered more serious, in part because it is not concerned with the everyday.

Housewife data, housewife studies, and housewife movements embody aspects of the everyday world; consequently, they are easily marginalized. A close relationship between thought, emotion, and action compounds the class and political disadvantages of women toxic waste activists, because the experts whom they challenge rest their authority on scientific rationality. As Smith notes:

> In relation to men (of the ruling class) women's consciousness did not, and most probably generally still does not, appear as an autonomous source of knowledge, experience, relevance, and imagination. Women's experience [does] not appear as the source of an authoritative general expression of the world. (1987, p. 51)

Women whose approach to knowledge is intensely or primarily subjective conform most closely to cultural stereotypes about female thinkers and thinking as subjective, intuitive, and nonrational. As Ruddick describes maternal action:

> Rather than separate reason from feeling, mothering makes reflective feeling one of the most difficult attainments of reason. In protective work, feeling,

thinking, and action are linked; feelings demand reflection, which is in turn tested by action, which is in turn tested by the feelings it provokes. (1989, p. 70)

In studying women's approaches and relationship to knowledge, Field Belenky et al. (1986) found that women often rejected science, seeing it as a form of alien expertise; nevertheless, toxic waste activists claim science and use it in their activism, even as they reformulate it into means of reaching their own goals. They deliberately use science as a tool for social justice, bringing their activism into the practice of science, thus going beyond pure knowledge seeking; indeed, their use of science is essential. The results of lay scientific studies of contamination in their communities confirm their intuitive and subjective knowledge, rooted in daily experience about the origins of their families' illnesses.

The consequences of living in proximity to a toxic hazard affect many members of the local community. Once residents learn that they share these risks, the consequences also become shared and viewed as a community-wide problem. The realization that each family is not alone with its problems is an important step toward forming a grassroots coalition, and it is a pivotal point in the involvement of blue-collar women in community activist groups. Learning of the commonality of their situation helps them realize that their health problems are tied to geography, and they begin to look for causes nearby. As national activist leader Gibbs reflected on a community meeting:

The meeting had one good effect: it brought people together. People who had been feuding because little Johnny hit little Billy are now talking to each other. They had air readings in common or a dead plant or a dead tree. They compared readings, saying, "Hey, this is what I've got. What have you got?" The word spread fast, and the community became close-knit. (1982, p. 26)

The sense of connection is important on a national or regional level as well as locally. When Hinds found out about the situation at Love Canal, she said the experience of toxic exposure she shared with residents there made it more possible to bear her own fears and moved her to work in her community: "I was so relieved. Not to see others suffering, but to realize there was someone else we could talk to about this who would know what we were talking about, who would understand how we felt" (Garland, 1988, p. 96).

Connectedness, the state of sharing, also makes the activists feel that they are accomplishing something real, and it bolsters their beliefs that their actions are right both morally and as a response to a real threat. In Los Angeles, white middle-class women came together with the poor minority activists to work against construction of an incinerator for toxic waste. As Charlotte Bullock explained, "We are making a difference.... When we come together as a whole and stick with it, we can win because we are right" (Hamilton, 1990, p. 219). The women relied on each other for support and encouragement, both despite and because of their ridicule by white male experts.

A key resource that women toxic waste activists often use in gaining the necessary specific technical expertise for their campaigns is involvement with one or more mentors. Some mentors are people who can guide them in media relations and public or political processes; some are professional scientists from whom the activists learn technical knowledge, especially science, including standards, methods, and strategies for using information. Mentoring is one of the traditional ways that people learn expert knowledge and the social relations in which to carry out that knowledge. Indeed, one of the gendered obstacles facing women in graduate training and academic positions is the lack of a mentor. Mentoring is harder to find outside of those locations where it is, at least ideally, part of the normal routine. Few of the grassroots toxic waste activists have scientific or professional training related to assessment of toxic hazards; however, in most cases, a relationship grows with someone who does, and this help is critical at the beginning of their activism in substantiating the activists' suspicions about the hazard and its consequences. Gibbs notes in the acknowledgments to her book, *Love Canal: My Story*:

> Wayne and Kathy Hadley are more than sister and brother-in-law to me. I will always remember them as the "teachers." They were both there at the start of Love Canal to help me found our organization, teach me politics, science, public speaking, and more. I will always be eternally grateful for the help and guidance they gave me to fight this awesome battle. (1982, p. ix)

For many mentors, especially those who work with the activist groups as professionals, working with and for a grassroots toxic waste campaign often has negative consequences for the mentor; these people are seen by supervisors and officials as transgressing against their interests and by peers as being out of line for taking part in lay efforts. In Gray, one staffer from the public health office was sympathetic to the toxic waste campaign and provided information to the group on what the issues were and what the state of ongoing investigations was. That staffer left the office under pressure, after allegations of embezzlement, as the protest movement became effective; similarly, in the Love Canal campaign, scientist Beverly Paigen (1982) was censored, restricted, and punished by her supervisor at the state health agency for working with residents on a study analyzing the site.

Traditional mentoring relationships—for example, among businessmen—have the purpose of strengthening the mentor's power base, influence, or prestige in the organization, as well as providing benefits to the person being mentored. Women's mentoring relationships often have a different basis. They are based on a greater reciprocity and, when women mentors were asked why they mentored, they said it was for the rewards of fostering development and growth in addition to more material gains (Lapsley, 1992). Among professional women, in particular, mentoring was a way of addressing obstacles, such as exclusion and disempowerment, that had hindered their own advancement. Although it is not yet clear how these distinctions may figure in mentoring among toxic waste activists, certainly the retribution and other negative consequences for the mentors leads us to ask what the rewards are for them. The guidance and concrete forms of assistance have distinct hallmarks of mentoring and, for the individual women who lead toxic waste protest groups, the relationship clearly serves them much like a conventional mentoring one. Furthermore, mentoring in toxic waste activism seems usually to take the form of mentoring to a group. That is, the mentor who has access to knowledge and techniques of science (or of politics) works together with a small group of the leaders of a toxic waste campaign, rather than in an intimate one-to-one relationship with an individual. Additionally, lay activists who have achieved knowledge through being mentored take on the responsibility of mentoring others in the group and in other groups.

Public Boundaries, Private Dilemmas

Prior to beginning their activist careers, most women toxic waste activists centered their lives in the private, or domestic, sphere of home and family. It is not only that they were not previously political activists; they had not been active participants in shaping the course of events in the public sphere. They approach the political realm as an area of power that has always excluded them. This encounter involves crossing the boundaries between the traditionally female private domain and the traditionally male public world of politics and policy determination.

Women become toxic waste activists when institutions of the public domain threaten the well-being of the more private household and local community sphere; yet, a community exists simultaneously in the public and the private domain. We do not mean to dichotomize the two rigidly, since they are simultaneous and overlapping and are experienced as such by individuals. Domain identity changes across time. Despite shifting boundaries, the ideological distinction between private and public, intimate and common, remains generally both constant and easily identified by participants and observers alike—women's activities are legitimized in the private domain, while men exercise authority in the public domain (Sanday, 1974).

A domain boundary is shaped less by the content of the activity in question (e.g., children, health, goods, or other services), but more by the location of the task within a matrix of characteristics. Public domain characteristics include primarily civic, industrial, professional, organizational, public, community, extra-domestic, bureaucratic, and formally rational dimensions; private domain characteristics are familial, intimate, informal, personal, nurturant or preservative, and household related. Each of these dimensions can be named as a feature of some entity or activity that may be clearly located in the other domain. Taken together as constituent parts of a whole, these characteristics shape the normative expectations of one domain or the other. When women do work that is located within a public domain, their authority and legitimate right to control over the work are often challenged (Sanday, 1974).

The work of women toxic waste activists illustrates this challenge. Their efforts at mobilizing local communities to combat toxic hazards are often dismissed initially as mere collections of "housewife data" gathered by "hysterical housewives." As housewives, they cannot do science (in the eyes of expert professionals), nor can they challenge the local political and corporate power structures. As their efforts gain support, however, both the activists themselves and the community at large usually redefine their activism as work appropriate for mother, thereby moving it conceptually into a normatively female domain. As Smith writes, "Experiences, concerns, needs, aims and interests arising among people in the everyday and working context of their living are given expression in forms that articulate them to the existing practice and social relations constituting its rule" (1987, p. 56). Even though the work may remain the same, as an extension of mothering it becomes normatively appropriate for housewives to be doing it. This work also represents a challenge along class lines to the authority to define appropriate kinds of behavior.

Institutions in the public domain are supposed to protect elements within the private sphere. When activists discover that local industry values its bottom line or national reputation more highly than it does the health of children in the community, this realization violates the trust that the women toxic waste activists have placed in the ideal of corporate good citizenship and governmental protection. It represents a breech in the implicit societal contract between the private world—their children's health and safety—

and the public world of governmental public health policy and industrial regulation. One of the mothers in Woburn phrased it this way:

> I think we all think, somehow we are all very comfortable thinking that industry just wouldn't do this to us, government wouldn't allow industry to do this to us even if industry wanted to, and it is a very difficult thing to grasp in the first place. (Brown and Mikkelsen, 1990, p. 60)

The problems occur inside the home, but they are consequences of actions taken outside of it.

Images and Ideologies of Motherhood

The images and reality of both social and biological mothering are basic to the experience of women toxic waste activists. These images are manipulated both by the activists and by their opposition. A politics of mothering is at the root of the activists' own claim to legitimacy, and to their rights to take action within the community. Mothering a sick child is most often the experience that motivates these activists to action. These women nearly always use their own maternal identities in public relations to gain attention to toxic waste issues. The image of a mother rising to the defense of her children is an extremely powerful one, and it is skillfully used to great advantage by women toxic waste activists.

Industry and government frequently use negative beliefs about mothers to discredit the activists' claims of connections between toxic waste and health consequences, since "housewife data" are not real science to the experts. Women who are upset by growing knowledge of the consequences of toxic waste exposure in their homes are easily labeled hysterical. The labeling gives government and industry permission not to take the activists' claims seriously, by playing on cultural beliefs about women's temperament and science. Each case study of a grassroots toxic waste fight led by women includes some anecdote about authorities and experts claiming that the activists could not possibly know what they were talking about because they were only housewives and mothers. For instance, Bullock recounted,

> I did not come to the fight against environmental problems as an intellectual, but rather as a concerned mother. People say, "But you're not a scientist. How do you know it's not safe?" I know if dioxin and mercury are going to come out of an incinerator stack, somebody's going to be affected. (Hamilton, 1990, p. 215)

The view of women as hysterical housewives is a contemporary example of a form of oppression that has been used throughout history to psychopathologize women for their particular forms of experience and perception. Just as "scientific" psychiatry has so often supported traditional gender roles and punished women who rebel, so too do organized professional and governmental bodies involved in environmental hazards. Activists turn this to their own use, as in Gibbs's frequent claim to be hysterical and proud of it. Otherwise, she adds, women's fear of publicizing personal toxic waste experiences will silence them. Capek's (1987) study of activists at the Jacksonville, Arkansas, Vertac site found that women experienced a tension between "issues of the head" and "issues of the heart." The issues of the head, manifested in sympathetic expert research and testimony, could be defeated in court and other official locations. "So," leader Patty Frase recounted, "we kept it an emotional issue because an emotional issue you can't beat."

Bound up in the strategic use of motherhood as a basis for activism is an ongoing concern with the reproductive consequences of toxic exposure. Reproductive concerns are central to women's toxic waste activism, and these issues are always present in the activists' accounts of their experiences. Women are particularly concerned with the reproductive hazards they face and with the impact on childhood health of familial exposure to toxic wastes. According to one activist, "One of the most sobering aspects of the ecological degradation we endure is the impact on our capacity to bear healthy children" (Nelson, 1990, p. 177). This impact leads to considerable worry about reproductive health, producing a very generalized anxiety. The centrality of women's power as the creators of human life is thus injured. This outcome no doubt is important in fostering the depression so widely identified at toxic waste sites.

Currently, much research on gender differences in attitudes toward toxic hazards has dealt with issues of social reproduction—that is, gender roles in social movement participation and in family caregiving roles. These issues of *social reproduction* are crucial, yet we must also understand the centrality of *biological reproduction*. In addition to the impact of environmental hazards on their children's health, problems in women's own reproductive lives—miscarriage, stillbirth, and birth defects—are often among the women's first clues to toxic health effects. Much of the environmental epidemiology literature had addressed reproductive outcomes, many of which are widely attributed to toxins (Phillips and Silbergeld, 1985).

Corporate and governmental responses to women in contaminated communities are comparable in many ways with responses toward women's organizing around reproductive hazards in industrial and occupational settings. For example, in the case of lead exposure, women are removed from high-paying jobs instead of the corporation working to clean up the facility where they work; in some cases, women are asked to be sterilized "voluntarily" as a condition of keeping their jobs (Randall and Short, 1983); similarly, some firms institute "fetal protection policies" in which they restrict the actions of mothers and potential mothers rather than clean up the work sites. These policies sometimes include surveillance of menstrual cycles and exclusion of fertile or married women from some occupations (Nelson, 1990). These approaches can be compared to the response by public health authorities at Love Canal, who allowed pregnant women and those with children under age 2 to be relocated to nearby motels, but not families with older children or women trying to conceive.

Issues in Marriage and Family Life

Working-class and lower middle-class women face greater hurdles in becoming activists than do more upper middle-class women. Despite economic necessity to the contrary, blue-collar families often view having a non-wage-earning wife as a symbol of achieving the American dream (Hochschild, 1989). Professional women have more freedom of movement, access to resources, and social acceptance of their activism. They have a history of activism through volunteer and reform work; indeed, the argument has been made that the rise of middle- and upper-class women's activism is a kind of social housekeeping (Ladd-Taylor, 1986). Much of this reform work, such as settlement houses, included the effort to change culture and customs within working-class (especially immigrant) communities and make them resemble upper- and middle-class life (Gordon,

1988). Immigrants often experienced such reform work as unwelcome interference and perceived women reformers as busybodies or worse.

While working-class women have been active in other community-based issues—most notably education—that have not required technical skills and knowledge, in toxic waste activism, women are taking on scientific concerns that transcend their domestic and neighborhood experience. Becoming expert in scientific matters relative to their husbands is perceived as a threat to the normal domestic balance of knowledge. Furthermore, the more time women spend on activism, the less they may be available for domestic labor in their homes, presenting an additional challenge to the family status quo. It is not surprising that activists in contaminated communities report high rates of conflict, separation, and divorce (Brown and Mikkelsen, 1990; Edelstein, 1988). The Citizens' Clearinghouse for Hazardous Waste receives a higher rate of reports of battering following protest activities (Krauss, 1993).

Although intense familial conflict does not always occur, Robin Cannon's account from the Los Angeles incinerator struggle is typical:

> My husband didn't take me seriously at first either. He just saw a whole lot of women meeting and assumed we wouldn't get anything done. I had to split my time. I'm the one who usually comes home from work, cooks, helps the kids with their homework, then I watch a little TV and go to bed. Now I rush home, cook, read my materials on LANCER [the incinerator project].... Now the kids are on their own, I had my own homework.... After about six months everyone finally took me seriously. My husband had to learn to allocate more time for babysitting. (Hamilton, 1990, p. 220)

It is ironic that women's involvement in toxic waste organizing is an outgrowth of one of their chief domestic responsibilities—family health. Women are the organizers of health surveillance and medical care for themselves and their children and often their husbands as well. Again, from Cannon:

> All these social issues as well as political and economic issues arc really intertwined. Before, I was concerned only about health and then I began to get into the politics, decision making, and so many things. (Hamilton, 1990, p. 221)

Since health concerns are the root of most toxic waste activism, women organizers are actually fulfilling an important traditional role, albeit in an untraditional and sometimes controversial fashion.

Can Social Movement Theories Explain Women's Toxic Waste Activism?

In their belief in the potential of lay science and in their emphasis on collectivity, women toxic waste activists have produced a movement similar to social movements in health care, where women also play leading roles. This is evident in the women's health movement (Ruzek, 1978), the mental patients' rights movement (Chamberlin, 1990), and the occupational health and safety movement (Chavkin, 1984), which have pointed out otherwise unidentified problems, worked to abolish the conditions giving rise to them, and educated citizens, public agencies, health care providers, officials, and institutions. These movements made it possible to learn of such hazards, diseases, and medical malfeasance as DES, Agent Orange, asbestos, pesticides, unnecessary hysterectomies, sterilization abuse, black lung, and brown lung. These movements oppose the use of professional hegemony to maintain power, prestige, and authority and to exercise control over service users. How can these kinds of movements be explained by social movement theories?

The existing social movement theories offer at best partial explanations for understanding the toxic waste movement, and even less for the matter of women's involvement. It might be tempting to consider resource mobilization theory as having the capacity to explain women's toxic waste activism in that knowledge is a resource for which activists contend. Resource mobilization theory, however, downplays the importance of grievances in the rise of social movements, preferring to focus on an economic cost/benefit model of participation (McCarthy and Zald, 1977). Klandermans considers that movement participation is a rational choice motivated by the "perceived instrumentality of the collective good" (1984, p. 585); however, women toxic waste activists start from a localized need rather than a broader collective goal. They often start from immediate rage or personal loss, without the likelihood of specific rewards.

Resource mobilization theory further views movements as coordinated by professionalized political organizers whose capabilities determine social movement success (McCarthy and Zald, 1977), but the toxic waste movement is not a professionalized movement: Organizations spring up quite spontaneously. Resource mobilization theory also emphasizes inter- and intraorganizational structure, to the neglect of how toxic waste activists learn "protest repertoires" (Capek, 1993) in a community setting. As case studies of toxic waste activists demonstrate, problem definition and social action occur in light of community sentiments, even if personal rage or loss provides the initial impulse. Activists focus much rage on how corporate culprits have harmed the community, and one of the main functions of collective action is to retrieve the sense of community (Couch and Kroll-Smith, 1991).

"New social movement" theory has challenged resource mobilization theory by focusing on a macro level of analysis—that is, large-scale political-economic and sociocultural trends. As Habermas (1983) points out, movements such as the women's, students', peace, and environmental movements differ from traditional redistributionist politics in that they emphasize quality of life, self-realization, more humane social relations and communications, and collective participation. They are also more likely to involve members of the new middle class, especially young, highly educated people.

These theorists argue that the new movements are unique in both form and ideology. Their form is typically decentralized, egalitarian, and tactically innovative. These movements put forth a critique of advanced state capitalism, especially its domination over rationality, nature, consciousness, and human relationships. New social movements center around concepts of autonomy and self-determination (Alario, 1989). In fact, new social movement theorists exaggerate this cultural uniqueness, many older movements also emphasized sharp cultural conflict. New social movement theorists may claim such uniqueness to distinguish themselves from orthodox Marxist views, but this opposition to equity politics unfortunately leaves existing inequality unchallenged (Plotke, 1990).

New social movement theory does not adequately explain the toxic waste movement, since toxic waste activists are largely working class and lower middle class, are not highly educated, and are not young. Nor are they part of an alternative culture, as are the new social movement adherents (including activists in the broader environmental movement). New social movement theorists err when they argue for a postmodern, postmaterialist, non-class-based politics. In fact, toxic waste activists direct their efforts at core political and economic institutions. Gender, class, and other forms of power are still central, and toxic waste activists are largely recruited from the ranks of the less powerful (poor, minority, and women). This is precisely why women and minority toxic waste activists have pursued the environmental justice approach.

New social movement theory may seem appealing because it takes on a more macro-level analysis and because it offers a cultural critique of traditional ways of knowing, but its drawbacks are large, and the appealing parts (e.g., the macro-level forces, personal transformation) are better found in the "political process model" and in "frame analysis." The political process model emphasizes favorable political opportunities such as the overall political climate, the stability of political alignments, economic change, the existence of other social movements, and the presence of allies among sympathetic political parties and government agencies that stems from a split in the ruling elites (McAdam, McCarthy, and Zald, 1989; Tarrow, 1989). Tarrow emphasizes political and institutional resources rather than the resource mobilization view of resources being organizational and entrepreneurial. The key function of protest is disruption of the normal range of political life, a phenomenon that occurs outside the boundaries of organizations; collective action itself is more central than are the actions of social movement organizations.

Frame analysis models also offer valuable insights into the alternative conceptual schemes we see in women activists. "Frame amplification" involves values amplification (e.g., pursuit of democratic principles) and belief amplification (e.g., belief in efficacy of collective action). "Frame transformation" involves perception of an "injustice frame" and a shift in how people attribute responsibility. Stemming from this is a transformation of domain-specific interpretive frames—how a domain of life previously taken for granted is reframed. At a larger level, there is a transformation of global interpretive frames—a thoroughgoing change of one's whole sense of the world and oneself (Gamson, Fireman, and Rytina, 1982; Snow et al., 1986).

This frame analysis model is quite congruent with the ways of knowing approach and helps us understand how toxic waste activism is embedded in deep personal change. Reframing is a cognitive, developmental process in which activists change their views of their personal capacities to know and act. At the level of knowing, reframing means trusting one's subjective knowledge, which stems from both personal experience and collective definition and holding to that knowledge in the face of mainstream scientific knowledge. Participants learn that knowledge is not a set of detached, completely knowable scientific truths, but a contextual application of both science and politics. Through these understandings, people transform a personal trouble into a social problem. At the level of action, reframing means sensing oneself as strong and capable of taking action in the face of counterpressures from powerful social institutions and agencies. This includes breaking through traditional marriage roles as well as taking on prevailing civic associations, corporations, and government agencies.

The political process and frame transformation models do not completely provide a sufficient theory but are useful as building blocks for a new theoretical framework—a new *global movement* perspective (Brown and Masterson-Allen, 1994) that can help explain both the toxic waste and the environmental movements. Toxic waste activism, like general environmentalism, gains much of its strength from a globally shared perception of problems and solutions. Like the women's movement, the environmental movement deeply affects nearly all parts of social structure and social action (e.g., food policy, population control, migration, war, community structure, economic development). This universalizing feature enables these movements to appeal to many who would not otherwise be drawn to the core of the movement. In addition, international treaties and agreements on environmental issues bring potential for international cooperation in the post-cold war period (Brown and Masterson-Allen, 1994).

This new global movement approach incorporates a political-economic approach to the global environmental crisis, reflecting a change of productive and consumption spheres in the world. An increase in hazardous processes, products, and byproducts has stemmed from profit-maximizing or bureaucratic criteria in both capitalist and state-capitalist societies. Both manufactured consumerism and personal desires for disposable commodities on the part of the public consumptive sphere have increased environmental degradation. Precisely because toxic waste is so central to global environmental crisis, the toxic waste movement is a unique component of the larger environmental movement. There is a growing awareness of environmental degradation on the part of corporations, governments, nongovernmental organizations, political parties, individuals, and social movements. Each has its own reasons for controlling the problem and, thus, the goals and interests of one sector often contradict those of another. The global movement perspective considers these contradictions to be central to the problem-defining and claims-making activities of the many parties. Rather than moving toward a unitary environmentalist conception, these parties will continue to have conflict with each other.

This global movement perspective is also interested in the attitudes, beliefs, and behaviors of individuals and categories of individuals (especially women, minorities, and poor people). The transformation of scientific knowledge is central here. Scientific abuses and excesses have led to growing dissatisfaction with science and technology, especially since toxic hazards are a key product of unbridled growth of science and technology. Economic growth and military activities are largely responsible for that growth; as toxic concerns mount, citizens more readily make connections between those core elements of the social order and the contaminated communities in which they live. Activists observe science in the service of corporate and governmental interests and, thus, seek a lay democratic science, which we see in popular epidemiology. As with other social movements in health, the toxic waste movement alters accepted scientific definitions of the problem and leads to significant advances in scientific knowledge. In so doing, it challenges normal routines of corporate power, political authority, and professionalism.

A global movement shows how the toxic waste movement is rapidly altering the environmentalist worldview as it imprints its concerns of "environmental justice" on an environmental movement that has too often failed to take up toxic waste issues at the local level and that has failed to see the excess burden of toxic hazards on minority and poor communities. This influence is strengthened by other movements. Ecofeminism sees many parallels between the domination of women and the domination of nature. This domination derives from a patriarchal attitude and practice of instrumental control over people and nature, as opposed to harmonious, egalitarian relationships. Ecosocialism and eco-Marxism examine the ways in which dominant political, economic, and social inequities lead to continued environmental degradation. Especially for ecosocialists, this may lead to participation in "green" political parties, which make environmental issues central to an overall political program.

Where does the toxic waste movement fit in this environmental transformation? It helps steer the environmental movement clear of an emphasis on personal solutions and recycling; it emphasizes the well-being of humans as a crucial issue, as opposed to wildlife and wilderness; it forces awareness of the Third World ramifications of toxic waste dumping; it introduces a different class and racial awareness based on the groups most affected; it places gender in a central role, due to the predominance of women activists and their particular approaches to social problems; and it offers political participation to many who would otherwise not be recruited, especially minorities and working-class people.

We believe that this new global movement perspective helps to understand women's toxic waste activism because this approach ties together the micro and macro levels through a focus on ways of knowing and social structural forces.

Conclusion

In this article, we have examined a significant phenomenon of our time—a burgeoning social movement concerned with toxic wastes and their known or feared health effects. We have seen that this movement is largely populated by women, especially by women without prior scientific and political experience. These women, as well as most of their men co-participants, share an approach to knowledge that is democratic, collective, and a synthesis of subjective and objective features. Attitude surveys and case studies have provided us with a picture of women's toxic waste activism that is concerned with local problems, especially when they threaten health. This differs from the larger environmental movement where women's attitudes are not necessarily more pro-environment than those of men and where women do not predominate as leaders and members.

The special characteristics of the toxic waste movement have also led us to examine how existing social movement theories could account for it. We showed the flaws in resource mobilization theory and new social movement theory and found solid contributions from frame analysis and political process theories. But the uniqueness of this movement begged for a unique approach to social movements, which we find in the nascent global movement perspective that incorporates personal transformation of knowledge, social structural causes of movement development, and the universalizing nature of toxic politics.

We do not wish to essentialize ways of knowing as only a women's phenomenon. Indeed, there is a large number of men activists who also take the popular epidemiology approach. Both men and women share in these efforts a dislike of the hierarchical and remote structure of most national environmental organizations that have failed to address environmental health effects. Both share a distrust of government and corporate structures that keep citizens at a distance. They have also experienced the failure of most mainstream science to come to their assistance. In response, they have relied on networks of trust, associations of collective research, and shared empathy over sickness and death of family and friends; hence, they have gravitated to a way of knowing and a method of action that centers on democratic participation and relational logic.

Yet, since women predominate in this activism, and since they play particular roles in relating to environmental health, we have focused on women's ways of knowing. Women toxic waste activists change their relationship to their known world—their families and communities and the corporate and political institutions that guide them—and, in the process, transform themselves as knowers. The changes reinforce one another. Women toxic waste activists experience knowledge and authority—especially scientific knowledge and political authority—in a way that is strongly determined by their gender as well as by social class. The activists' relationship to knowledge and their belief in themselves as knowers play a central role in their experience of activism.

Such belief in themselves and in the moral, as well as scientific, foundation of their activism is hard won. Women toxic waste activists create within themselves the abilities to organize and to lead the process of change in their communities, despite opposition or evasion by local public health authorities, by industry experts, and often even by their neighbors and families. This is a remarkable transformation, and they deserve recognition not only for their efforts to protect their families and communities, but also for the courage to change themselves in the process.

References

Alario, Margaret. 1989. "The state crisis and new social movements: The emergence of a political paradigm." Paper presented at the annual meeting of the American Sociological Association, San Francisco, August.

Bale, Anthony. 1990. Women's toxic experience. In *Women, health, and medicine in America: A historical handbook*, edited by Rima Apple. New York: Garland.

Belenky, Mary Field, Blythe McVicker Clinchy, Nancy Rule Goldberger, and Jill Mattuck Tarule. 1986. *Women's ways of knowing: The development of self, voice, and mind*. New York: Basic Books.

Blocker, Jean, and Douglas Lee Eckberg. 1989. Environmental issues as women's issues: General concerns and local hazards. *Social Science Quarterly* 70:58–93.

Brown, Phil, and Susan Masterson-Allen. 1994. Citizen action on toxic waste contamination: A new type of social movement. *Society and Natural Resources* 7:269–286.

Brown, Phil, and Edwin J. Mikkelsen. 1990. *No safe place: Toxic waste, leukemia, and community action*. Berkeley: University of California Press.

Bullard, Robert D. 1993. Anatomy of environmental racism and the environmental justice movement. In *Confronting environmental racism: Voices from the grassroots*, edited by Robert D. Bullard. Boston: South End.

Cable, Sherry. 1992. Women's social movement involvement: The role of structural availability in recruitment and participation processes. *Sociological Quarterly* 33:35–47.

———. 1994. Personal communication, 5 April.

Capek, Stella M. 1987. *Toxic hazards in Arkansas: Emerging coalitions*. Paper presented at annual meeting of Society for the Study of Social Problems, Chicago, August.

———. 1993. The environmental justice-frame: A conceptual discussion and an application. *Social Problems* 40:5–24.

Chamberlin, Judi. 1990. The ex-patients' movement: Where we've been and where we're going. *Journal of Mind and Behavior* 11:323–336.

Chavkin, Wendy. 1984. *Double exposure*. New York: Monthly Review.

Couch, Stephen R., and J. Stephen Kroll-Smith, eds. 1991. *Communities at risk: Collective responses to technological hazards*. New York: Peter Lang.

Dunlap, Riley E., and Angela G. Mertig. 1992. The evolution of the U.S. environmental movement from 1970 to 1990: An overview. In *American environmentalism: The U.S. environmental movement, 1970–1990*, edited by Riley E. Dunlap and Angela G. Mertig. Philadelphia: Taylor & Francis.

Edelstein, Michael R. 1988. *Contaminated communities: The social and psychological impacts of residential toxic exposure*. Boulder, CO: Westview.

Freudenberg, Nicholas. 1984. Citizen action for environmental health: Report on a survey of community organizations. *American Journal of Public Health* 74:444–448.

Freudenberg, Nicholas, and Carol Steinsapir. 1992. Not in our backyards: The grassroots environmental movement. In *American environmentalism: The U.S. environmental movement, 1970–1990*, edited by Riley E. Dunlap and Angela G. Mertig. Philadelphia: Taylor & Francis.

Gamson, William, Bruce Fireman, and Steven Rytina. 1982. *Encounters with unjust authority*. Homewood, IL: Dorsey.

Garland, Anne Witte. 1988. *Women activists: Challenging the abuse of power*. New York: Feminist Press.

Gibbs, Lois Marie. 1982. *Love Canal: My story.* Albany: State University of New York Press.
Gilligan, Carol. 1982. *In a different voice: Psychological theory and women's development.* Cambridge, MA: Harvard University Press.
Gordon, Linda. 1988. *Heroes of their own lives: The politics and history of family violence.* New York: Viking.
Habermas, Jurgen. 1983. New social movements. *Telos* 44:33–37.
Hamilton, Cynthia. 1990. Women, home, and community: The struggle in an urban environment. In *Reweaving the world: The emergence of ecofeminism,* edited by Irene Diamond and Gloria Feman Orenstein. San Francisco: Sierra Club Books.
Hamilton, Lawrence C. 1985a. Concern about toxic wastes: Three demographic predictors. *Sociological Perspectives* 28:463–486.
———. 1985b. Who cares about water pollution? Opinions in a small-town crisis. *Sociological Inquiry* 55:170–181.
Harding, Sandra. 1986. *The science question in feminism.* Ithaca, NY: Cornell University Press.
Hochschild, Arlie (with Anne Machung). 1989. *The second shift: Working parents and the revolution at home.* New York: Viking Penguin.
Howe, Holly L. 1990. Public concern about chemicals in the environment: Regional differences based on threat potential. *Public Health Reports* 105:18–96.
Keller, Evelyn Fox. 1986. Gender and science: Why is it so hard for us to count past two? *Berkshire Review* 21:7–21.
Kelly, M. Patricia Fernandez. 1990. Delicate transaction: Gender, home and employment among Hispanic women. In *Uncertain terms: Negotiating gender in American culture,* edited by Faye Ginsburg and Anna Lowenhaupt Tsing. Boston: Beacon.
Klandermans, Bert. 1984. Mobilization and participation: Social-psychological expansions of resource mobilization theory. *American Sociological Review* 489:583–600.
Krauss, Celene. 1993. Women and toxic waste protests: Race, class and gender as resources of resistance. *Qualitative Sociology* 16:247–262.
———. 1994. Personal communication, 2 April.
Kroll-Smith, J. Stephen, and Stephen R. Couch. 1990. *The real disaster is above ground: A mine fire and social conflict.* Lexington: University Press of Kentucky.
Ladd-Taylor, Molly. 1986. "Mother-work: Ideology, public policy, and the mother's movement, 189–1930." Ph.D. dissertation, Yale University.
Lapsley, Hillary. 1992. "Mentoring in women's lives." Lecture given at the Center for Research on Women, Wellesley College, February.
Levine, Adeline Gordon. 1982. *Love Canal: Science, politics, and people.* Lexington, MA: Lexington Books.
McAdam, Doug, John D. McCarthy, and Mayer N. Zald. 1989. Social movements. In *Handbook of sociology,* edited by Neal Smelser. Thousand Oaks, CA: Sage.
McCarthy, John D., and Mayer N. Zald. 1977. Resource mobilization and social movements: A partial theory. *American Journal of Sociology* 82:1212–1241.
McStay, Jan R., and Riley L. Dunlap. 1983. Male-female differences in concern for environmental quality. International *Journal of Women's Studies* 6:291–301.
Merchant, Carolyn. 1980. *The death of nature: Women, ecology, and the scientific revolution.* San Francisco: Harper & Row.
Mohai, Paul. 1992. Men, women, and the environment: An examination of the gender gap in environmental concern and activism. *Society and Natural Resources* 5:1–19.
Morgen, Sandra. 1988. "It's the whole power of the city against us!": The development of political consciousness in a women's health care coalition. In *Women and the politics of empowerment,* edited by Ann Bookman and Sandra Morgen. Philadelphia: Temple University Press.
Nelson, Lin. 1990. The place of women in polluted places. In *Reweaving the world: The emergence of ecofeminism,* edited by Irene Diamond and Gloria Feman Orenstein. San Francisco: Sierra Club Books.
Paigen, Beverly. 1982. Controversy at Love Canal. *Hastings Corner Reports* 12:29–37.

Phillips, A., and Ellen Silbergeld. 1985. Health effects of exposure from hazardous waste sites: Where are we today? *American Journal of Industrial Medicine* 8:1–7.

Plotke, David. 1990. What's so new about new social movements? *Socialist Review* 20(February/March): 81–100.

Portney, Kent E. 1991. *Siting hazardous waste treatment facilities: The NIMBY Syndrome.* Boston: Auburn House.

Price-King, Linda. 1994. Interview, 20 April.

Randall, Donna M., and James F. Short. 1983. Women in toxic work environments: A case study of social problem development. *Social Problems* 30:411–424.

Ruddick, Sara. 1989. *Maternal thinking: Towards a politics of peace.* Boston: Beacon.

Ruzek, Sheryl Burt. 1978. *The women's, health movement: Feminist alternatives to medical control.* New York: Praeger.

Sanday, Peggy R. 1974. Female power in the public domain. In *Women, culture, and society,* edited by Michele Rosaldo and Louise Lamphere. Palo Allo, CA: Stanford University Press.

Schahn, Joachim, and Erwin Holzer. 1990. Studies of individual environmental concern: The role of knowledge, gender, and background variables. *Environment and Behavior* 22:767–786.

Sherkat, Darren E., and T. Jean Blocker. 1991. "Environmental activism in the protest generation: Differentiating sixties activists." Paper presented at the annual meeting of the American Sociological Association, Cincinnati, OH, August.

Smith, Dorothy E. 1987. *The everyday world as problematic: A feminist sociology.* Boston: Northeastern University Press.

Snow, David A., E. Burke Rocheford, Jr., Steven K. Wordern, and Robert D. Benford. 1986. Frame alignment processes, micromobilization, and movement participation. *American Sociological Review* 51:464–481.

Solomon, Lawrence S., Donald Tomaskovic-Devey, and Barbara J. Risman. 1987. "The gender gap and nuclear power: Attitudes in a politicized environment." Unpublished manuscript, North Carolina State University, Department of Sociology.

Stetanko, Michael, and Jordan Horowitz. 1989. Attitudinal effects associated with an environmental hazard. *Population and Environment* 11:43–57.

Tarrow, Sidney. 1989. *Struggle, politics, and reform: Collective action, social movements, and cycles of protest.* Ithaca, NY: Cornell University Center for International Studies.

Taylor, Dorceta E.1989. Blacks and the environment: Toward an explanation of the concern and action gap between Blacks and whites. *Environment and Behavior* 21:175–205.

———. 1993. Environmentalism and the politics of inclusion. In *Confronting environmental racism: Voices from the grassroots,* edited by Robert D. Bullard. Boston: South End.

Taylor, Verta. 1992. "'Watching for vibes': Bringing emotions into the study of feminist organizations." Paper presented at the 1992 annual meeting of the American Sociological Association, Pittsburgh, PA, August.

Wright. Stuart A. 1992. "Grassroots environmentalism, knowledge, and toxic waste siting: A study of the 'Nimby' Syndrome." Paper presented at the annual meeting of the American Sociological Association, Pittsburgh, PA, August.

PART FIVE

POLICY ALTERNATIVES

Photograph courtesy of Carol Ireson.

Chapter 15

Policy Review

Women and International Forestry Development

AUGUSTA MOLNAR

Environment Division, Asia Technical Department
The World Bank
1818 H Street NW
Washington, DC 20433
USA

Introduction

Women's involvement in forestry is gradually coming into its own as an important policy concern in the developing world, in the forestry sector, and as a precondition for sustainable economic development in general. It has been documented for a growing number of geographic regions that women are key actors in the forestry sector as (1) collectors of a wide range of products, (2) repositories of knowledge regarding forest product use and growing patterns, (3) producers of trees and sale items manufactured from forest products, (4) decision makers regarding the management of forest resources, and (5) farmers whose agriculture-cum-livestock production systems depend on the availability of forest products. There is a growing body of experience regarding concrete strategies and approaches that can be applied in forestry programs and development in related sectors to increase as well as qualitatively improve women's involvement in forestry.

This review examines some of the evidence from evaluations of projects involving forestry and natural resource management and from literature on women and development. It also draws heavily upon the author's own experience working on women and development and natural resource management in the World Bank Asia region.

Interest in women and forestry has arisen from a new emphasis, increasingly promoted by governments and development agencies, on forestry strategies including community or social forestry that meet the local demands of the rural and urban population for forestry products and that allow for sustainable forestry growth in the face of increased population pressure on areas once designated primarily for commercial forestry production and extraction. This has combined with a focus in agriculture on those areas that depend mainly on rainfed farming, many of which are arid or semi-arid ecozones, where agriculture productivity and household income-generating strategies often depend on the availability of forest products for soil replenishment, animal fodder, raw materials for local industries, wild foodstuffs, fuels, and marketable items for commercial sale. Recognition of women's involvement in forestry and agriculture has also been a result of increased women-in-development–focused research into the extent of women's participation in the productive sectors. A focus on women from women-in-development (WID) specialists has created new areas of emphasis in forestry as well, including examination of the role of forestry in household food security and the present and potential economic value of nontimber forest products.

Ongoing Development Work on Women and Forestry

A wide variety of international and local organizations are carrying out research and concrete programs that have meeting the needs of women in forestry and increasing participation by women as one aim (Fortmann and Bruce, 1988; Kumar, 1988; Williams, 1986). The United Nations Food and Agricultural Organization's (FAO) Forest Department has taken a lead policy role on this topic (FAO and SIDA, 1987), and the International Labor Organization (ILO) has done policy reviews of women and energy (Cecelski, 1985). The multilateral and bilateral donor agencies have all helped in the financing of government forestry programs that are increasingly cognizant of women (Molnar and Schreiber, 1989). Environmental research and policy institutes are building their expertise in this area. The International Union for the Conservation of Nature has recently hired an officer to build a program on women and natural resource management; World Resources Institute has been examining women's roles in the context of the Tropical Forestry Action Plan initiative. Both of these programs focus most strongly on Africa and the Caribbean region. In addition, a large number of international and local nongovernmental organizations (NGOs) are involved in field projects that test strategies and approaches, such as Catholic Relief Services (CARE) Majia Valley Worldbreak project; the Save the Children—Sahel (SOS) assisted program in Sudan; Mahiti's project in Gujarat, India; Proshika's program in Bangladesh, which has received funds from the Ford Foundation (Khan and Khan, 1988); and Environment Development Activities (ENDA)-Zimbabwe's program to rehabilitate village lands.

Some local NGOs have initiated or supported popular movements that highlight women's involvement in planting trees and managing of forest resources. Some, like the Chipko movement in India, have emerged as an angry response from village women, heavily dependent on the forest, to the unrestrained commercial exploitation of the forest resource. Others, such as the Proshika program in Bangladesh, have developed as means to provide income, assets, and employment to the landless poor, many of whom are women. Still others are environmental movements, such as the Green Belt movement, which draw on existing women's political organizations (Maathai, 1988). Some of the spread of these movements and their success in mobilizing women can be attributed to the importance of forest lands and trees to rural women. At the same time, organizing around forestry has provided women with a consciousness-raising and political forum that benefits them in other ways as well. These movements have influenced government forestry agencies to develop more socially responsive agendas and have fostered splinter movements elsewhere in their countries of origin.

How Women Participate in Forestry

The information currently available documents the wide range of women's participation in forestry. It is commonly recognized that rural women are primary collectors of fuelwood for domestic cooking and heating and that the sale of fuelwood and charcoal to urban and industrial markets is an important source of income for the rural poor, particularly women (Cecelski, 1985; DeBeer and McDermott, 1989; FAO and SIDA, 1987; Kaur, 1988). Less recognized is the extent to which women use fuelwood for home-based industries and the extent to which women's food security strategies are dependent not only on their access to fuelwood but on their collecting and processing a range of foodstuffs from trees and forests to complement the agricultural and animal products consumed in the household (FAO, 1989). Nor is it recognized that the primary responsibility (and labor burden) is borne by women in many cultures to provide fodder for livestock, largely from trees and forest lands (Acharya and Bennett, 1982; Sarin, 1990; World Bank, 1990b).

On-farm, women in different regions of the world play major roles in tree planting and tree management. In Southeast Asia, multistory home gardens managed in large part by women may provide 50% of a household's total food supply and income (Stoler, 1981). In other regions, fruit and fodder trees planted in the homestead or on field boundaries are often the responsibility of women. Although women in some regions may be less involved in on-farm block planting of fast-growing trees than men, their planting and management of freestanding trees used for multiple purposes may result in equally substantial and more varied production. Recent studies demonstrate that one freestanding tree managed over a long term may generate much more leaf and wood biomass than a row of fast-growing trees, as well as producing fruits, fibers, or seeds useful for food or income (Banerjee, 1989; Robinson and Thompson, 1989). In general, women's interest in on-farm tree planting has been greatly underestimated in forestry programs, yet when women have been provided the opportunities to establish nurseries or to grow seedlings through targeted extension, NGO programs, or support to local women's groups, they have been active participants (FAO and SIDA, 1987; Fortmann and Rocheleau, 1985; Sarin, 1990). In aeras of Africa and Latin America where urban migration is extensive, women in a large proportion of households have the primary responsibility for agriculture, and with it, responsibility for tree-growing (Rocheleau, 1988).

Off-farm, a tendency in earlier analyses of women's roles in forestry was to focus on their exploitation of limited forest resources to obtain the subsistence and commercial products needed for their survival in the face of land degradation and deforestation. Now, however, there are a growing number of cases where women have actively managed forest resources off-farm being documented in the developing countries. In the hills of Nepal and Himalayan India, women have traditionally been central as decision makers when forests are closed for regeneration and in decisions about community rules of harvesting and production extraction. In tribal areas, women whose livelihood depends largely on the collection, sale, and processing of nontimber forest products (NTFP) have historically made decisions about harvesting techniques that sustain production of diverse products and about rotational use of different parts of the forest and grasslands (Chen, 1990). In Zimbabwe, a successful program that draws upon women's initiative to foster community regeneration of local woodlands is being expanded.

In fact, women often have a better knowledge base than men in the same locality about the qualities, growing patterns, and potential uses of forest species and grasses. A survey in Sierre Leone revealed that women could name 31 products they gathered or made from the nearby bush, while men only named 8 (Hoskins, 1988). A study in Karnataka, India, of women's NTFP enterprises documented the employment of more than 6000 women in collection and sale of a condiment that most local foresters had not even heard about (Campbell, in press; FAO, 1989).

Forestry Projects and Women's Participation

There is a range of forestry development activities in which projects have actively tried to encourage women's participation or to capitalize on their traditional roles in the sector. These include plantation forestry, to increase the production and supply of timber and wood products for both commercial and domestic needs; agroforestry or farm forestry on private lands to generate domestic products and household income and return nutrients to the soil; development of wood-processing industries; rehabilitation of degraded forests and grasslands for environmental protection and production of forest products; promotion of fuelwood saving devices, such as improved cookstoves; nursery development to supply seedlings; and establishing shelter belts to reclaim and stabilize land. Projects that have made attempts to include women have faced a number of difficult constraints caused by the socioeconomic situation of women in the society and the institutional orientation of foresters and other technical extension staff undertaking the projects. Constraints include

(1) Women's restricted access to productive resources, such as credit, land, information, training, inputs, and marketing channels
(2) Women's restricted roles in public decision making outside (and sometimes within) the household
(3) Restrictions on women's participation in the labor market due to conflicting demands of child care and daily household tasks, including food preparation and fodder, water, and fuel collection
(4) Lack of skills or incentives for project personnel to incorporate women's knowledge, needs, and preferences into the design of interventions and the menu of technical recommendations.

Restricted Access to Productive Resources

Women's more restricted access to resources can be a constraint on successful projects in a number of ways. Women may not automatically receive information about the availability of a government planting or nursery program, as has happened in South Asian forestry programs. They often have less access to credit for forestry activities, because this is often contingent upon possession of land as collateral or evidence of an income-earning capacity that many women, particularly divorced or widowed women, lack, as has occurred in Kenya (Chavangi, 1988). Women have reduced access to markets because of restricted mobility, lack of capital, and greater labor burdens, and they have reduced access to land due to traditional tenure systems. Credit access has an important effect on a woman's ability to market forest products. Some case studies document that as nontimber forest products become commercialized, women lose control over their sale because the women lack access to the capital to do their own marketing (Campbell, in press). In contrast, a study carried out in Indonesia in Kalimantan documents women's ability to retain control when markets change (Colfer, 1981).

Insecurity of tenure is a particular issue in forestry programs that must be dealt with within a specific sociocultural context (Bruce, 1989; Fortmann and Bruce, 1988). In parts of Africa where women have customary use rights over land, such as in the Gambia or Senegal, formal entitlement programs have given title to male household heads and have reduced women's interest in planting trees on lands over which they no longer have secure rights (Dickerman, 1988). Elsewhere in Africa, such as in Kenya or Cameroon, tree planting by women is discouraged because it can help to establish long-term claims over lands customarily held by men. Women often have rights to produce from trees planted on family lands only with permission of male landowners. As the values of trees and tree products change, customary rights of women to household trees and tree products may change. In the Dominican Republic, women have abandoned traditional palm fiber handicrafts because men have diverted family palm trees to a cash crop, rather than allowing women to collect the palm fronds (FAO and SIDA, 1987; Rocheleau, 1988). *Acacia nilotica* pods that were formerly fed to cattle by village women on village commons in Gujarat are now auctioned under government social forestry schemes to the highest male bidder for the village development fund.

The restrictions on women's decision making on tree product disposal can stem from a combination of women's limited land and tree tenure and women's traditional lack of decision making power regarding the disposal of cash crops in commercial markets. In West Bengal, women have forgone potential fuelwood benefits from eucalyptus plantations, because men sell the trees in blocks to maximize profits, generating more wood at once than can be stored in the homestead plots. Women do not make suggestions to harvest the trees in rotation, because they do not have enough involvement in and knowledge about the market situation to know if this is feasible (Molnar, 1986; J. Gabriel Campbell, pers. comm., 1988).

Where land is communally held, as in some of the community lands and corporately held lands in India or Africa, women's lack of participation in public institutions may limit their decision-making authority in the use of such lands. Projects have inadvertently cleared shrub areas valuable to women for their wild foodstuffs or medicinal plants, for fast-growing tree plantations, without any concern being voiced by the local male villagers consulted (Williams, 1985).

Restrictions on Decision Making Outside the Household

Forest projects that have encouraged group action for afforestation or forest rehabilitation often have not introduced mechanisms for women's participation in group decision making. In some societies, the decision-making spheres of men and women may be so distinct that the only workable solutions are to form separate groups of women and men. This has been tried successfully in an NGO-assisted project in Sudan that allocated a portion of the village woodlot to a women's group for establishment and management (FAO, 1988). This also has been a positive strategy of some of the NGOs working with poor women in South Asia, because they have little voice in community decision-making forums. So far, there are not well-tested models of this type that have been successfully applied on a broad scale. Some of the pitfalls with these models experienced in, for example, India or Bangladesh result when male-dominated groups try to regain control of the plantations once the lands become productive. Women do not get adequate legal support from the courts or government personnel to defend their claims, and forestry extension staff do not have the requisite skills to provide needed technical and logistic support to the women's groups.

There has been progress, however, with the involvement of women in management decisions for rehabilitation of forestry in a number of countries where women are traditionally more outspoken, for example in Nepal, Zimbabwe, Rwanda, and northern India. A model developed in a project assisted by ENDA in Zimbabwe for rehabilitation of village lands by women-dominated village groups is to be expanded under a World Bank–funded forestry project. Women's forest committees have emerged in some villages in Nepal to manage the village forests. In West Bengal, women have encouraged men to form forest protection committees for forests they use to collect nontimber forest products (Chen, 1990). *Mahila Mandals,* local women's groups, have undertaken afforestation of plots of state forest land in Himachal Pradesh in India (Sarin, 1990).

Project planners may also make the mistake of failing to analyze properly the specific differences in decision-making responsibilities by gender within and outside the household. Although women may not attend public meetings, they may traditionally make informal decisions regarding utilization or closure of a village forest resource or common land, particularly when they are the ones responsible for grazing or collection of fuel and fodder from that area. In such cases, it does not require a change in women's social status to involve them in forest management decisions, only appropriate extension and information dissemination. NGOs with a rural development presence can serve this role in government programs, as is being tried in India. Some programs hire women foresters to play such an information dissemination role. Women and men may also have different responsibilities for types of products from the same trees; women may lop trees for fruit, leaves, and branches, while men harvest them for timber (FAO and SIDA, 1987). Women are often the driving force in programs to rehabilitate degraded lands. Once they are convinced and involved, they are the ones to convince men in the community that forest rehabilitation is worth their time and effort (Kumar, 1988).

Restrictions on Women's Participation in the Labor Market

Developers of broad-scale forestry projects have often lost valuable opportunities to involve women as a result of failing to identify and address restrictions on women's participation resulting from conflicting demands on their labor. The FAO documented a project in Kenya in which trees died because project staff failed to recognize that women were traditionally responsible for watering trees but did not have time to bring water the long distance required to keep the trees alive (FAO and SIDA, 1987). Projects often schedule meetings to present information about planned development activities at times when women are too busy to attend or provide forestry inputs at centers that are too distant from women's workplaces to allow them to collect the inputs. Training sessions may be too long for women to take the whole course, or there may not be the provision for child care needed to allow younger women to attend.

It is only recently that how much traditional forestry sector employment is undertaken by women has been documented. Forestry plantation schemes used by governments to generate employment for rural poor or in off-season or drought periods are often undertaken with female labor in parts of Africa or India, because men can find higher-paying employment farther from the home. A large proportion of labor for commercial forestry in planting, pitting, and even harvesting and grading is done by women in different countries in Africa and Asia. Where forestry projects generate sizable employment for women, there is seldom an analysis of what this implies for women's work burden. In contrast, women may assume agricultural or animal husbandry tasks traditionally allocated to men, along with their traditional work burden, to allow men to earn extra income (Chen, 1990). This can have a negative effect on their health status and on the time available to rear their children (Kumar and Hotchkiss, 1988). It may also impel them to keep children out of school. Positive action can be taken, if the constraints are understood. Employers can provide child-care facilities, introduce time-saving devices for other work done by women, or provide lighter-weight implements that are more easily (and productively) used by female laborers.

Women's lack of interest in programs may stem from planners' poor understanding of their needs and time demands. Successful improved-stove programs, for example, are those that take a more holistic view of the cooking issues than just the need to conserve fuel supplies or save money. In fact, there is a growing concern that many of the stoves touted as successful fuel-savers do not in fact save much fuel. For nearly every positive evaluation, researchers seem able to cite a negative evaluation of the same stove model drawing on a different sample or using a different study methodology. The United Nations Development Program (UNDP) Bank funded Energy Sector Management Assistance Program (ESMAP) has found in its evaluation of stoves that there are great difficulties in preventing stove users from telling surveyors what they think they wish to hear, regardless of the stoves' genuine performance. There is a clear need for more and better evaluations of these technologies throughout the developing countries.

For this reason, stove designs that concentrate on shorter cooking times for meals and that address health issues such as smoke pollution in the house may be more popular among women. For instance, the Lorena models used in several South Asian stove programs have not proven as fuel-efficient as expected from laboratory-field testing, but women continue to use them because of the high value of smokelessness (Pandey, 1987; Sarin, 1990). Elsewhere, energy-saving programs teach women better kitchen management to save fuel in existing stoves. Stoves and other women-oriented program components may also have positive effects on women's productivity by improving their own and their families' health and freeing time for other productive activities. Government programs that establish plant nurseries for households or small groups can give women participants an added benefit: access to the nurseries' water systems for domestic use or for generating additional income from fruit or vegetable gardening (using the wastewater).

Lack of Skills and Incentives for Foresters and Project Staff to Involve Women

Most forest services are heavily male-dominated, and training programs do not impart information about women's roles in forestry or practical ways to reach women. There is increasing emphasis on the recruitment of female extension agents and female foresters. These tend to be long-term solutions, however, and may be seen as overburdening the government with recurrent staff costs. For example, in several state forestry projects in India, governments are recruiting more female staff, and the number of women foresters is increasing. It is still an open question how intensive the use of female extension agents should be. There is much that has been accomplished by training male staff to work with women and to be sensitive to women's needs and interests, along with seeking to recruit more women as staff. NGOs and some governmental programs have also tapped into existing female extension services in health, education, and agriculture to get information to women about forestry. This is cost-effective and sustainable. Women are most effectively involved through groups. Local women's organizations can be a target for extension advice from technical personnel, both male and female, rather than trying to reach large numbers of women as individuals (Molnar and Schreiber, 1989; World Bank, 1990a).

Reorienting forestry toward women results in new types of forestry. As mentioned previously, there is increasing documentation of the value of nontimber forest products in many parts of the world, much of which is collected and marketed by women. A publication by FAO documents the importance of forestry in food-security strategies of households around the world, particularly in areas that are prone to drought or flooding

(Falconer, 1987; Falconer and Arnold, 1989). Projects rehabilitating forests or establishing new plantations can greatly increase the economic and social returns by promoting technical forestry models that generate a much wider range of products than timber and fuelwood. Research is increasing on multicanopy plantation or forest management models that produce long-rotation timber, yet include grasses, intermediate rows of fuel-producing shrubs, fruit- or nut-bearing species, oils, resins, fibers for local handicrafts, medicines, or leaves for fodder and industries (Campbell, in press; DeBeer and McDermott, 1989). Women in communities traditionally dependent on forest resources often have indigenous technical knowledge about fruiting and flowering requirements of trees, seasonal variations, and systems of controlled lopping for fuel or leaf fodder that are least damaging over the long term to forests (Molnar, 1989; Robinson and Thompson, 1989). Tapping this information can lead to a new direction for forestry research if there is sufficient commitment. On-farm, too little attention has been paid to the multipurpose uses of trees grown. There has been an assumption that households will grow trees exclusively for fuelwood, when local patterns of tree growing may be ones in which fuelwood is only a by-product of trees grown ostensibly for other purposes.

Gaps in Knowledge about Women and Forestry

There are a number of areas in which more research and analysis is needed to involve women further in forestry. As yet, there is little good economic analysis of the potential returns from forestry models that benefit women relative to other, more female-excluding forestry models. One case study in Nepal documents the decline in agricultural productivity resulting from women's increased labor burden in collecting fuel and fodder, but this is unusual (Kumar and Hotchkiss, 1988). There has been little conclusive research of the effects of fuel shortages on dietary practices, fertility patterns, or women's and children's health (Anne Fleuret, pers. comm., 1988). There is also not enough research on the decline in women's income-earning potential from loss of raw materials for handicrafts or shortages of fuel-based industries. Where male migration is high, agroforestry may result in increased labor burdens for women without commensurate benefits when they do not control use of final produce.

The impacts of forestry projects on women are often assumed rather than measured. Many projects throughout Africa and Asia promote stall-feeding of animals and introduction of improved breeds for higher productivity. These do not evaluate the costs to women, particularly poor women, of spending more time in cutting and carrying fodder or whether the stronger emphasis on cattle rather than sheep and goats reduces women's personal income because they and old people are often the ones with greatest control over income from *small* animals (Ki-Zerbo, 1981). Increased numbers of trees may not also imply increased fuelwood supplies for women, as the project in India, cited earlier, documents (Molnar, 1986). Projects in Kenya have documented that men do not traditionally grow trees for fuelwood; they view it as women's responsibility to provide fuel from gathered sources. Unless their priorities change, as has happened in one project due to extension work, women may not benefit (Chavangi, 1988).

Conclusions

Women's involvement in forestry has emerged as an important policy concern for forestry and sustainable-resource management in developing countries. Considerable experience has been gained in forestry and forestry-related programs undertaken by government, international donors, and nongovernmental organizations. Considerable progress has been made by taking a more cross-sectional look at women's roles in forestry and identifying the positive contributions that their participation makes in enhancing the returns from forestry and introducing more sustainable management systems, as well as meeting their varied needs. The experience is new, however, and there has not yet been time to evaluate properly the advantages of particular strategies and the impacts on women over the longer life of forest plantations. Innovative approaches have been tried on a small scale, but their applicability to broad-scale programs and other regions needs further testing. New research directions in forest management and development for nontimber forest products should generate new technical models. Central to their successful involvement is making institutional changes in forest services to recruit more women professionals and improve the skills for reaching rural women and drawing on their own indigenous know-how.

References

Acharya, M. and L. Bennett. 1982. *Women's roles in the subsistence sector.* World Bank Working Paper No. 526. Washington, DC: World Bank.

Banerjee, A. K. 1989. *Shrubs in tropical forest ecosystems.* World Bank Technical Paper No.103. Washington, DC: World Bank.

Bruce, J. 1989. *Rapid appraisal of tree and land tenure for the design of community forestry initiatives.* Community Forestry Note No. 5. Rome: FAO Forest Department.

Campbell, J., ed. In press. *Dynamics in forest-based small scale enterprises. Women's roles: Two employees from India.* Rome: FAO.

Cecelski, E. 1985. *The rural energy crisis: Women's work and basic needs, perspectives, and approaches to action.* ILO World Employment Programme; Research, Rural Employment Policy Research Programme, Geneva.

Chavangi, N. 1988. *Case study of women's participation in forestry activities in Kenya.* Draft prepared for FAO Forest Department, Rome.

Chen, M. 1990, May. *Women and wasteland development; ILO policy paper.* Discussion Draft.

Colfer, C. 1981. Women, men, and time in the forests of East Kalimantan. *Borneo Research Bulletin* 13(2):883–919.

DeBeer, J. and M. McDermott. 1989, July. *The economic value of nontimber forest products in South East Asia.* Netherlands Committee for IUCN.

Dickerman, C. 1988. Security of tenure and land registration in Africa: Literature review and synthesis. Land Tenure Paper 137. Madison, WI: Land Tenure Center.

Falconer, J. 1987. *Forestry and nutrition: A reference manual.* Rome: FAO Forest Department.

Falconer, J., and J. E. M. Arnold. 1989. *Household food security and forestry: An analyzing of the socioeconomic issues.* Rome: FAO Forest Department.

FAO (Food and Agricultural Organization of the UN). 1988. *Women and community forestry in Sudan.* (Filmstrip and text.) Rome: FAO.

FAO (Food and Agriculture Organization of the UN). 1989. Women in community forestry: Field guide for project design and implementation. Draft. Rome: Forest Department.

FAO and SIDA (Food and Agriculture Organization of the UN and Swedish International Development Authority). 1987. *Restoring the balance: Women and forest resource.* Rome: FAO Forest Department.

Fortmann, L., and J. Bruce, eds. 1988. *Whose trees?* Boulder, CO: Westview.

Fortmann, L., and D. Rocheleau. 1985. Women and agroforestry: Four myths and three case studies. *Agroforestry Systems* 2:253–272.

Hoskins, M. 1988. Rural women, forest outputs and forest products. Discussion draft. Rome: FAO Forest Department.

Kaur, R. 1988, September. Women's role in forestry in India. Background paper for Indian Country Study of Gender and Poverty, World Bank, PHRWD, and India Country Department.

Khan, I., and A. Khan. 1988. *Agroforestry for the poor: The Proshika experience.* Dhakka: National Workshop on Homestead Plantations and Agroforestry.

Ki-Zerbo, J. 1981. Women and the energy crisis in the Sahel. *Unasylva* 33(133):5–10.

Kumar, N. 1988, August. The role of women in forestry and natural resource management. Consultant Report to the World Bank, PHRWD, Washington, DC.

Kumar, S. K., and D. Hotckhiss. 1988. *Consequences of deforestation for women's time allocation, agricultural production, and nutrition in hill areas of Nepal.* Research Report 69. Washington, DC: International Food Policy Research Institute.

Maathai, W. 1988. *The Green Belt movement: Sharing the experience and the approach.* Nairobi, Kenya: Environmental Liaison Center International.

Molnar, A. 1989. Forest conservation in Nepal, encouraging women's participation. In *SEEDS: Supporting women's work in the Third World,* ed. A. Leonard, pp. 98–122. New York: Feminist Press.

Molnar, A. 1986. Social forestry experience in India and Nepal. Consultant Report to the World Bank, India Country Department, Washington, DC.

Molnar, A., and G. Schreiber. 1989, May. *Women and forestry: Operational issues.* Working Paper 184, Policy Planning and Research Complex, World Bank.

Pandey, S. 1987. *Women in Hattisunde forest management in Dhading District, Nepal.* MPE Series No. 9. Kathmandu, Nepal: International Center for Integrated Mountain Development.

Robinson, P., and I. Thompson. 1989. *Fodder trees, nurseries and their central role in the hill farming systems of Nepal.* Social Forestry Network Paper 9a. London: Overseas Development Institute.

Rocheleau, D. E. 1988. Gender, resource management and the rural landscape: Implications for agroforestry and farming systems research. In *Gender issues in farming systems research and extension,* ed. S. Poats et al., pp. 149–170. Boulder, CO: Westview.

Sarin, R. 1990. Participation of Mahila Mandals in social forestry: A case study of Himachal Pradesh social forestry program. Report submitted to the World Bank, Women and Development Division, Washington, DC.

Stoler, A. 1981. Garden use and household economy. In *Agriculture and rural development in Indonesia,* ed. G. Hansen. Boulder, CO: Westview Press.

Williams, P. 1985, January. *Women's participation in forestry activities in Burkina Faso.* Hanover, NH: Institute of Current World Affairs.

Williams, P. 1986. Women's participation in forestry activities in Africa. Project Summary (mimeograph).

World Bank. 1990a. *Gender and poverty in India: Issues and opportunities for women in the Indian economy.* Washington, DC: India Country Department Study.

World Bank. 1990b, July. *World development report—poverty.* Washington, DC: World Bank.

Chapter 16

Women and Community Forestry in Nepal: Expectations and Realities

IRENE TINKER

Department of City & Regional Planning and
Department of Women's Studies
University of California, Berkeley
Berkeley, California, USA

Community forestry programs in Nepal, based on the erroneous assumption that farming families are the major cause of deforestation, ignore other critical causes. Recent emphasis on the inclusion of women in these programs, based on the recognition of women's subsistence activities and understanding of nature, minimizes fundamental constraints embedded in Nepalese society. A review of community forestry programs funded by international agencies and nongovernmental organizations (NGOs) indicates increasing participation of both women and men at the community level. Exaggerated expectations of the potential of community forestry in solving Nepal's ecological crisis could undermine the fragile gains of Nepali women.

Keywords community, cookstoves, development, donors, forestry, Nepal, NGOs, women

Increasing deforestation and forest degradation in the Himalayan kingdom of Nepal is of vital concern for the development and prosperity of both the country and the region. The extent of the problem was gradually documented after communications with the outside world were expanded following the overthrow of the hereditary Rana prime minister rule in 1950. Survival activities of the subsistence farm family were presumed to be the major cause of this environmental stress. To reach families in the remote areas of the country, the Nepalese government, in conjunction with foreign aid donors, has over the last two decades instituted forestry programs that involve the community itself in protecting and upgrading public forests. More recently, as the critical role women play in using and maintaining forests has been recognized, women have been explicitly included in community forestry.

Much enthusiasm has been generated by these imaginative programs that seem to promise so much: they combine current development planning preferences for community participation with a recognition of women's utilization of natural resources. Unfortunately, like so many panaceas promoted by development agencies, the expectations raised are often unrealistic. Because the causes of ecological stress in Nepal are now perceived as multifaceted, focusing the "solution" on a single cause is hazardous. Disappointing outcomes of community forestry programs could be too easily assigned to women's inclusion, thus promoting a backlash that would undermine women's struggles for empowerment in this fragile transitional society.

Two Problematic Assumptions

The high expectations for the women and forestry programs are fueled by two problematic assumptions that inform current forestry programs: that the major culprit causing forest degradation is the poor subsistence farming family, and that women are closer to nature than are men. These assumptions are perceived as problematic not because they are completely wrong, but because they are only part of the truth; yet the programs based on these assumptions have pushed aside the complexities and oversimplified both the problem and the solution. By designating women as the guardians of community forests and designating community forestry as the primary approach to forest policy, planners and women alike are investing their hopes and expectations in the success of programs that ignore the reality of family and class power relationships.

Causes of Deforestation

Conventional wisdom that assumes subsistence farming families are the major cause of land degradation was greatly reinforced by the first oil crisis in 1973. Until then, research on traditional household fuel consumption was minuscule. The realization that a majority of households in developing countries relied on biomass products led many observers to conclude that firewood consumption by subsistence farmers was the cause of the rapid deforestation and desertification observed in both semiarid and mountainous regions. Overnight, crash programs were funded, in response to this perceived root cause of environmental degradation, that would increase the supply of trees and reduce the demand for firewood. Improved cookstoves were invented and tested by appropriate technologists worldwide; social forestry was established as a subfield that supported forest use by people rather than emphasizing commercial interests (Agarwal, 1986; Dankelman & Davidson, 1988; Tinker, 1987).

After a decade of effort it is clear that neither improved cookstoves (Smith, 1987) nor social forestry programs (Fortmann, 1988) have lived up to their expectations. Many failures in both programs were the result of top-down technological "solutions" that neglected local needs and social constraints, particularly pressures on women's time to tend new plantings or to alter cooking practices (Cecelski, 1985; Hoskins, 1989; Islam et al., 1984). Trees selected for social forestry programs overlooked the multifold subsistence uses of forest products for food, fiber, and fodder (Falconer & Arnold, 1991; Fortmann & Rocheleau, 1985; Hoskins, 1983).

Critics not only question this technological response to deforestation, but also challenge the fundamental policy assumption that environmental degradation is primarily caused by subsistence farming families (Agarwal, 1989; Sarin, 1989). Brower (1991) goes even further to argue that "hasty application in the Himalaya of experience and practice derived from Western relations with environment and resources" has become part of the problem (p. 154). Gilmour (1991) suggests that the crisis scenario for Nepal was based on short-term observations that lacked historical perspective. He notes that deforestation is not a recent phenomenon: a century ago, forests were cleared to mine iron and copper while the wood was used for smelting. Nor has deforestation increased markedly in the last three decades although a decline in the density of national forests is apparent.

Three significant alternative causes of the environmental difficulties in Nepal are often overlooked: the geological characteristics of the region, illegal timber harvesting, and rapid modernization that is being stimulated by the development process itself. The Himalayan mountains are continuously being pushed upward as the tectonic plate under the Indian subcontinent moves northward. This geological destabilization is exacerbated by monsoon rains that concentrate 80% of all rainfall in 4 months of the year. Surface erosion caused by these forces accounts for less than one-sixth of the total with "most sediment resulting from mass wasting process . . . (that are) largely outside man's control" (Wallace, 1988, p. 3).

The negative environmental implications of rapid modernization, particularly urbanization and infrastructure projects, tend to be ignored by governments intent on economic development. Brick kilns in Kathmandu Valley not only consume vast amounts of wood, but also contribute to the air pollution that obscures the glorious mountain views throughout much of the year. Population densities exploded in the Terai, the lowlands that border India, once malaria was controlled; the government policies, including community forestry, continue to be framed in terms of the hills and so neglect profound regional variations (Soussan et al., 1991). The sole exception to government inaction has been the environmental impact of increased trekkers in the mountains: fuel for campfires must be brought in and trash removed.

Despite the expectation that urbanization would promote an energy transition to more efficient fuels (Leach, 1990; Sathaye & Meyers, 1990), urban demand for firewood remains strong in Nepal. A 1986 study in Bhaktapur, a town in the Kathmandu Valley with a population of about 53,000, enumerated 6,000 backloaders from 1423 households engaged in supplying firewood (Paudyal, 1986). In Kathmandu the price for kerosene in 1988 was lower than for firewood, but kerosene supplies were unreliable and kerosene stoves expensive so few families switched. Of the households surveyed in Chhampi, a degraded forest area near Kathmandu, 35% sell firewood to urban markets as their main source of income (Shrestha, 1989).

Road expansion in the hill areas is another major cause of environmental degradation. Visually, road cuts appear as wounds on the fragile hills of Nepal, with soil oozing down the slopes. During construction, trees on both sides of the road are used for heating asphalt for surfacing and for cooking. The Swiss Agency for Technical Assistance (SATA) has switched to using cold bitumen emulsion on new roads, arguing that the bitumen process is cheaper when the cost of replacing trees is added to asphalting costs.

A serious cause of forest depletion is the favoritism and corruption that have been an intrinsic part of granting rights to cut timber. Officials in the Ministry of Forestry, from local forest guards to the ministerial level, have been linked to the illegal sale of privileges in every recent regime; the 1980 referendum and the 1991 election campaign added to deforestation when politicians promised forest lands in return for votes (Shoumatoff, 1991; Wallace, 1988). Political will to reduce corruption and address the environmental impacts of urban fuel demand and road building does not yet seem to exist within the power structure and so cannot be addressed by the donor community. In contrast, forestry projects aimed at poor subsistence farmers are politically possible and, if funding is provided, economically feasible.

Women and Nature

Women's extensive use of forest products for subsistence and their concomitant knowledge of seeds and species has only recently been documented (Dankelman & Davidson, 1988; Hoskins, 1983). In Nepal indigenous knowledge is devalued by westernized government foresters and foreign experts alike (Brower, 1991; Gilmour, 1991), making it difficult for either the development community or the local patriarchal society to accept the importance of inherited learning among women (Stamp, 1989). Ignoring women's knowledge is consistent with the imposition of western stereotypes of women's roles through development programs that discount women's economic roles and undermine their status within both society and the family (Tinker, 1990).

Ecofeminists celebrate women's affinity with nature and draw a parallel between the desire of industrialization to conquer nature and the desire of men to dominate women. The feminist perspective counters aggression and celebrates "the interconnectedness and diversity of nature," writes Shiva (1989, p. 14), a leader of this movement. Such an appeal to the superior nature of women resonates with many feminists who believe that women are more peaceful (Boulding, 1977), more moral (Gilligan, 1982), or more nurturing (Sen, 1988). The dramatic emergence of the Chipko Andolan, when women in the Indian Himalaya hugged trees to prevent commercial cutting, is emblematic of the exploitation of women and of nature by development. So romanticized has this movement become that the current needs of these women living in the hills of Uttar Pradesh have been obscured, according to a young Indian scholar who insists that "we must disentangle the realities from the myths" (Mehta, 1991, p. 13). Nonetheless, this championing of poor women over development has made Shiva something of a cult figure among environmentalists worldwide. Increasingly, environmentalists in the South have joined with ecofeminists to question the mainstream environmentalist constituency of the North, with their technocratic views, and to support those grassroots movements in the South concerned with social justice.

In Nepal, dependent as it is on the largess of foreign aid, the argument for including women in forestry programs is based on efficiency, not social justice. Development practitioners support the inclusion of women in forestry projects because women are major users of forest products (Acharya & Bennett, 1982; Molnar, 1991; Molnar & Schreiber, 1989) and are likely to continue using the forest illegally unless alternatives for fuelwood, fodder, or leaves are provided. For them, women's productive roles, not their needs, provide the justification for including them in forestry programs. (ICIMOD, 1988; Peluso, 1991). Yet the influence of ecofeminism penetrates the program objectives and reflects claims of superior knowledge by women (Acharya, 1992; Shrestha & Gurung, 1991).

Community Forestry in Nepal

Community forestry programming was developed in response to a perception of increasing deforestation and forest degradation in Nepal. For many years the preferred explanation for this deterioration was the impact of the 1957 Private Forests (Nationalization) Act, which placed all forests under government control. The 1957 act was meant to reduce large forest holdings granted by the former feudal overlords, known as Ranas, to their supporters; but the lack of cadastral surveys defining clear boundaries to national forests made it possible for individuals to clearcut forest land for agriculture use and then register the land as private property (Wallace, 1987, 1988). Forest degradation was attributed to villagers who, it was assumed, no longer had vested interest in forest protection or upkeep and so overexploited the existing forests (Gautam, 1987; Regmi, 1992). Despite widespread rejection of this assumption, it was incorporated into the 1989 Master Plan for the Forestry Sector, 1989–2010, funded by the Asian Development Bank and the Finnish government, and thus remains the basis for current policy (Finlayson, 1987).

Subsequent legislation under the 1957 act set up a system of permits for use of forest products, regulations difficult for the small forest service to impose over the extensive nationalized forests. This legislation turned forest officers and rangers into policemen, an unpalatable role according to former Nepali forester Tej Mahat. Assigned to the Chautara Forest Division of the Middle Hills in 1973, Mahat began experimenting with ways of using the consumers of forest products in forest protection; his pilot project was influential in shaping community forestry in Nepal (Mahat, 1987). Like Mahat, most development experts considered the forest nationalization approach not only wrong but unenforceable, given the topography of the country and an understaffed forest service: "in the hills of Nepal *there is generally no alternative* to community forestry—and that its essence is a *real transfer of responsibility* for forest protection and management, from the central government to the users" (Finlayson, 1987, his emphasis).

The National Forestry Plan of 1976 had as its central tenet the transfer to village communities, then known as *panchayats*, of the right to manage local forests for their own benefit. Religious institutions such as monasteries and temples could also acquire forests, while a category of leasehold forests allowed other institutions or individuals similar rights. The forest available was immense. Wallace (1988) estimated that local villages "could have formal responsibility for managing more than 2.5 million hectares—over two-fifths—of existing forests" (p. 17). The implications for the forest service were staggering. Instead of a centralized cadre of foresters trained in commercial forestry and charged with protecting the forests, the service was to reinvent itself as a rural development agency capable of working with poor villagers not only to protect and use existing forests but also to plant noncommercial trees for fuelwood and fodder. The forestry institutes would have to drastically alter their courses; students continue to be trained as rangers who display "little empathy towards villagers and their problems. . . . The task of reversing this situation and creating a cadre of people with a new 'world view' (implies) a generational change" (King et al., 1988, pp. 17–18).

The general outline of community forestry in Nepal was not affected by the return to multiparty elective government in May 1990, with no substantial change beyond substituting the word "community" for "panchayat" wherever its appeared (Tuladhar, 1991). The 1989 Master Plan for the Forestry Sector, 1989–2010, was endorsed by both the interim and elected governments and formed the basis of the Forest Act of 1992 that was passed in September and signed by the king in January 1993. Although this legislation repeals earlier laws that emphasized central state control of the forests, its brevity leaves many of these recommendations for further legislation or regulation.

International Influence

Much of the impetus for emphasizing community forestry came from the donor community. Their influence was derived not only from their funding but also from the influence of the large number of professionals brought into Nepal as advisors and managers. As early as 1954, a forestry expert from the Food and Agricultural Organization (FAO) observed that "deforestation frequently assumes disastrous proportions" (Robbe in Wallace, 1987, p. 9). Australia was asked by the Nepalese government to plant eucalyptus near Kathmandu in 1962; in 1978 the Nepal-Australia Forestry Project (NAFP) began working in the Chautara Forest Division and took the lead in implementing the community forest concept (Griffin, 1987; NAFP, 1985). USAID funded the first aerial survey in 1964.

By 1987, 38 international agencies and organizations plus 3 Nepali groups were running projects and collecting data related to forestry, usually as a component of broader development programs: 8 multilateral organizations or international banks, 15 bilateral agencies, and 15 international nongovernmental organizations (NGOs); locally, a mass youth organization and two consulting organizations.[1] These 40 various agencies and organizations along with five semigovernmental corporations, two international research groups, and the International Centre for Integrated Mountain Development (ICIMOD), plus 22 different governmental departments or boards, were, in 1987, involved in 68 discrete forestry programs.[2] More than half of these projects included some sort of community forestry component such as tree planting, forest protection, or introducing improved cookstoves; those without a community participation element included commercial timber programs, research into forest products used as medicine or for food or crafts, environmental and topographical surveys, and conservation and resource policy and planning (Carter, 1987). Among the 35 programs with some type of participatory activities there were six major afforestation projects, 15 watershed projects, and 14 small forestry projects run by NGOs.

The interplay between donors and the government is obvious from the number of such projects. The financial implications are immense; for example, the Tinau Watershed Project in Palpa District, jointly financed by Swiss and German aid, supported 70% of activities carried out by the district forest office. The CARE project at Begnas Lake near Pokhara hired villagers to build stone fences around plantations to encourage their cooperation. Not everyone was happy with this trend of coopting villagers with income. Gautam and Roche (1987), a Nepali forest officer and a Swiss aid official, note that the Integrated Hill Development Project in Dolakha District pays "the villagers to grow trees in the nursery . . . then pays the villagers to plant and . . . protect the trees from themselves"; not only is the cost extremely high, but the process has made the people dependent on the project and reduced their previous involvement in forest protection (p. 12).

Implementing Community Forestry

The Panchayat Forest Rules and Panchayat Protected Rules, promulgated in 1978, created two categories of community land, both based on the lowest level political and administrative unit, known then as panchayat. Nepal listed 2,913 village and 23 town panchayats; population of village panchayats varied from 2,000 to 5,000. Degraded forests, enclosed so that growth could rejuvenate, were termed Panchayat protected forests and are now called natural forests. Overgrazed and barren land was to be turned into village plantations, called Panchayat forests, using seedlings supplied through the forest service; today they are called forestry plantations. Income from the sale of forest products from the plantations, but not from the forest, went to the village panchayat.

To receive title of these forest lands, the panchayat had to produce a management plan in consultation with the district forestry officer (DFO). The plan would then move through several higher levels of the bureaucracy, a procedure that "considerably slowed down the process" (Kayastha & Karmacharya, 1987, p. 5). Difficulties in developing an acceptable panchayat management plan encouraged some local forestry officials to proceed informally, a choice that may have made the agreements more durable after the end of the panchayat system in 1990.

Because the community forest program assumed that years of chaos and instability in the rural areas had undermined any traditional organizations that controlled access to and use of forests, an important feature of the management plan was the establishment of village-level user committees, who were expected to implement and continuously manage the plan. From the beginning there was debate over the composition of the committees: Should they consist of locally elected village officers, or of actual forest users? Would village elite allow members from all groups? Increasingly, as the continued strength of both caste and gender hierarchies was recognized, the question became: Is there a *community* at all? Hobley (1985) argues that the term implies "that local communities are an homogeneous entity, united for common action by their need for firewood and fodder . . . (and ignores) the differential access to both natural and political resources within the village dependent upon . . . caste, class and gender" (p. 2). Hamlets within a village, known as wards, often reflect ethnically different settlements. Furthermore, panchayat boundaries are often arbitrary: geographically related areas may be in different panchayats while wards in one village may be close on a map but many hours away across a deep ravine. Because of the terrain, several panchayats, even in different districts, may have customary rights to the same forest lands, indicating the need for multi-village committees (Campbell et al., 1987). This difficulty of reconciling the interests of local user groups with those of the larger panchayat was mentioned frequently by panchayat officials during my field research.

The original assumption that fuelwood was the primary need was quickly adjusted as the role of livestock in hill agriculture became better understood: animal manure is essential as fertilizer, and 35% of animal feed comes from trees. This farming system requires one large animal per person per family of five or six; thus, for each hectare of agricultural land, 1.3 ha of forest land are needed to supply fodder (Applegate & Gilmour, 1987). Although forest-farming linkages are the basic characteristic of the hills, foresters had not been trained to grow fodder trees. Griffin (1987) describes the NAFP as starting from an "almost unbounded ignorance" (p. 8). Through trial and error, the project workers learned a technique, now widely copied, for growing fodder trees that had proved difficult to propagate: pine trees are first planted on denuded land; their fast growth provides shelter for the natural regeneration of the fodder trees that the villagers prefer.

Forestry programs have been much slower to adapt their original concepts regarding the introduction of theoretically improved cookstoves (Joshee, 1986). Some 35,000 improved and highly subsidized cookstoves were distributed by 11 different projects funded by six independent agencies through forestry programs between 1981 and 1987. An assessment in 1987 found that the stove model being disseminated was "inappropriate for a large segment of the population" and recommended redirecting research toward less expensive models possibly based on widely used indigenous stove types (Wood, 1987, p. 17). The insistence on pairing cookstoves with forestry projects requires forestry workers in many projects to promote both stoves and community forestry at the same time (Acharya, 1992).

Including Women

Women fetch fodder daily to feed the animals; in contrast, fuelwood can be stored, so men often harvest trees at some distance a few times a year. As more men seek work away from the hills, women take over even more of the farming chores. The development agencies realized that community forestry could not succeed unless women supported the efforts to protect new plantations and degraded forest; cookstoves would not be used if they did not harmonize with women's work schedules (Kumar & Hotchkiss, 1988).

During the 1980s, most forestry projects began using professional women as project and field staff, but located in Kathmandu. NAFP appointed a Nepali woman as Women's Coordinator, whose tasks were to interview NAFP staff and villagers about women in the project, distribute improved cookstoves, and offer scholarships to girls (Shrestha, 1989). The coordinator found that women in the project area were aware of the need to protect the forest but gravely concerned that if cutting fodder in the forest were banned those women without trees on their own land would suffer (Kharel, 1987).

This worry was well founded. In Banskharka, the authoritarian village chief had formed his own forest committees and closed the forests, forcing many near-landless lower castes to break the law or to migrate to India. Women-headed households were particularly vulnerable; in one ward five of eight households had women heads because of male outmigration. The village chief allowed the male-headed households to register common pastures as their own personal property, thus usurping the forestry rights of the women even though *all* of the households had contributed equally to buying the forest 40 years before. Accusations of witchcraft against some of the women were used to justify the actions (Hobley, 1987).

At project sites, local women were hired as nursery managers and forest watchers. Although gender hierarchies remain strong in the Hills, there is greater equality among the lower castes, who are most likely to take these jobs, than there is among village Brahmins and Chettris (Kharel, 1987). The Tinau Watershed Project hired six female motivators to visit women's user groups and encourage cultivation of improved grasses in silvapastoral plots, according to the DFO Diwakar Pathak. With alternative fodder from these grasses the women could accept closing of the degraded forest areas so that natural regeneration was possible. Because many poor women were surviving by selling firewood in the towns, the project also launched an income-generating effort. One widow had begun to make orange and lemon squash in her rural home, bottle it, and then sell it to roadside snack shops. Collecting and selling seeds from the forest was another income activity.

Women's membership on forest user committees was more problematic. Traditionally, two classes of villagers use the forest: women, children, and some men who collected fodder, bedding, and fuelwood from the forests, and men from the traditional elite who decide rules for forest access and exploitation of timber. Since user committees were expected to facilitate the planting and weeding of the panchayat plantations and the controlling access to panchayat forests for collecting fuelwood, all women's activities, men could accept women members, especially when the suggestion came from the district forest controller (DFC) or the local development office (LDO).

Whether to accomplish women's participation by separate women's committees or by ensuring female membership in male forest committees was the subject of some debate. The NAFP opted for integrated committees. Although Australian aid has been active in Nepal since 1966, women's roles in forestry were first mentioned in the project document outlining goals for the third phase that commenced in 1985 (NAFP, 1985). The objective is to raise "the social and economic status of women in the project area by involving them directly in the community project"; the idea of a separate women's component was rejected "because the role of women is integral to almost every one of (the Project's) aspects" (NAFP, 1985, pp. 16–17).

Evaluation of women's participation in Thapagaon, in the NAFP area, indicates the difficulties in changing local attitudes. Apparently women were put on the forestry committee and one elected as vice-chairman to please project personnel, but "were actively discouraged from attending the meetings. As one woman said, 'we are only invited to meetings when foreigners will be present, otherwise we are completely excluded'" (Hobley, 1987, p. 9). Nor did their husbands or fathers, who did attend, inform them of decisions made at the meetings (King et al., 1988). The NAFP coordinator commented that most village women are illiterate and have a very localized view of life, while the "male members of society . . . may not be very keen to give power and authority to the women as it will threaten the *status quo*" (Kharel, 1987, p. 3).

Such powerlessness supports the need for separate rather than integrated programs, according to many Nepali feminists. Indira Shrestha harshly criticized the NAFP women's programs in an evaluation report:

> It is disquieting to find that . . . there is no awareness of the need to make any special provisions to incorporate women into community forestry, either in terms of forest officials, field staff, or the rural population. . . . Unless there is a distinct component for women . . . and training programmes for women to build up a cadre of female field staff, the Nepal Australia Forestry Project cannot achieve its envisaged goal. (Shrestha, 1989, p. 56)

NAFP showed no signs of changing; it funded a training manual that reiterates that women in forestry must be part of all programs and that male extension workers must learn to work with women (Siddiqi, 1989).

An integrated approach to forming forest committees was also tried in phase two of the USAID-funded Integrated Rural Development Program in the Rapti Zone, an area covering five districts in western Nepal. The forestry officer was charged with setting up forest committees in two districts, with women as 25% of the membership. These committees were restricted to users of forest products, and panchayat officers could not be members. This arrangement allowed women to have a say in daily forest use while leaving the preparation of the management plan with the traditional male village elite.

Darchula District, in the far western part of Nepal, "has the oldest and greatest number of women's forestry committees" (Inserra, 1988, p. 4), a situation encouraged by an enthusiastic DFC and a woman VSO (Volunteer Service Overseas, British). At first women were not consulted about membership but simply appointed to the committee by the male village leadership. While the majority of members are from upper castes, lower caste Tamang and Gurung women are bold and independent members who draw on their experience gained from trading local goods, especially the wool they spin themselves, in India (Prasai et al., 1987). Women members are reported to "enjoy belonging to a group and 'improving' themselves by taking on new tasks and responsibilities" (Inserra, 1988, p. 9).

Two panchayats eventually decided that committees composed only of women were not reasonable and elected men to join (Prasai et al., 1987). Men also frequently attended meetings of the all-women committees; while they often dominated the discussions, at other times the women argued freely with the men. In one village the women asked men to form an advisory group to assist them. However, when women in Pipalchaurie, Darchula District, tried to discipline some men who were illegally cutting wood in the government forest, the women got no support because men thought that they were "getting above themselves" in trying to discipline men. According to Inserra (1988), "Support from local men is very important for female membership on forestry committees. . . . Women are taking on a new role by joining a committee, a role that has traditionally been filled by men" (p. 12).

A workshop on "Women's Participation in Forest Management" was held in April 1987, sponsored by Winrock International, to provide an opportunity for field workers, 16 Nepalese and 13 expatriates, to exchange information and to make policy recommendations. Two conclusions seem to undermine the entire concept of women in community forestry. First, the group rejected separate components for women, arguing that "The issue is not 'how to involve women' but 'how to involve motivated community members who will ensure that community interests are met in the management of the resource.'" (Tisch, 1987, p. 3). Second, the workshop rejected the idea that only female staff could reach village women: waiting 20 years for sufficient women foresters was unreasonable. "The emphasis should be on making current personnel effective. . . ." (Tisch, 1987, p. 5).

These ideas stand in stark contrast with recommendations made at the "International Workshop on Women, Development, and Mountain Resources," held at ICIMOD in November 1988. Women at this workshop were well-known feminist researchers and writers rather than field workers; their views were endorsed by the men attending. Their report emphasizes the importance of including "gender as an analytical category" in research, policy, and program implementation. Looking at gender variables should allow greater understanding of women's decision-making roles in resource management, a necessity if women are to be treated as more than laborers. Women's limited land ownership rights, a topic usually avoided because of its deep cultural roots, was identified as a major deterrence to women's ability to decide resource use. The report also calls for more women to be hired as development professionals (ICIMOD, 1988). Subsequently, the Institute of Forestry began offering 15% of all seats to women (Bloom & Luche, 1992).

These contrasting views reflect, first, the difficulties encountered in the field when trying to implement policies designed in Kathmandu and exacerbated by the cultural distance between educated urban women and the women villagers. Furthermore, only a limited number of trained women (and men) are willing to work outside the capital. Second, the contrasting policy recommendations reflect the ongoing debate among women in development proponents: Should programs for women be separate or integrated? Experience has shown that neither approach works alone. Program design that incorporates women's concerns along with those of men provides a stronger conceptual framework, enhanced funding, and improved bureaucratic support; program implementation usually functions better when women work together in their own organizations (Tinker, 1990).

While the two workshops disagreed on project design and staffing, both emphasized the diversity of women's activities and called for closer cooperation among various development projects operating at the village level. Even here the contrast in perspectives is evident. Winrock participants called for training women in skills needed for community participation; ICIMOD participants spoke of involving women in the market economy to improve their status and economic security. The two views respond to women situated at opposite ends of the transition that defines their identity: from being embedded in community and family to functioning as an independent person within the larger society.

Current Issues

Nepal's return to multiparty elective government in May 1990 created an atmosphere in the country that encourages greater citizen participation. In rural areas, the new climate allows "more space and leeway for . . . creative and innovative grassroot development experiments" (Tamang, 1990). Indeed, evidence suggests that even under the previous regime many remote communities continued to monitor and manage their own forests (Fisher, 1989; Gilmour, 1989; Shrestha, 1987), substantiating the argument that "Common property regulation is likely to be the most effective response to scarcity" (McGranahan, 1991, p. 1276). Writing specifically about Nepal, Hobley (1985) insists that common property has clearly defined rights and use patterns that sustain natural resources and contends that overuse occurs only when resources are "open to unregulated access by individuals" (p. 663). Fisher (1989) has analyzed existing evidence of current management systems and concludes that they are not simply copies of ancient practices but, rather, are "dynamic responses to new situations (that) sometimes adopt some of the formal features of earlier systems" (p. 1).

Increased community involvement in forests was observed by Fox (1993), who writes of his surprise when he returned, in 1990, to the same area of central Nepal where he had studied forest use practices 10 years earlier. He found the forests in *better* condition than they had been in 1980, primarily because there were fewer livestock to be fed and cared for, a change caused by a chain of local events. A USAID project had developed a plantation on degraded land and hired forest guards to protect the trees; grazing and cutting of the trees for fuel or fodder were prohibited, although collection of dry fuelwood was allowed. Villagers were forced to reduce their livestock holdings, an action facilitated by the increased availability of chemical fertilizers to replace manure for agricultural use due to a recently completed motorable road to Gorkha, an hour's walk away. To feed the remaining livestock, farmers planted trees on their own lands while forming an informal users committee to protect the plantation after the project withdrew. The informal nature of the committees allowed their continuation after the end of panchayat government; but the lack of a management plan has brought controversy over dividing proceeds from sales (Fox, 1993).

Ancillary effects of these changes have benefitted both women and children. Fewer livestock to tend meant that more children were in school. Less fodder to collect meant that women had time to respond to activities initiated by two NGOs that have begun work in the village. Fox (1993) found it ironic that among the income activities suggested was raising livestock, although buying an improved breed of cattle on credit has long been part of women's programs, whether run by government or by NGOs. Finding such efforts in conflict with reduced fodder use, Fox (1993) suggests the need for coordination among projects in a single area.

The new atmosphere has encouraged greater activity among women who participated in unprecedented numbers during anti-panchayat demonstrations. The 1990 Constitution proclaims equal rights for women and repeals all previous laws that discriminate against them. However, the 1950 Constitution also provided greater inheritance and marital rights for women, but the changes were never introduced because implementation of such laws requires fundamental changes in attitudes toward women. Already the parties that supported women's rights during the writing of the new constitution "went back to tradition when these values came in conflict with traditionally entrenched practices" (Acharya, 1993, p. 23).

More women have become active in forest user groups. A recent evaluation of women in community forestry in Gorkha run by Save the Children, USA, credits their training programs in the raising and planting of seedlings, and in forest management, with increasing production of local nurseries run by women's groups. The profits have been used to provide a revolving credit fund for their membership (Acharya, 1992).

Aware of decreasing forest resources, more and more villages are taking advantage of community forestry legislation to register their own forests and restrict access of neighboring villagers; in turn, the affected villages rush to register their own forests, creating a domino effect. In Gorkha District, 84 community forestry areas had applied for rights by the end of 1992 and 32 had received them. Four of the five user groups sponsored by Save the Children, USA, in *ilaka* (subdistrict) 1 of Gorkha District were women's groups. In one instance, three separate women's groups combined their small plantations to apply for recognition and increase the pool of households able to patrol the forests. In another, the women registered as the official group but set up a separate male advisory committee. Nearby, women and men together established a ward plantation, in that instance only men were listed as official users. The evaluator emphasized that "The practice of listing only the women in the user group . . . does not mean that women and men are opposed to each other. It would have been impossible to make the forestry program a success without the cooperation of the men" (Acharya, 1992, p. 19). While both women and men take leadership roles in community forestry, a division of labor exists: women tend to manage plantations while men manage natural forests. These trends in Gorkha are significant but are not necessarily predictive of how other districts might proceed.

Administrative changes are slowly making an impact. Although the panchayat system has been abolished, the same administrative divisions (except for zones) remain, so that the issues of community, and of locally entrenched hierarchical authority patterns, persist. Village development committees have replaced the panchayats at the community level, but DFOs are allowed to bypass the community and register user groups at the ward level. Because these units are small, ward user groups from the same village development committee or even from neighboring village committees may join into a more logical geographic unit to manage natural forests and/or forest plantations.

The conflicting roles of forest rangers are being addressed, forestry planning staff informed me, by creating three distinct cadres and training them accordingly. Rangers assigned to the hills will be trained as community foresters to work with women and men to sustain the forest. Rangers in the Terai will fulfill a more technical role, advising industries that are growing their own materials. Special conservation units will manage national forests surrounding the highest mountains. Private interests in forestry are to be granted recognition through greater use of leasehold rights, which will transfer management to "various classes of community" (Bonita & Kanel, 1987, p. 79).

Summation

In this article I have argued that community forestry programs based on the erroneous assumption that the subsistence farm family is largely responsible for forest degradation in Nepal create unrealistic expectations. On the positive side, recognition that villagers must have control over the forestry resources they use if they are to maintain them has altered rural dynamics and allowed more participation throughout the country.

It is an equally unrealistic expectation that women, with their special affinity to nature, can solve deforestation through forestry or cookstove programs. This assumption makes them particularly vulnerable to being identified as the cause of failure when program makes goals are not achieved. On the other hand, the inclusion of women in forestry committees sends a signal of legitimacy and approval from government that challenges the status quo. Simply raising the issue has forced both planners and administrators to confront the sexual division of both labor and decision making at the village level. Evaluations of pilot programs reflect the complexity of social change and suggest caution about importing assumptions to Nepal either about women and nature or about gender relationships.

The history of community forestry in Nepal and of the efforts to include women in forestry programs illustrates the complexity of economic development and its inevitable corollary, social change. Easy solutions offered by conventional wisdom need to be set aside and the entire development program in Nepal reconsidered in light of the realities portrayed in recent reports. Identification of the broad array of programs and lifestyle changes that are at the root of deforestation in the Himalayan hills is the first step toward addressing environmental degradation in Nepal.

Notes

1. Multilateral agencies: FAO, United Nations Development Programme, United Nations Environmental Programme, UNICEF, and UNESCO; plus the World Bank, the Asian Development Bank (ADB) and the European Economic Community. Bilateral agencies: Australia, Canada, China, Finland, Germany, India, Netherlands, New Zealand, Norway, Sweden, Switzerland, United Kingdom, U.S.S.R., and United States. Japan forestry funds flowed through ADB. International

NGOs: with own projects—Action Aid (European), CARE, International Planned Parenthood Federation, King Mahendra Trust for Nature Conservation, Lutheran World Federation, Save the Children Federation (USA), United Mission to Nepal, and World Neighbors; sole funders—Helvetas (Swiss), Asian Community Trust (Japan), Red Cross, and Himalayan Trust (NZ); research—Association for Research, Exploration, and Aid (Australia) and Roughtans & Partners (UK). Nepali groups: Youth Activities Coordination Council, New Era, Forest Services (Carter, 1987).

2. In the spring of 1988, in Nepal as a Fulbright fellow, I interviewed more than 50 foresters, scholars, and consultants involved with planning, supervising, or evaluating these programs, both in Kathmandu and in the field, where I also talked with panchayat and ward members.

References

Acharya, H. 1992. Participation of women in the management of community forests in Gorkha, Nepal. Report to the Nepal Field Office, Save the Children Federation, USA in Kathmandu, Nepal.

Acharya, M. 1993. Political participation of women in Nepal. In *Women in politics worldwide,* eds. B. Nelson and N. Chaudhary. New Haven, CT: Yale University Press.

Acharya, M., and L. Bennett. 1982. Women's roles in the subsistence sector, World Bank working paper no. 526. Washington, DC: World Bank.

Agarwal, B. 1986. *Cold hearths and barren slopes: The woodfuel crisis in the third world.* London: Zed Books.

Agarwal, B. 1989. Rural women, poverty and natural resources: Sustenance, sustainability and struggle for change. *Economic and Political Weekly,* October 28.

Applegate, G. B., and D. A. Gilmour. 1987. Operational experiences in forest management development in the hills of Nepal. Occasional paper no. 6. Kathmandu, Nepal: International Centre for Integrated Mountain Development.

Bloom, G., and J. Luche. 1992. Enhancing USAID/Nepal's commitment to gender/WID considerations with a focus on the democracy programs. Prepared for USAID by GENESYS. Washington, DC: Futures Group.

Bonita, M., and K. Kanel. 1987. Some views of the master plan for forestry sector project on community forestry. *Banko Janakari* 1(4):76–81.

Boulding, E. 1977. Integration into what? Reflections on development planning for women. In *Women and technological change in developing countries,* eds. R. Dauber and M. Cain. Boulder, CO: Westview Press.

Brower, B. 1991. Crisis and conservation in Sagarmatha National Park. *Society and Natural Resources* 4(2):151–163.

Campbell, J. G., R. P. Shrestha, and F. Euphrat. 1987. Socio-economic factors in traditional forest use and management: Preliminary results from a study of community forest management in Nepal. *Banko Janakari* 1(4):45–54.

Carter, J. 1987. Organizations concerned with forestry in Nepal. Occasional paper 2/87. Kathmandu, Nepal: Forestry Research and Information Center.

Cecelski, E. 1985. *The rural energy crisis, women's work and family welfare: Perspectives and approaches to action.* Geneva: ILO.

Dankelman, I., and J. Davidson. 1988. *Women and environment in the third world: Alliance for the future.* London: Earthscan.

Falconer, J., and J. E. M. Arnold. 1991. Household food security and forestry: An analysis of socioeconomic issues. Community forestry note #1. Rome: Food and Agriculture Organization.

Finlayson, W., 1987. ed. *Banko Janakari* 1(4): Issue on Community Forestry Management.

Fisher, R. J. 1989. Indigenous systems of common property forest management in Nepal. Working paper no. 18. Honolulu, HI: East-West Center.

Fortmann, L. 1988. Great planting disasters: Pitfalls in technical assistance in forestry. *Agriculture and Human Values.* Winter/Spring 49–60.

Fortmann, L., and D. Rocheleau. 1985. Women and agroforestry: Four myths and three case studies. *Agroforestry Systems* 2:253–72.

Fox, J. 1993. Forest resources in a Nepali village in 1980 and 1990: The positive influence of population growth. *Mountain Resources and Development* 13(1):89–98.

Gautam, K. H. 1987. Legal authority of forest user groups. *Banko Janakari* 1(4):1–4.

Gautam, K. H., and N. H. Roche. 1987. The community forestry experience in Dolakha District. *Banko Janakari* 1(4):12–15.

Gilligan, C. 1982. *In a different voice.* Cambridge, MA: Harvard University Press.

Gilmour, D. A. 1989. Forestry resources and indigenous management in Nepal. Working paper no. 17. Honolulu, HI: East-West Center.

Gilmour, D. A. 1991. Trends in forest resources and management in the middle mountains of Nepal. Proceedings of a workshop on soil fertility and erosion issues in the middle mountains of Nepal, pp. 32–46, IDRC, Kathmandu and Ottawa.

Griffin, D. M. 1987. *Innocents abroad in the forests of Nepal: An account of Australian aid to Nepalese forestry.* Kathmandu, Nepal: Nepal-Australia Forestry Project.

Hobley, M. 1985. Common property does not cause deforestation. *Journal of Forestry* 83(11):663–664.

Hobley, M. 1987. Involving the poor in forest management: Can it be done? The Nepal-Australia project experience. ODI social forestry network. Network paper. London: Overseas Development Institute, Regent's College.

Hoskins, M. 1983. *Rural women, forest outputs, and forestry projects.* Rome: Food and Agriculture Organization.

Hoskins, M. 1989. Activities of the community forestry unit of FAO relating to domestic energy. *Nordic seminar on domestic energy in developing countries,* pp. 19–21. Lund, Sweden: Lund University, Lund Centre for Habitat Studies.

ICIMOD. 1988. Women in mountain development. Report of the international workshop on women, development and mountain resources: Approaches to internalising gender perspectives. Kathmandu, Nepal: International Centre for Integrated Mountain Development.

Inserra, A. E. 1988. Women's participation in community foretry in Nepal: An analysis for the forestry development project. Kathmandu, Nepal: USAID.

Islam, M. N., R. Morse, M. H. Soesastro, eds. 1984. *Rural energy to meet development needs: Asian village approaches.* Boulder, CO: Westview Press.

Joshee, B. R. 1986. Improved stoves in minimization of fuelwood consumption in Nepal. Forestry research paper series no. 2. Kathmandu, Nepal: Winrock International Institute for Agricultural Development.

Kayastha, B. P., and S. Karmacharya. 1987. Legislation in community forest management. *Banko Janakari* 1(4):5–8.

Kharel, S. 1987. Women in forestry. Nepal-Australia Forestry Project discussion paper. Kathmandu, Nepal: Nepal-Australia Forestry Project.

King, G. C., M. Hobley, and D. A. Gilmour. 1988. Management of forests for local use in the hills of Nepal. Kathmandu, Nepal: Nepal-Australia Forestry Project.

Kumar, S., and D. Hotchkiss. 1988. *Consequences of deforestation for women's time allocation, agricultural production, and nutrition in hill areas of Nepal.* Washington, DC: International Food Policy Research Institute.

Leach, G. 1990. The energy transition. *Energy Policy* 20(2):116–123.

Mahat, T. 1987. *Forestry-farming linkages in the mountains.* Occasional paper no. 7. Kathmandu, India: International Centre for Integrated Mountain Development.

McGranahan, G. 1991. Fuelwood, subsistence foraging, and the decline of common property. *World Development* 19(10):1275–1287.

Mehta, M. 1991. The invisible female: Women of the UP Hills. *Himal* Sep/Oct:13–15.

Molnar, A. 1991. Women and international forestry development. *Society and Natural Resources* 4(1):81–90.

Molnar, A., and G. Schreiber. 1989. *Women and forestry: Operational issues.* Washington, DC: Women in Development, World Bank.

Nepal-Australia Forestry Project (NAFP). 1985. Project document phase 3. Kathmandu, Nepal: Nepal-Australia Forestry Project.

Paudyal, K. R. 1986. Noncommercial cooking energy in urban areas of Nepal. Forestry research paper no. 4. Kathmandu, Nepal: Winrock International Institute for Agricultural Development.

Peluso, N. 1991. Women and natural resources in developing countries. *Society and Natural Resources* 4(1):1–3.

Prasai, Y., J. Gronow, U. R. Bhuja. 1987. Women's participation in forest committees: A case study. Forestry research paper no. 11. Kathmandu, Nepal: Winrock International Institute for Agricultural Development.

Regmi, S. C. 1992. Women in forestry: Study of a woman's forest committee in a Nepalese village. Research report series no. 2. Kathmandu, Nepal: Winrock International Institute for Agricultural Development.

Sarin, M. 1989. Improved stoves, women and domestic energy: The need for a holistic perspective. In *Nordic seminar on domestic energy in developing countries.* Lund, Sweden: Lund University, Lund Centre for Habitat Studies.

Sathaye, J., and S. Meyers. 1990. Urban energy use in developing countries: A review. In *Patterns of energy use in developing countries,* ed. A. V. Desai. New Delhi, India: Wiley Eastern.

Sen, G. 1988. Ethics in third world development: A feminist view. Rama Mehta lecture. Radcliffe College, Cambridge, MA.

Shiva, V. 1989. *Staying alive: Women, ecology and development.* London: Zed Books.

Shoumatoff, A. 1991. The mountain is rising. *Conde Nast Traveler.* August.

Shrestha, B. K. 1987. The impact of decentralization on community forest management in Nepal. *Banko Janakari* 1(4):9–15.

Shrestha, G. R., and S. M. Gurung. 1991. Sub-sector review paper on the role of women in fuel, fodder, and forest management. Kathmandu, Nepal: Center for Rural Technology.

Shrestha, I. 1989. *Women in development in Nepal: An analytical perspective.* Kathmandu, Nepal: Integrated Development Systems.

Siddiqi, N. 1989. Towards effective participation: A guide for working with women in forestry. Technical note 1. Kathmandu, Nepal: Nepal-Australia Forestry Project.

Smith, K. 1987. The biofuel transition. *Pacific and Asian Journal of Energy* 1(1):13.

Soussan, J., E. Gevers, K. Ghimire, P. O'Keefe. 1991. Planning for sustainability: Access to fuelwood in Dhanusha District, Nepal. *World Development* 19(10):1299–1314.

Stamp, P. 1989. *Technology, gender, and power in Africa.* Ottawa, Canada: International Development Research Centre.

Tamang, D. 1990. *Lessons from participatory action research and rural development extension in the K-BIRD areas.* Kathmandu, Nepal: Service Extension and Action Research for Communities in the Hills (SEARCH).

Tinker, I. 1987. The real rural energy crisis: Women's time. *Energy Journal* 8:125–146. Longer version in Desai, A. V., ed. 1990. *Human energy.* New Delhi, India: Wiley Eastern.

Tinker, I., ed. 1990. The making of a field: Advocates, practitioners, and scholars. In *Persistent inequalities: Women and world development.* New York: Oxford University Press.

Tisch, S. T. 1987. Final recommendations from the workshop on women's participation in forest management. 7 pp. mimeo. Kathmandu, Nepal: Winrock International Institute for Agricultural Development.

Tuladhar, A. R. 1991. Forestry in an accountable democracy: Confusion, conflicts and choices. *Himal* Sep/Oct:33–36.

Wallace, M. B. 1987. Community forestry in Nepal: Too little, too late? Research report series no. 5. Kathmandu, Nepal: Winrock International Institute for Agricultural Development.

Wallace, M. B. 1988. Forest degradation in Nepal: Institutional context and policy alternatives. Research report series no. 6. Kathmandu, Nepal: Winrock International Institute for Agricultural Development.

Wood, T. S. 1987. *Assessment of cookstove programs in Nepal.* Arlington, VA: Volunteers in Technical Assistance (VITA).

Chapter 17

Integrating Gender Diverse and Interdisciplinary Professionals into Traditional U.S. Department of Agriculture-Forest Service Culture

JAMES J. KENNEDY

College of Natural Resources
Utah State University
Logan, UT 84322-5215
USA

Abstract *In compliance with environmental legislation and affirmative action policies of the 1970s, a generation of interdisciplinary and gender diverse professionals have been hired by the U.S. Department of Agriculture-Forest Service. Two studies examine the first wave of these new employees (half of whom were women), finding most committed to their profession, involved with mentors, and integrating into the agency culture. When career development difficulties did occur, they were usually associated with type of profession rather than gender. Wildlife and fisheries biologists, new to the Forest Service, had more difficulty accepting and becoming committed to the agency than their forester or range manager colleagues. There were several important gender differences, with women usually the more satisfied about their current status and future Forest Service career prospects.*

Keywords career development, mentors, natural resources or environmental policy, natural resources or environmental professionalism, natural resources or environmental values, organizational behavior.

Introduction

In the 1960s federal resource management agencies in the United States, such as the Soil Conservation Service or Forest Service, were proud, cohesive, and successful organizations (Clarke and McCool, 1985; Kaufman, 1960). They were also professional and gender monocultures with utilitarian conservation ethics, soon to be confronted with the social policies and environmental values of an urban, postindustrial U.S. society (Kennedy, 1988). The biggest agency change vector was the National Environmental Policy Act (NEPA) of 1969 (U.S. Department of Agriculture-Forest Service, 1983).

This article examines the first generation of this new and more diverse U.S. Department of Agriculture-Forest Service (USFS) professional work force in the entry-level stages of their careers. Two studies investigate the reasons for selecting a natural resource profession and the USFS, the ability to understand and adapt to agency culture, the job satisfaction, and any mentor relationships—with special attention to gender and professional differences. The conceptual base is the career development theories of organizational behaviorists such as Schein (1978) and Van Maanen (1977).

Two Samples of Female and Male USFS Natural Resource Professionals

The first study (Kennedy and Mincolla, 1982) is based on a 50% sample of the foresters, range managers, and wildlife/fisheries biologists (81% return rate; $N = 109$; randomly selected within regional and professional strata). All were hired in USFS Regions 4 and 6 between 1978 and 1981. Region 4 includes Nevada, Utah, eastern Wyoming, and southern Idaho; Region 6 consists of Oregon and Washington. Together these regions employ about 27% of these three types of agency professionals (Akin, Glazier, and Marine, 1981). The young employees in this sample were representative of professional and gender diversity (50% were women and only 40% were foresters) of the entry-level USFS population in the late 1970s. This study is referred to as the R4/R6 study.

Although the R4/R6 study anticipated many gender-related differences, there were more pronounced differences associated with type of profession, especially for wildlife/fisheries biologists. In 1979, wildlife and fisheries biologists were the most common new type of post–NEPA professional in the USFS (2.5% of the USFS professional work force then, they increased to about 8.5% by 1990; Akin et al., 1981; Martin, Lynch, and Berry, 1990). These newcomers, regardless of gender, often differed from their forester and range manager colleagues in professional allegiance, acceptance of agency values, and perception of USFS career barriers.

Results of the R4/R6 study prompted the USFS Wildlife/Fisheries Division (Washington, DC) to fund a survey of all entry-level wildlife/fisheries (WL/F) managers. This will be referred to as the WL/F-MGR study (Kennedy and Mincolla, 1985)—a 43% sample (86% return rate; $N = 99$; proportionally distributed among all USFS regions) of WL/F managers with 1-6 years in the agency. About 60% of this sample (and population) were wildlife and 30% were fisheries biologists (with few career development differences between the two professional types). Half the sample were women, and there were more career development similarities between men and women than differences. When differences did occur, however, women WL/F managers always felt better than their male colleagues about the current status and future possibilities of their USFS careers.

Women in the R4/R6 study had a mean age of 26 years and were about 1.5 years younger than the men (nonsignificant at $p \leq .05$). Twenty-nine percent of women and 36% of men had master's of science (M.S.) degrees. Reflecting the eagerness of the USFS to realign gender deficiencies, women in the WL/F-MGR study had a mean age of 29 years (about 3 years younger than their male colleagues). They also had less graduate education; 40% of women and 61% of men WL/F managers had M.S. degrees, χ^2 ($N = 99$) $p < .01$. The mean USFS permanent job tenure was about 2 years for the R4/R6 study and 4 years for the WL/F-MGR study respondents.

These are the first career development studies of entry-level natural resource professionals in the USFS (Kennedy and Roper, 1989), and they depict the initial post–NEPA generation of WL/F biologists and affirmative action candidates. The sampling strategies and sample sizes well represent the career development of that new generation of USFS professionals in these regions (e.g., a 10% test–retest reliability, 6 months after original mailings, had a mean Pearson's correlation of .80 for the WL/F-MGR study). However, there is no way of knowing if they represent other types of professionals in the USFS (e.g., archaeologists or engineers), represent entry-level colleagues who resigned before the surveys, or are comparable with other natural resource agencies.

The results of the R4/R6 and the WL/F-MGR studies are presented in a narrative manner, as they relate to specific organizational development and socialization tasks (Schein, 1968). These are tasks that men and women normally encounter in adapting to their professions and to the early stages of a USFS career (Dalton et al., 1977).

Becoming Committed to a Natural Resources Profession and to the USFS

Men and women in the R4/R6 study differed significantly ($p = .04$) in reasons for selecting a natural resource position in the USFS. The most frequent motivation for men (62%) was one that fit a geographic preference category (e.g., with replies such as, "I wanted to work as a Forest Service forester in the west for quality of life, especially hunting and fishing"). Women cited geographic reasons third in frequency (31%) and usually because friends and family were there, not that the hunting or fishing was great. The most often-cited motivation for women (second for men) was concern for resources and the environment. Even within this category, men and women expressed their relationships differently. Women tended to use verbs such as "care for," "love of," "concern for" natural resources, wildlife, or the environment. Men generally used "work in," "work with," or "manage"; they rarely used "care" and never "love." Both cited professional and USFS job fulfillment or satisfaction (third in frequency for men, second for women).

Developing a Professional Commitment

Eighty percent of respondents in the R4/R6 study were past the initial exploration stages of their careers (Super et al., 1957) and stated a long-term commitment to stay in their professions. Only 2% made this decision in high school, but by college graduation 60% had done so. The remaining 20% required 1-3 years of job experience before becoming convinced that they had made the right professional choice.

Fewer women (75%) than men (92%) in the R4/R6 study had a long-term commitment to their profession ($p = .04$). Recall that these women were slightly younger and had less graduate education (and socialization) than male colleagues. Women were also more uncertain about the future effects of family status or spouse on their professional career plans. In contrast, there were no significant professional commitment differences between men and women in the WL/F-MGR study, perhaps because these respondents were a few years older, had more experience in their profession, and had been in permanent USFS employment twice as long as the R4/R6 sample.

Adapting to USFS Culture and Developing an Agency Commitment

Van Maanen (1977, p. 18) described recruits in an organization as, "standing at the entrance point of an organizational career, the individual is essentially a stranger. . . ." As a stranger to USFS, entry-level professionals must understand the power structures, values, communication systems, and professional cliques before they can successfully adapt to them (Miller, 1967; Schein, 1968, 1978).

Peer Support and Role Models. Men and women in both studies soon identified USFS superiors and peers whom they respected and turned to for advice. It was type of profession, not gender, that seemed to influence the selection of these significant people. For example, biologists in the R4/R6 study rarely named a forester as an important peer. Because most USFS professionals are foresters (about two-thirds in this area), it would have been easier for them to find three to four forester peers at their field-level job location. Yet they sought out other WL/F biologists. Range managers and WL/F biologists often had to seek peers in other districts to locate colleagues in their profession. Although distance inhibited daily contact, WL/F biologists and range managers cited similar high peer support and friendship bonds as did foresters.

Women in the R4/R6 study rated (on a five-point scale) the support "as a person and professional" by their primary peers significantly higher ($p = .04$) than did male colleagues. Such supportive peer relationships of women in our studies are consistent with Chodorow (1978) and Gilligan's (1982) findings that women are more motivated and successful in establishing and sustaining human relationships.

Recruits in the R4/R6 study were also asked, "In your current position, whose praise, compliments, or criticism would have the greatest effect on you?" Most of the respected people described were at their job location. With the exception of range managers, however, only one-fourth were immediate supervisors. Professional respect (rather than rank or gender) determined the esteem and influence of these special people. Once again, there was a "professional clan" association: foresters were most influenced by foresters, WL/F biologists by other biologists.

Adapting to the USFS Value System. Along with getting the respect and support of others, professional recruits must understand and adapt to USFS value systems to bond with the organization. This important and complex issue was approached from several perspectives.

Professional recruits in the R4/R6 study were asked to describe the three values or attitudes most rewarded by USFS status and promotion and whether they agreed with the importance the agency placed on them. Open-ended replies consistently described the agency value system (in descending priority) as (a) being loyal to the USFS; (b) having a strong work ethic and being productive; and (c) getting along with people and in teams. Fewer than half (45%) believed that all three values ought to be so highly rewarded by their agency: 19% disagreed with one, 18% with two, and 18% rejected all three of these values. Perceptions and acceptance of rewarded USFS values did not vary by gender, but WL/F biologists expressed more disagreement with the importance placed on USFS loyalty and productivity than did foresters or range managers. For example, a significantly higher proportion of WL/F biologists (25%, $p = .05$) disagreed with all three rewarded USFS values than did foresters (12%) or range managers (16%).

Respondents in the WL/F-MGR study were asked a slightly different value question: "What are the two most important attitudes/values to be a successful WL/F manager in the Forest Service?" Fifty-one percent described values associated with cooperation and teamwork, 32% cited attitudes falling into the professional knowledge/competency category, 28% mentioned learning how to compromise/give-and-take (another teamwork attitude and skill), 20% noted a strong work ethic, and 18% listed loyalty to USFS. These replies focus on essential attitudes to be a successful WL/F specialist in the multiple-use, interdisciplinary, public involvement culture of the USFS, rather than those directly associated with promotion.

Entry-level professionals in the WL/F-MGR study were then asked what was "the biggest attitude/value change, if any, you had to make in the first year or two as a permanent employee?" All cited important (and often stressing) attitude changes: 42% were in the "learning to get along with people/teams" category, 27% in "accepting wildlife/fisheries are low priority in USFS," and 20% in "accepting many decisions are political."

Recruits often find it difficult and painful that the organizations they join are more complex and political than the relatively simple world of college. Accepting a second-class status for one's wildlife/fisheries profession is not easy either. Even for female WL/F managers, this perceived USFS resource bias (i.e., timber, range, or fire-fighter chauvinism), rather than gender bias, was usually their biggest entry-level conflict.

Personal Life, Career, and the Agency. Most entry-level men in the R4/R6 study were married (64%). All lived with their spouses and none had ever been divorced or widowed. In contrast, only 34% of their women colleagues were married (about one-quarter of this group lived separate from their husbands), 10% were widowed or divorced, and 56% (vs. 36% of men) had never married ($p = .001$). Men in the WL/F-MGR study were also much more likely to be married (70%) than were women (45%; $p = .02$). Correction for age still displayed significant differences.

Women in the R4/R6 study were more likely to perceive the USFS as insensitive to dual-career problems (68%) than were their male colleagues (44%). All married women in the sample reported that they had experienced dual-career conflicts, compared with only half the married men ($p < .01$).

Developing a Long-Term Commitment to the USFS. Recall that the majority of respondents to both surveys were committed to their natural resource management professions when reporting to their first permanent USFS assignment. Although the USFS was the first or second choice for permanent employment for most (82% in the WL/F-MGR study), developing a long-term commitment to the agency like that to the profession takes time and positive socialization. As indicated earlier, this socialization may also involve some struggle between professional and USFS values.

When surveyed, about 33% of both samples had decided to stay in the USFS and 20% had decided to leave. It is not surprising that the majority of entry-level professionals in both studies were still unsure if they wanted a USFS career. Commitment did not differ by gender, but WL/F biologists were consistently less committed to a long-term USFS career than their forester or range manager colleagues.

Another question in both studies was, "If you could start your professional career over again, with what organization would you want to work?" Eighty-one percent of foresters in the R4/R6 study, versus 63% of range managers and 57% of WL/F biologists, would choose the USFS again ($p = .01$). In response to the same question in the WL/F-MGR study, the majority of entry-level WL/F-managers (54%) would not choose to work for the USFS again. This indication of agency commitment varied by professional type, not by gender.

Satisfaction with USFS Job and Career

The R4/R6 study focused on initial job satisfaction, whereas the WL/F-MGR study monitored general USFS job attitudes after 4 years (mean) employment and one or more assignments.

Women in the R4/R6 study initially experienced less overall first permanent job satisfaction than did their male colleagues ($p = .01$). This gender difference disappeared, however, after a transfer or with two or more years at their first assignment. There was no gender-related difference on the impact of that first job on commitment to profession, but first job impact on commitment to the agency was significantly lower ($p = .02$) for women than for men. Further analysis (Mincolla and Kennedy, 1985) found that a woman's first job supervisor had a much greater impact on her development of USFS commitment than the first supervisor did for men. Here, as with peer support (discussed earlier) and later in mentor relationships, women in our studies seem more dependent on and vulnerable to people. This can be both an asset (the usual case with the first job supervisor) and a liability. Men may have weighted the liability possibilities greater, because they tended to perceive and behave more independently and self-sufficiently (in a mythic male hero mode), keeping their on-the-job personal relationships more at arms length. This probably protected them somewhat but may have also foreclosed many career (and human) development options.

Another reason for such first permanent job variation between men and women may be differences in expectations. Women in the R4/R6 study consistently had higher first job expectations than men about the job challenge, professional prestige, or group morale that they would experience in the USFS. Such optimism caused a greater disparity for women when they encountered first job realities—a phenomena that Hughes (1958) called "reality shock." There were two t-test differences (significant at $p \leq .05$) between women's greater first job expectations, in "group morale" and in "opportunity to pursue my personal career goals" than what they found on the job.

These women may have had less realistic USFS job and career expectations, because many (45%) had no previous temporary USFS job experience (vs. 20% of men; $p = .01$). Women were also less precommitted to a USFS career while in college: Twice the proportion of women than men said the possibility of future USFS employment was not an important consideration when selecting a natural resource major ($p = .01$).

Based on these R4/R6 study findings, job satisfaction for women and men was expected to be similar in the WL/F-MGR sample. This was generally true, with little difference in "liking their job," "knowing what's expected," having their "professional advice accepted" by the USFS, or being "assisted with training/career development." But in other important areas in which differences in satisfaction occurred, women were always more satisfied with current jobs and more optimistic about future USFS assignments.

Table 1 illustrates five job/career satisfaction attitudes of entry men and women in the WL/F-MGR study. Women expressed significantly higher χ^2 ($p < .05$) job satisfaction than their male colleagues in all cases. Further analysis of variance tests indicated that gender explains why women rated their jobs as more challenging, $F(1, 95, p = .04)$, and why they were more satisfied with future promotion prospects, $F(1, 95, p = .02)$. Less graduate education, $F(1, 95, p = .03)$, is associated with women finding their jobs more interesting than men (the first time in this study where education had a significant, $p \leq .05$, impact). Gender and educational factors were both involved in satisfaction with current rank. That is, women with bachelor of science (B.S.) degrees were the most content with their rank, followed by women with a graduate degree, men with a B.S. and, finally, men with a graduate degree. In contrast, women's optimism with promotion prospects was only associated with gender.

Table 1
Job and Career Satisfaction Controlled for Gender and Graduate Education

	Analysis of Variance (p-Value for F-Test)		
Job Satisfaction Variable	Gender	Educational Level	Total
1. What I do is challenging ($p = .04$)[a]	.04	.47	.16
2. My work is interesting ($p = .05$)	.20	.03	.04
3. Forest Service generally treats me as a valuable employee ($p = .02$)	.11	.44	.25
4. Satisfied with current rank ($p = .01$)	.02	.006	.001
5. Satisfied with promotion prospects ($p = .03$)	.02	.26	.02

[a]p-Values in parentheses, chi-square when the five-point Likert scale is collapsed to three (*agree very much* and *agree* combined; *neutral*; and *disagree very much* and *disagree* combined).
(Source: Kennedy and Mincolla, 1985.)

Career Counseling and Mentoring

Developing a written and approved training plan is an annual event in the USFS. Formal career counseling is not so specified, promised, or delivered. Even though career goals are the logical ends to which training is a means, entry-level respondents in both studies did not believe that they received reasonable formal career counseling to develop adequate career goals and training strategies.

The majority (80%) of entry-level natural resource managers in the R4/R6 study said that career counseling was important, yet 51% received no formal career counseling (i.e., scheduled time to discuss career options). Most career counseling was obtained informally and from peers (only 7% received no informal counseling). Few in the R4/R6 study were satisfied (25%) with the overall quality of career counseling. Women were less satisfied than men ($p < .05$).

The Mentor Process

Career counseling and much other career development assistance can be provided by mentors. After an average of 2 years in permanent USFS employment, 62% of R4/R6 study participants reported a person who had a "great positive professional and career impact" and whom they considered a mentor. There were no differences in the proportion of men and women who stated that they had a mentor. As with important peers, mentor's profession seemed to be a bigger selection variable than gender; foresters usually selected other foresters as mentors, WL/F biologists selected biologists (especially college professors), and range managers selected other range managers. In the time between the R4/R6 study and the WL/F-MGR study, much was learned from the mentor literature (e.g., Kram, 1985; Rawlins and Rawlins, 1983; Shapiro, Haseltine, and Rowe, 1978), so that the second survey contained more and better designed mentor questions.

Respondents in the WL/F-MGR study were more likely to have mentors (80%) than were those in the R4/R6 sample (62%). Recall that the WL/F-MGR respondents were in the USFS about 2 years longer than the R4/R6 sample, and 18% began their mentor relationship in the last 2 years. Most of their mentors were other WL/F biologists (68%), 90% were male, and 60% were USFS employees (25% were college professors). As with the R4/R6 and other studies (e.g., Burke, 1984; Henderson, 1985; Kanter, 1977), women in the WL/F-MGR study were as likely to report having a mentor as their male colleagues. Neither of our studies confirm Ragins's (1989) concerns that women professionals are handicapped in finding and developing supportive mentor relationships in traditionally male organizations or professions.

Entry-level WL/F managers evaluated nine potential roles their mentors played, on a five-point scale ranging from *very great* (1) to *very low* (5). Respondents then ranked, from 1 to 9, the greatest personal or professional needs they had for a mentor at that time. Greater mentor needs (in declining order of ranked frequency) were:

(1) encouragement and advice in identifying and achieving my career potentials,
(2) help in forming my values and ethics,
(3) being a role model,
(4) sponsoring my USFS career, and
(5) teaching me "how to make it" in the USFS.

Fortunately, the majority of respondents felt that their mentors actually filled these important needs (68–83% gave their mentors favorable ratings). When asked to recall any needs that their mentors failed to fulfill, 83% wrote "none."

Although 60% of those with mentors in the WL/F-MGR study judged that their mentor was a great or very great friend, this was ranked as a second-to-last need. Apparently friendship and "teach me technical skills" (also ranked low) can be provided by more casual and less intimate sources, such as peers or technical journals. What these young people needed of their mentors was more personal, intimate, and influential—someone to help them establish a professional and agency identity (including values and ethics), understand the agency culture, develop career strategies, and then provide agency sponsorship in achieving career goals.

Mentoring Differences Between Male and Female Entry-Level WL/F Managers

Women in the WL/F-MGR study reported different roles and better mentor support than did male colleagues, even though 94% of their mentors were men.

Mentors had similar friendship and role model impacts on men and women WL/F managers. However, in two critical roles, many more women rated as great (or very great) the role their mentors played, in "teaching me how to make it in the USFS" (81%) and "sponsoring my USFS career" (85%). The comparable favorable ratings by men were 54% and 49%, respectively ($p < .01$ in both cases).

Analysis of variance clarified the relationships between gender, education, and the support provided to WL/F managers by their mentors (Table 2). The greater "teach me how to make it" support of women's mentors was mostly associated with gender ($p < .01$) rather than level of education. Sponsorship of women's advancement was also highly associated with gender ($p < .01$), but educational level was also significant ($p < .02$). Consistent with most differences between men and women controlled for education, women WL/F managers with B.S. degrees felt best about their current and future

Table 2
Women's Mentor Impacts Controlled for Gender and Educational Level

Mentor Relationship	Analysis of Variance (p-Value for F-Tests)	
	Gender	B.S. or M.S. degree
My mentor taught me how to make it in the USFS ($p = .007$)[a]	.01	.26
My mentor sponsored and supported my USFS advancement ($p = .003$)	<.01	.02

Note. B.S. or M.S. = bachelor's or master's of science degree.
[a] p-Values in parentheses, chi-square when the five-point scale is collapsed to three (*agree very much* and *agree* combined; *neutral*; and *disagree very much* and *disagree* combined).
(Source: Kennedy and Mincolla, 1985.)

USFS careers, followed by women with graduate degrees, men with B.S. degrees, and, finally, men with graduate degrees. This is educational effect. Cause was not as directly addressed or statistically established.

Based on general observations of these data and of the changing USFS culture, the best hypothesis is that educational and mentor effects may interact in several ways. First, those in the WL/F-MGR study with graduate degrees were more likely to be mentored by college professors, most of whom appeared to be inappropriate role models for success in the multiple-use, interdisciplinary, public involvement world of the USFS. In addition, WL/F biologists with M.S. degrees stated a greater research (vs. field management) orientation. They were also more "species or population" (vs. "habitat") specialized than colleagues with no graduate degree. This may have inhibited adaptation to the more practical, habitat management culture of the USFS.

The relationship between education and professional specialty may frustrate those searching for simple relationships between gender and USFS job satisfaction or career development (Kennedy, 1982). For example, an unmarried, female, nongame wildlife biologist on her first permanent USFS assignment in rural Utah may associate much job and life dissatisfaction with her gender. Yet her marital status, a nongame and wildlife specialty, and the religious-cultural variables in an isolated, conservative community may provide all of the handicaps any young woman (or man) should have to endure in adapting to their first permanent assignment. This is not to suggest that entry-level women professionals do not experience subtle and overt sexual prejudice in the USFS or local communities. The Region 5 (California) class action suit on USFS sexual discrimination (Briggs, 1982), Enarson's (1984) book, and many articles from *Women in Forestry* (now *Women in Natural Resources*) document sexual discrimination in the forestry profession and in the USFS. However, our studies found the uniqueness of wildlife/fisheries values and professional identity to be more influential in career development conflicts than gender.

Conclusions

Unlike white male foresters, willingly hired in the first part of this century, the USFS was legislatively directed to employ many of the WL/F managers and women in our studies. Laws such as the NEPA (1969) encouraged the employment of different specialists (e.g., archaeologists, soil scientists, and WL/F biologists) to incorporate their sensitivities, values, and skills into an organization dominated by one professional culture (Kaufman, 1960). As such, WL/F managers and other specialists were to be "change-agents" in such federal agencies (Bennis, 1966; Kanter, 1983; Kennedy, 1988). At the same time, equal employment opportunity and affirmative action executive orders sought a more balanced work force and often put women in similar roles. Because over half the USFS WL/F biologists hired in the 1970s were women, many of these new recruits found themselves in multiple and unanticipated change-agent roles, for which they were usually unprepared (in both attitudes and skills).

It is not easy for a proud, successful, and cohesive agency to smoothly absorb change-agents and to quickly adjust its organizational character and style (especially when the motivation to change is largely externally imposed). Changing an organizational culture requires time and patience—and usually some blood, sweat, and tears. Yet Table 1 displays evidence that women WL/F managers believe affirmative action and equal employment opportunity policies are achieving desired effects in their lives and on the USFS culture. Many women WL/F biologists also feel that the agency has changed its attitudes and behavior toward their gender in the 1970s more rapidly and appropriately than toward their wildlife/fisheries profession and its values.

There was another complication in USFS and WL/F biologist relationships during the initial post-NEPA (1969) decade. Few female or male WL/F managers (and other specialists) selected their profession or were trained in college to be persistent, skillful, and successful organizational change-agents. Most WL/F managers in our studies selected their profession to understand and work with animals and fish, out in the field, away from people and bureaucracies, where seldom would be heard a discouraging word. Few were educated to understand and manage people or complex organizational cultures (Cutler, 1982; Kennedy, 1985b). Biologists would have majored in law or social work if they wanted to be societal change-agents; as such, they would have received better coursework and role modeling in college and summer jobs to succeed in that endeavor.

Therefore, one should not be surprised or necessarily disturbed to discover some stress in the USFS culture and in the careers of new specialists we have studied—be they men or women. NEPA (1969) and affirmative action policies initiated a "shotgun marriage" dynamic in which neither the USFS nor nontraditional professionals volunteered for or were well prepared to adapt easily. Both change-agents and the agency had to learn to accommodate one another on the job, usually by trial-and-error, and sometimes with pain.

Our studies present some personal and professional "snap-shots" of new USFS recruits in this initial post-NEPA era (1970–1980). The picture could be considerably different in the 1980s and 1990s. WL/F biologists have been the most rapidly growing USFS profession in the 1980s (expanding from about 2% to 8.5% of the professional work force; Martin et al., 1990). Many WL/F biologists (a high portion being women) are now entering USFS line positions, and their agency reputation is shifting from reactive critics of USFS programs to proactive and essential team members (e.g., note the high-priority emphasis on wildlife/fisheries values in the agency's *Long-term strategic plan,* U.S. Department of Agriculture–Forest Service, 1990).

The pioneering role WL/F biologists played in USFS cultural change in the 1970s has now expanded to incorporate a new wave of archaeologists in the 1980s and botanists in the 1990s (e.g., USFS Region 6, Oregon/Washington, now has one to three botanists in all its national forests). Although these new nontraditional professionals probably have been no better educated and role-modeled to be effective agency change-agents than were WL/F biologists, it is hoped that the agency has matured in the decades since NEPA (1969). In addition, these new recruits might find veteran women and men change-agents of the 1970s to mentor and guide them and their agency in adapting to the social values of an urban, postindustrial American society (Kennedy, 1985a).

Acknowledgment

This research was sponsored by the Utah State University Agricultural Experiment Station, MacIntire-Stennis Project No. 712 (Journal Paper No. 3194), and cooperative grants from the U.S. Department of Agriculture–Forest Service.

References

Akin, B., M. Glazier, and T. Marine. 1981. *Workforce planning data book.* Washington, DC: U.S. Department of Agriculture–Forest Service.

Bennis, W. 1966. *Beyond bureaucracy.* New York: McGraw-Hill.

Briggs, S. P. 1982. Final report to the Regional Forester, Region 5, and to the Director of Pacific Southwest Forest and Range Experiment Station. Unpublished manuscript, Urban Management Consultants of San Francisco.

Burke, R. J. 1984. Mentors in organizations. *Group and Organizational Studies* 9(3):353–372.

Chodorow, N. 1978. *The reproduction of mothering.* Berkeley: University of California Press.

Clarke, J. N., and D. McCool. 1985. *Staking out the terrain: Power differentials among natural resource management agencies.* Albany, NY: State University of New York Press.

Cutler, M. R. 1982. What kind of wildlifers will be needed in the 1980s? *Wildlife Society Bulletin,* 10(1):75–79.

Dalton, G. W., P. H. Thompson, and R. L. Price. 1977. The four stages of professional careers. *Organizational Dynamics,* 6(1):19–42.

Enarson, E. P. 1984. *Woods-working women: Sexual integration in the U.S. Forest Service.* University: University of Alabama Press.

Gilligan, C. 1982. *In a different voice: Psychological theory and women's development.* Cambridge, MA: Harvard University Press.

Henderson, D. W. 1985. Enlightened mentoring: Characteristic of public management professionalism. *Public Administration Review,* 45(6):857–863.

Hughes, E. C. 1958. *Men and their work.* Glencoe, IL: Free Press.

Kanter, R. M. 1977. *Men and women of the corporation.* New York: Basic Books.

Kanter, R. M. 1983. *The change masters.* New York: Simon and Schuster.

Kaufman, H. 1960. *The forest ranger.* Baltimore, MD: Johns Hopkins Press.

Kennedy, J. J. 1982. Dealing with masculine/feminine gender labels in natural resource professions. In *Women in natural resources: An international perspective.* ed. M. Stock, J. E. Force, and D. Ehrenreich. Moscow, ID: University of Idaho Press.

Kennedy, J. J. 1985a. Conceiving forest management as providing for current and future social values. *Forest Ecology and Management,* 13(4):121–132.

Kennedy, J. J. 1985b. Viewing wildlife managers as a unique professional culture. *Wildlife Society Bulletin,* 13(4):571–579.

Kennedy, J. J. 1988. Legislative confrontation of groupthink in U.S. natural resource agencies. *Environmental Conservation,* 15(2):123–128.

Kennedy, J. J., and J. A. Mincolla. 1982. Career evolution of young 400-series U.S. Forest Service professionals. Unpublished manuscript, College of Natural Resources, Utah State University, Logan, UT.

Kennedy, J. J., and J. A. Mincolla, 1985. Career development and training needs of entry-level wildlife/fisheries managers in the USDA Forest Service. Unpublished manuscript, College of Natural Resources, Utah State University, Logan, UT.

Kennedy, J. J., and B. B. Roper, 1989. Status of and need for career development research in natural resource agencies: A Forest Service example. *Transcripts of the North American Wildlife and Natural Resources Conference,* 54:432–438. Washington, DC: Wildlife Management Institute.

Kram, K. G. 1985. *Mentoring at work.* Glenview, IL: Scott, Foresman and Co.

Martin, T., M. Lynch, and C. Berry. 1990. *Work force data book 1989–1990.* Washington, DC: U.S. Department of Agriculture–Forest Service.

Miller, G. A. 1967. Professionals in bureaucracy: Alienation among industrial scientists and engineers. *American Sociological Review,* 32(4):755–767.

Mincolla, J. A., and J. J. Kennedy. 1985. Early career development processes of women and men resource managers in the USDA Forest Service. Unpublished manuscript, College of Natural Resources, Utah State University, Logan, UT.

Ragins, B. R. 1989. Barriers to mentoring: The female manager's dilemma. *Human Relations,* 42(1):1–22.

Rawlins, M. E., and L. Rawlins. 1983. Mentoring and networking for helping professionals. *Personnel and Guidance Journal,* 62(2):116–118.

Schein, E. H. 1968. Organizational socialization and the profession of management. *Industrial Management Review,* 9(2):1–16.

Schein, E. H. 1978. *Career dynamics: Matching individual and organizational needs.* Reading, MA: Addison Wesley.

Shapiro, E. C., F. P. Haseltine, and M. P. Rowe. 1978. Moving up: Role models, mentors, and the patron system. *Sloan Management Review,* 19(3):51–58.

Super, D. E., J. Crites, R. Hummel, H. Moser, P. Overstreet, and C. Warnath. 1957. *Vocational development: A framework for research.* New York: Teachers College Press.

U.S. Department of Agriculture–Forest Service. 1983. *The principal laws relating to Forest Service activities* (Agriculture Handbook 453). Washington, DC: U.S. Government Printing Office.

U.S. Department of Agriculture–Forest Service. 1990. *The Forest Service program for forest and rangeland resources: A long-term strategic plan.* Washington, DC: U.S. Government Printing Office.

Van Maanen, J., ed. 1977. *Organizational careers: Some new perspectives.* New York: Wiley and Sons.

Epilogue

Our career development studies of USDA-Forest Service (USFS) employees began about 1980, in the revolutionary post–National Environmental Policy Act and affirmative action era of U.S. and USFS change (Kennedy and Mincolla, 1985). As pioneering scholars in this area, we anticipated (and secretly hoped for) dramatic research results that illustrated clear, profound differences between men and women professionals. We were confounded, humbled, and educated by the complexity and ambiguity deserving of the new generation of USFS recruits and a complex, evolving U.S. and agency culture. It has been cognitive and value education for many researchers.

Since the early 1980s, there have been many studies of the USFS organization and its employees. A dramatic illustration of progress is the summer 1995 issue of *Policy Studies Journal*, with eight articles and many citations devoted to this subject. Mohai's (1995) lead article cites our and several other USFS employee surveys and observes their motivation: "interest in examining attitude changes in the Forest Service has stemmed from the anticipation that changes in attitudes will result in changes in behavior, and ultimately in management and policy decisions in the agency" (pp. 248–249). That is a utilitarian assumption worthy of the traditional USFS perception of trees and forests being useful primarily as timber. Actually, I began career development research as a curious and concerned university professor, interested in how entry-level foresters, range managers, and wildlife/fisheries biologists were making the transition from classrooms to field-level challenges in a proud, traditional, evolving multiple-use agency. We were more interested in the success and anguish of these young people themselves, rather than as a means for heroic inference to agency policy decisions. Just as prosaically, our professional and gender diversity interests (little ethnic diversity resulted from our samples) were in how those employee characteristics were associated with the essential, early career development tasks of mastering and enjoying their job and agency expectations; finding relatable peers, role models, or mentors; and adapting their personal and professional values to the agency reward system. We did not attempt to measure national or USFS diversity objective achievement, and only tangentially speculated on employee diversity impacts on agency cultural dynamics or national forest management (although we were young and eager, and the temptation to be profound was often too great; e.g., Kennedy, 1988; Kennedy and Thomas, 1991).

A good example of recent benchmark monitoring of USFS achievement in employee diversity objectives is Thomas and Mohai (1995). However, like our findings, theirs cannot support dramatic, obvious conclusions. They document, for example, that the biologist professional series grew 46% between 1983 and 1992 (from about 5 to 8% of the USFS workforce). Yet foresters still dominate the agency's workforce (at 40%), even though that is close to a 50% decline since Kaufman's (1960) study—and even though the acting director of the Bureau of Land Management and chief of the USFS (circa 1995) are biologists with Ph.D.s, no less. Likewise, the proportion of women increased in all job categories since 1983, but white males still dominate leadership positions as well as forester/range professional series, from which line officers traditionally have been selected. How optimistic or pessimistic should one be about such change?

To further confound heroic inference, the first broad value survey of USFS employees (Kennedy et al., 1992) found many more value similarities than differences between ranks, professional series, or gender—and more challenges to common stereotypes than support (see also Brown and Harris, 1993; Mohai, Stillman, and Liggett, 1994). For example, recent USFS forester recruits scored as high on established environmentalism scales as their biologist and ecologist colleagues, and well above business, engineer, or range manager recruits. Also, most line officers agreed with the majority of USFS professional employees that the agency puts too much emphasis on timber and grazing values for their own and perceived public preferences. They believed that timber and range budget priorities should be reduced and other multiple uses emphasized (Kennedy, 1991; Kennedy and Quigley, 1993).

A subsequent study (Cramer et al., 1994) compared "new" male and female district rangers (most female district rangers were "new," with 1 to 3 years as rangers) versus experienced (mostly men) rangers. Searching for dramatic male–female differences in this historic leadership shift, we again found more similarities than differences. For example, the majority of new or experienced, male or female district rangers believe the USFS most rewards the following values (ordinally ranked): (1) achieve targets, (2) be loyal to the agency, (3) work hard, (4) stay within budgets, and (5) work well within teams; whereas, they believe the USFS system should support: (1) care for healthy ecosystems, (2) concern for employees, and (3) concern for future generations. Among the new district rangers, women scored higher on environmentalism scales than their male colleagues.

Once again, these results suggest evolutionary (versus revolutionary) change in USFS employee attitudes and behaviors as women, ethnic minorities, or other new types of employees become line officers (Kennedy and Quigley, 1993) and as these younger employees usually better represent the values of a more recent urban, postindustrial America. How that will affect individual and organizational behavior will be even more challenging to measure and predict. These studies might also support Twight and Lyden's (1989) belief in the tenacity of the proud, traditional USFS culture, plus recognition of the power of budgets developed in congressional committees, which generally have been dominated by Western representatives and traditional commodity interests since the 1950s (Alston, 1972; Faraham, 1995).

Yet with all the invitations for pessimism in the USFS becoming a more open, responsive agency with better balanced environmental values, the cumulative effect of U.S. and USFS evolution since 1970 (that we researchers continue to measure with mixed results) seems to be emerging and focusing more on environmental values (USDA-Forest Service, 1994) and an "ecosystem management" paradigm shift (Kennedy and Dombeck, 1995) for a new century of national forest and grassland (Kennedy, Fox, and Osen, 1995) management. The direction of this urban post-industrial and environmental evolution in agency workforce and resources management appears set, regardless of the research snapshots we take and attempt to interpret along the path.

Additional References

Alston, R. M., 1972. *FOREST—Goals and Decision-Making in the Forest Service* (INT-128). Ogden, UT: USDA-Forest Service, Forest and Range Experiment Station.
Brown, G., and C. C. Harris. 1993. The implications of workforce diversification in the US Forest Service. *Administration and Society*, 25(1):85–113.
Cramer, L. A., J. J. Kennedy, R. S. Krannich, and T. S. Quigley. 1994. *Attitudes and values of new and experienced Forest Service district rangers*. Paper presented at the 5th International Symposium on Society and Natural Resource Management, Ft. Collins, Colorado.
Faraham, T. J. 1995. Forest Service budget requests and appropriations: What do analyses of trends reveal? *Policy Studies Journal*, 23(2):253–267.
Kaufman, H. 1960. *The Forest Ranger*. Baltimore, MD: Johns Hopkins University Press.

Kennedy, J. J. 1988. Legislative confrontation of groupthink in the US natural resources agencies. *Environmental Conservation*, 15(2):123–128.

Kennedy, J. J. 1991. How employees view Forest Service values and the reward system. Testimony before the House Subcommittee of Civil Service (101st Congress). In *Forest Resource Management and Personnel Practices: Values in Conflict*, pp. 90–107. Washington, DC: U.S. Government Printing Office.

Kennedy, J. J., and M. P. Dombeck. 1995. The evolution of public agency beliefs and behavior toward ecosystem-based stewardship. In *Toward a Scientific and Social Framework for Ecologically-Based Stewardship*. Washington, DC: USDA-Forest Service.

Kennedy, J. J., B. L. Fox, and T. D. Osen. 1995. Changing social values and images of public rangeland management. *Rangelands*, 17(4):127–132.

Kennedy, J. J., R. S. Krannich, T. M. Quigley, and L. A. Cramer. 1992. *How Employees View the USDA-Forest Service Value and Reward System*. Logan: Utah State University, Department of Forest Resources.

Kennedy, J. J., and J. A. Mincolla. 1985. Early career development of fisheries and wildlife biologists in two Forest Service regions. *Transactions 50th North American Wildlife and Natural Resources Conference, 50*, pp. 425–435. Washington, DC: Wildlife Management Institute.

Kennedy, J. J., and T. M. Quigley. 1993. Evolution of Forest Service organizational culture and adaptation issues in embracing ecosystem management. In *Eastside Forest Ecosystem Health Assessment, Vol. 2*, eds. M. E. Jensen and P. S. Bourgeron, pp. 19–29. Washington, DC: USDA-Forest Service.

Kennedy, J. J., and J. W. Thomas. 1991. Exit, voice and loyalty of wildlife biologists in public natural resource/environmental agencies. In *American Fish and Wildlife Policy: The Human Dimension*, ed. W. R. Mangun, pp. 221–238. Carbondale: Southern Illinois University Press.

Mohai, P. 1995. The Forest Service since the National Forest Management Act. *Policy Studies Journal*, 23(2):247–252.

Mohai, P., P. Jakes Stillman, and C. Liggett. 1994. *Change in the USDA Forest Service: Are We Heading in the Right Direction?* General Tech. Report. St. Paul, MN: USDA-Forest Service, North Central Forest Experiment Station.

Thomas, J. C., and P. Mohai. 1995. Racial, gender, and professional diversification in the Forest Service from 1983 to 1992. *Policies Studies Journal*, 23(2):296–309.

Twight, B. W., and F. J. Lyden. 1989. Measuring Forest Service bias. *Journal of Forestry*, 87(5):35–41.

USDA-Forest Service (USDA-FS). 1994. *The Forest Service Ethics and Course to the Future* (FS-567). Washington, DC: Author.

Index

Activism. *See* Environmental concern/
    activism; Toxic waste activism
Agha Khan Rural Support
    Programme (AKRSP), 173
Agricultural societies. *See also*
    Irrigation; Sharecropping
  cultivator households in, 113–115
  division of labor in, 108–109
  fishing societies vs., 74–76
  in India, 72–74, 108–115
  labor process in, 110–113
  single women and. *See* Amuesha;
    Single women
  subsistence among Amuesha, 139–
    151. *See also* Amuesha
  women's workloads in, 109–110
Agriculture. *See also* Crop diversity;
    Irrigation; Seed saving; Uncultivated land
  feminization of, 150, 151
  interdependence between livestock
    husbandry and, 157–158,
    162–164. *See also* Livestock
    husbandry; Pakistan
  role of women in, 194
Agroforestry
  Chipko Movement in India as
    study in, 200–202
  gender differences in involvement
    in, 205–207
  KENGO Workshops in Kenya as
    study in, 202–205
  Plan Sierra Development Project in
    Dominican Republic as study
    in, 196–200

Amuesha
  location of, 141
  overview of, 139, 140
  role of single women among, 140,
    142, 149–151
  single women and subsistence
    agriculture among, 142–149
Anderson, Anne, 250
Anthropology, 50–51
Appalachia, coal mining in. *See* Coal
    mining
Attitudes, to environmental hazards,
    246

Biodiversity
  crop diversity and, 178–179
  decline in, 177–178
  levels of, 178
Brazil, 75
Bullock, Charlotte, 252, 255
Bun, 58–59

Cannon, Robin, 257
Capitalism, 35–36
Caribbean, 75
Caste system
  in Indian agricultural communities, 105–106
  in Indian fishing communities, 66,
    67, 71–72
Catholic Relief Services (CARE),
    268, 282
Chambri, 58–59

309

Chipko Movement (India), 116, 200–202
Class
    activism and, 256–257
    explanation of, 104
    irrigation and, 92–93, 123–124
    sharecropping and, 128, 134
    in south India, 105
Coal Employment Project (CEP), 35
Coal mining
    conclusions regarding women in, 46–47
    labor process of, 32–34
    methodology for study on, 37–38
    physical demands and dangers of, 38–41
    rural women in, 31
    sexualization of work relations and workplace and, 36–37, 43–46
    social aspects of women in, 41–43
    theoretical framework for women in, 35–37
    women's entry into male culture of, 34–35
Columbus exchange, 181
Community forestry
    current issues in, 286–288
    development of, 280–281
    implementation of, 282–283
    international influence for, 281–282
    role of women in, 283–286, 288
Constructed knowing, 250
Consultative Group on International Agricultural Research (CGIAR), 182
Crop diversity
    biodiversity and, 178–179
    developments in, 180–181, 189–190
    distribution of, 182
    exchange methods and, 181–182
    in United States, 183, 184
Cultivator households, 113–115

Decision-making authority
    of Pakistani women, 172–173
    restrictions on, 271–272
Deforestation. *See also* Forest work; Forestry
    causes of, 278–279
    in Nepal, 277–279
Division of labor
    in fishing communities, 55, 57
    in irrigation, 87, 108–109, 122
    overview of gender, 6–7
    pastoral, 158, 159
Division of power, 55–56
Dominican Republic, 196–200, 271

Ecofeminism
    criticisms of, 2–3
    cultural, 2
    explanation of, 2
    liberal, 2
    social, 2
EHN, 244, 247
Energy Sector Management Assistance Program (ESMAP), 273
Environmental concern/activism
    conclusions regarding gender differences in, 230–231
    data and methods used to examine, 220–223
    explanations for gender differences in, 9, 216–220
    results of study of gender differences in, 223–230
    role of women in, 244–248, 257–258
    studies of gender differences in, 215–216
    toxic waste, 241–262. *See also* Toxic waste activism
Environmental Development Activities (ENDA) - Zimbabwe program, 268, 271
Ethiopia, 179–180

Fishing communities
    conclusions regarding women in, 59, 76–77
    data on women in, 52–53
    frameworks for gender analysis in, 51
    in India, 65–81. *See also* Indian fishing communities
    overview of women in, 49–50
    political economy and, 53–56
    studies of women in various, 56–59, 75–76

Fodder. *See also* Livestock husbandry; Pakistan
    explanation of, 157, 167
    from irrigated land, 165–167
    from uncultivated land, 164–165, 167, 168
    women and, 158, 160–161, 172, 173

Forest work study
    expansion of logging following, 29
    forest gathering results of, 22–26
    operationalization of variables for, 20–21
    research design and procedures for, 18–19
    sampled women and households in, 21
    village selection and description for, 19–20

Forestry. *See also* Agroforestry; United States Department of Agriculture-Forest Service (USFS)
    access to productive resources in, 270–271
    assumptions regarding women in, 278–280
    decision-making restrictions related to, 271–272
    gaps in knowledge about women and, 274
    Nepalese community, 277–288. *See also* Community forestry
    ongoing development work on women in, 268
    recruitment of women in, 273–274
    restrictions on women's participation in, 272–273
    role of women in, 193–194, 267–268, 274–275
    types of participation by women in, 268–270
    women as heads of households and role in, 195
    women as users of products of, 194–195

Frame analysis models, 259

Gender
    anthropology and, 50–51
    classifications of sex vs., 50
    divisions of labor and. *See* Division of labor
    knowledge and strategies for sustainability and, 8–9
    natural resource policy and, 9–10
    personal approaches to, 56–59
    political economy and, 53–56
    property rights and, 7–8
    sharecropping and, 128, 133–135
    water resource policy and, 124–125

Gender differences
    in agroforestry involvement, 205–207
    in environmental and social activism, 9
    in environmental concern, 246. *See also* Toxic waste activism
    in environmental concern/activism, 9, 215–231
    social status and, 80–81

Gender theory, in anthropology, 50–51
Gene banks, 182, 183
Genetic conservation, 182–183. *See also* Seed saving
Gibbs, Lois Marie, 249, 250, 252, 253
Global movement approach, 259–260
GRAIN, 183
Green revolution, 180

Heads of households, 195. *See also* Single women; Widows
High-yielding varieties (HYVs), 180–181
Hinds, Cathy, 242, 249–252
Home gardens
  of single Amuesha women, 147–148
  in Southeast Asia, 269
Hostility, 41–43

India
  caste system in, 66, 67, 71–72
  Chipko Movement in, 116, 200–202
  connections with environment in, 3–4
  dairy projects in, 174
  forestry in, 200–202, 271, 272
  livestock husbandry in, 159, 161, 173–174
  patriarchy in, 105
  role of women in agriculture in, 194
  status of fishing vs. farming in, 76
  women's movement in, 116
Indian fishing communities
  changes following study of, 80–81
  comparative analysis of, 72–76
  distinguishing features of, 65
  factors related to role of women in, 76–77
  family life and daily activities in, 69–70
  leisure and friendship in, 70
  marriage and kinship networks in, 70–71
  Minakuppam as example of, 66–67
  women's economic activities in, 67–69
  women's place in, 71–72
Indian irrigation
  conclusions and implications for, 115–117
  gendered transformations in, 121–125
  overview of, 103
  patriarchy, 103–104
  research context and, 105–106
  as state intervention, 106–107
  women in agricultural labor households and, 108–113
  women in cultivator households and, 113–115
Integrated Hill Development Project (Nepal), 282
International Board for Plant Genetic Resources (IBPGR), 182, 183
International Centre for Integrated Mountain Development (ICIMOD), 281–282, 286
International Labor Organization (ILO), 268
International policies, 9–10
International Union for the Conservation of Nature, 268
International Workshop on Women, Development, and Mountain Resources, 286
Irrigation
  changes for women in, 101–102, 121–125
  class issues and, 92–93

division of labor in, 87, 108–109, 122
fodder production and, 165–167
in India, 103–125. *See also* Indian Irrigation
livestock husbandry and, 157–158. *See also* Livestock husbandry; Pakistan
in Peru, 85–102. *See also* Peruvian irrigation
politics and, 94–96
state-sponsored, 106–107
system construction and, 90–91
system operation and maintenance and, 91–92
water management and, 92
women and, 87–90

Japan, 76

KENGO Tree-Planting and Agroforestry Workshops (Kenya), 202–205
Kenya, 202–205, 272, 274
Kinship networks, 70–71
Knowledge
of activists, 248–251
experience and, 251–253
various perceptions of claims of, 5

Labor. *See* Division of labor
Laos
background and setting of, 17–18
sources of women's influence in, 15–17
women's economic activities in, 15
Laos forest work study
expansion of logging following, 29
forest gathering results of, 22–26
operationalization of variables for, 20–21
research design and procedures for, 18–19
sampled women and households in, 21
village selection and description for, 19–20
Latin America. *See also specific countries*
analysis of single women in, 139–140. *See also* Amuesha; Single women
feminization of farming in, 150, 151
role of women in agriculture in, 194
women in fishing and agricultural communities in, 75–76
Livestock husbandry. *See also* Pakistan
in India, 159, 161, 173–174
interdependence between agriculture and, 157–158, 162–164
role of Latin American women in, 194
women and, 158–160
Love Canal, 249, 253

Madagascar, 128
Madagascar sharecropping
conclusions regarding, 135–136, 138
gender and, 133–135
overview of, 128–130
resource access and, 127–128, 136
women and, 130–133
Majia Valley Worldbreak project, 268
Malaysia
agricultural communities in, 75
fishing communities in, 58
status of fishing in, 76

Males
- coal mining culture and, 34–35
- participation in mentoring by, 300–301
- relationship to nature and, 1–2
- as USFS natural resource professionals, 294–295

Maritime communities. *See* Fishing communities

Marriage, 70–71

Menopause studies, 57

Mentoring
- activist, 252–253
- male vs. female participation in, 300–301
- professional, 299–300

Minakuppam, India, 66–67. *See also* Indian fishing communities

Motherhood, 255–256

National policies, resource use, gender and, 9–10

Native Seeds (New Mexico), 184

Natural resource professionals
- career counseling and mentoring and, 299–301
- commitment of, 295–297
- female and male USFS, 294–295
- job and career satisfaction of, 297–299

Nature
- female relationship to, 1–2, 279–280
- male relationship to, 2

Nepal
- community forestry in, 280–288. *See also* Community forestry
- dairy projects in, 174
- deforestation in, 277–279
- problems for women in, 174
- role of women in agriculture in, 194

Nepal-Australia Forestry Project (NAFP), 281, 283–285

New Social movement theory, 258–259

Newfoundland, 56–57

Nova Scotia, 54–55

Pakistan
- interdependence of livestock husbandry and agriculture in, 162–164
- irrigated farming areas in, 165–167
- rainfed farming areas in, 164–165
- relationships between women and livestock in, 157–158, 167–168
- women and fodder in, 160–161, 174
- women and livestock husbandry in, 158–160
- women and uncultivated land in, 161–162
- women's decision-making authority in, 172–173

Pakistan Agricultural Research Council (PARC), 172, 173

Patriarchy
- in India, 105
- resolution between forces of capitalism and, 35–36

Peru
- irrigation in, 85–102. *See also* Peruvian irrigation
- seed saving in, 187–190
- subsistence and single Amuesha women in. *See* Amuesha
- subsistence and single women in, 139–151
- women in local government in, 94–95
- women's roles in, 85–86

Peruvian irrigation. *See also* Agricultural societies; Irrigation
   bureaucratic transition in, 93–95
   class issues and participation in, 92–93
   conclusions regarding women in, 97–98
   epilogue of study of women in, 101–102
   overview of, 85, 86
   politics of, 95–96
   system construction and, 90–91
   system operation and maintenance and, 91–92
   water management and, 92
   women and, 86–90
Plan Sierra Development Project, Dominican Republic, 196–200
Plant Variety Protection Act, 183
Plowing, 131
Political economy, 53–56
Political participation
   gender differences in, 218–219
   of Peruvian women, 94–95
   role of emotions in, 249
Politics, of irrigation, 95–96, 124–125
Popular epidemiology
   explanation of, 243
   toxic hazards activism and use of, 247
Postmodernism
   feminist critiques of, 5–6
   shifts related to, 4–6
   Third World women and, 138
Power. *See* Division of power
Procedural knowing, 250
Professionals. *See* Natural resource professionals
Property rights, 7–8

Reproduction, toxic exposure and, 256

Resource access, 270–271
Resource mobilization theory, 258
Rice sharecropping. *See* Madagascar sharecropping
Ruddick, Sara, 251–252
Rural Advancement Fund International (RAFI), 183

Save the Children, 268, 287
Seed Savers Exchange, 183, 184
Seed saving
   methods of, 182–183, 189–190
   in Peru, 187–189
   renewed interest in, 184
   in United States, 183–187, 190
   women and, 179–180, 190
Sex, classifications of gender vs., 50
Sexualization, of work relations and workplace, 36–37, 43–46
Sharecropping. *See also* Madagascar sharecropping
   conclusions regarding, 135–136, 138
   gender and, 133–135
   overview of, 127–128
   women and, 130–133
Shrestha, Indira, 284–285
Single women. *See also* Widows; Women
   agricultural tasks of, 149
   in agriculture and forestry, 195
   Amuesha, 139–151. *See also* Amuesha
Smith, Dorothy, 251, 254
Social closure theory, 36
Social movement theories, 258–261
Socialist ecofeminism, 2
Species diversity, 178
State, 104
Subsistence farming. *See also* Agricultural societies; Agriculture

Subsistence farming (*Cont.*)
  among Amuesha, 139–151. *See also* Amuesha
  land degradation and, 278
Survey of the Public's Attitudes toward Soil, Water, and Renewable Resources Conservation Policy, 220
Sustainability, 8–9

Taiwan, 76
Taylor, Dorceta, 249
Thailand, 75, 76
Third World. *See also specific countries*
  impact of Western development on women in, 3
  toxic waste dumping in, 260
  women as "other" in, 5
Tikopia, 58
Tinau Watershed Project (Nepal), 282
Toxic waste, 243
Toxic waste activism
  boundaries issues and, 25255
  conclusions regarding, 261–262
  experience and knowledge issues and, 251–253
  gender differences in attitude and, 246
  gender issues in, 244–246
  influence of gendered position of women on, 246–248
  marriage and family life issues and, 256–257
  motherhood issues and, 255–256
  origins of knowledge and, 248–251
  overview of, 241–244
  social movement theories and, 257–261
Toxic waste movement, 243
Tropical Forestry Action Plan, 268

Uncultivated land
  fodder production in, 164–165, 167, 168
  women and, 161–162
United Mine Workers of America (UMWA), 32, 34, 44
United Nations Development Program (UNDP) Bank, 273
United Nations Food and Agricultural Organization (FAO), 268, 281
United States, seed saving in, 183–187, 190
United States Department of Agriculture-Forest Service (USFS)
  background of, 293–294, 301, 305
  career counseling and mentoring and, 299–301
  commitment to natural resources profession and, 295–297
  job and career satisfaction within, 297–299
  recent development in, 305–306
  samples of female and male professionals in, 294–295
USFS. *See* United States Department of Agriculture-Forest Service (USFS)

Wage rates, in south Indian agricultural community, 111–113
Water management. *See also* Irrigation
  moral economy of Peruvian, 87
  women and, 92
Widows, 67, 69–71. *See also* Heads of households; Single women
Women
  as activists, 244–248
  decision-making authority of, 172–173
  effects of occupational status on, 76–77

gathering and use of forest products by, 194–195. *See also* Agroforestry; Forestry
as heads of households, 195
maritime settings and position of, 52–53
relationship to nature, 1–2
as seed savers, 179–180, 190
socialization of, 216, 219
Women Strike for Peace, 242
Women's groups. *See also* Chipko Movement (India)
impact of, 202
in Kenya, 203–204
role in community organization of, 195–196, 201
Women's movements
effects of, 259–260
in India, 116
Women's Participation in Forest Management workshop, 285
Work, 87
Work relations, 36–37, 43–46
Workplace, sexualization of, 36–37, 43–46
World Resources Institute, 268